Advanced Series in Agricultural Sciences 23

Advanced Series in Agricultural Sciences

Volumes already published in the series

Volume 10
Saline and Sodic Soils (1982)
E. Bresler, B. L. McNeal, D. L. Carter

Volume 11
A Theory of Feeding and Growth of Animals (1982)
J. R. Parks

Volume 12
Fertilization of Dryland and Irrigated Soils (1982)
J. Hagin, B. Tucker

Volume 13
Agricultural Soil Mechanics (1983)
A. J. Koolen, H. Kuipers

Volume 14
Energy and Agriculture (1984)
G. Stanhill (Ed.)

Volume 15
The Rhizosphere (1986)
E. A. Curl, B. Truelove

Volume 16
Population Genetics: Basic Principles (1987)
D. P. Doolittle

Volume 17
Irrigation with Treated Sewage Effluent (1991)
A. Feigin, I. Ravina, J. Shalhevet

Volume 18
Advances in Statistical Methods for Genetic Improvement of Livestock (1990)
D. Gianola, K. Hammond (Eds.)

Volume 19
Electromagnetic Radiations in Food Science (1992)
I. Rosenthal

Volume 20
Water Flow and Solute Transport in Soils (1993)
D. Russo, G. Dagan (Eds.)

Volume 21
Production of Sugarcane: Theory and Practice (1993)
G. Hunsigi

Volume 22
Management of Water Use in Agriculture (1994)
K. K. Tanji, B. Yaron (Eds.)

Volume 23
Woody Plants in Agro-Ecosystems of Semi-Arid Regions (1995)
H. Breman, J.-J. Kessler

Volume 24
Soilless Culture Management (1995)
M. Schwarz

Henk Breman Jan-Joost Kessler

Woody Plants in Agro-Ecosystems of Semi-Arid Regions

with an Emphasis on the Sahelian Countries

With 55 Figures and 75 Tables

Springer-Verlag
Berlin Heidelberg New York
London Paris Tokyo
Hong Kong Barcelona
Budapest

Dr. HENK BREMAN
Ir. JAN-JOOST KESSLER

Research Institute for Agrobiology and Soil Fertility
Ministry of Agriculture, Nature Management and Fisheries
P.O. Box 14, 6700 AA Wageningen
The Netherlands

ISBN-13: 978-3-642-79209-0 e-ISBN-13: 978-3-642-79207-6
DOI: 10.1007/978-3-642-79207-6

CIP-Data applied for

Softcover reprint of the hardcover 1st edition 1995

Typesetting: Best-set Typesetter Ltd., Hong Kong

SPIN: 10135483 31/3130/SPS – 5 4 3 2 1 0 – Printed on acid-free paper

This book is dedicated to Paul Sciarone, a graduate in law and in physics, whose insight into the value of the studies on rangeland dynamics in the Sahel region led to the initiative of providing funds for subsequent studies on the role and potential of woody plants in this region.

Preface

This book provides a quantitative analysis of the role of woody plants in semi-arid regions, for the assessment of their benefits in agrosylvopastoral land-use systems with productive and sustainability objectives. The insights presented and conclusions drawn allow the additional benefits of woody plants for specific climatic and physical site conditions and land-use systems to be estimated. The Sahel and Sudan zones in West Africa, on which the book focusses, represent resource-poor conditions, whose ecological dynamics have been relatively well studied. The role of woody plants in this region, as assessed in this book, is extrapolated to other semi-arid regions, leading to general conclusions on agroforestry's potential as an option for sustainable land use in semi-arid regions.

The origins of this book go back to 1982, when the Club du Sahel requested that available data on woody plants in the Sahel region be synthesised, to provide basic information to enable better attention to be given to woody plants in rural development programmes. We are grateful to the Club du Sahel for this challenge.

Various people contributed to studies used in this book. The preliminary inventory of the data available was made by Frits Ohler; later his work was continued by Franciska Dekker. Useful contributions were made by Chris Geerling and Frits Penning de Vries. We thank Arnoud Budelman, Sjaak Conijn and Meine van Noordwijk for their constructive criticism of drafts of some chapters, Pieter Oostra for spending many hours in correcting the list of references, and we are very grateful to Peter Uithol for doing the simulation studies and the layout. The Wageningen Agricultural University, Forestry Department is acknowledged as having enabled the second author to finalise this book.

The work involved in creating this book was supported by the Research Institute for Agrobiology and soil Fertility (AB-DLO) in Wageningen and, indirectly, by the Netherlands

Ministry of Foreign Affairs, through their support to the AB-DLO for research in Sahelian countries.

Wageningen, The Netherlands
December 1994

H. Breman
J.-J. Kessler

Contents

Abbreviations

a, b, c	Parameters, locally used with different values
C_3; C_4	characterisation of photosynthetic pathway
CC	canopy cover
CEC	cation exchange capacity
DR	diffuse radiation
DOM	digestibility of organic matter
FP	foliage production (dry matter)
f	see f_h, f_{hw} and f_w
f_h	average fraction of nitrogen in the aboveground herbaceous biomass that is lost annually
f_{hw}	see f_h; biomass of herbaceous and woody plants
f_w	see f_h; biomass of woody plants
I	amount of infiltrated water (mm)
La	valleys and depressions, (loamy-)clay soils
LI	litter input
M	temporary lakes
N_h	nitrogen annually available for the aboveground production of herbs in an equilibrium situation between nitrogen inputs and outputs
N_{hw}	see N_h; production of herbaceous and woody plants
Os	undulating dunes, sandy soils
PAR	photosynthetic active radiation
PET	potential evapotranspiration
Pl	plains, loamy soils
Ps	gently sloping plains, sandy soils
R	annual rainfall (mm)
r	radius of vertical tree crown projection (average horizontal distance from trunk to extremities of branches)
r_d	litter decomposition rate
r_h	little humification rate
r_t	litter turn-over rate
RG	plant species with rapidly germinating seeds
SG	plant species with slowly germinating seeds
SOM	soil organic matter

Sq	shallow, poorly developed loamy soils
SSWA	Sahel and Sudan zones of West Africa
n	number
TLU	tropical livestock unit
TR	transmitted radiation
IR	incoming radiation
WP	wood production (dry matter)
WUE	water-use efficiency
X, Y, Z	units, locally used with different significations

Ia, Ib, IIa, etc. situations of plant production (see Sect. 5.2 and Fig. 5.1)

1 Introduction

Paradoxically, although agroforestry is a well-known traditional land-use system, it has only recently been taken on board by development experts as a new technology. Whereas in most western countries trees are generally considered to be incompatible with cropping or grazing, agroforestry is currently attracting much interest because of environmental awareness of the ecological problems related to monocultures and of the food and fuelwood shortages in many developing countries.

The concept of agroforestry became scientifically recognised in the late 1970s. One of the first definitions stated that agroforestry is a sustainable land management system (King and Chandler 1978). Later, this claim was modified by stating that the rationale for developing agroforestry is to obtain a higher, more diversified and/or a more sustainable production of the combined tree and other agricultural and/or livestock resources than is possible with other forms of land-use (Lundgren 1982). Even though it is currently recognised that agroforestry systems are not a priori sustainable, agroforestry is still commonly embraced as "the ideal sustainable development technology, providing fuelwood, increased agricultural productivity, helping small-scale farmers and generating employment and income" (e.g. Winterbottom and Hazelwood 1987).

The claim of increased and/or more sustainable agricultural production is particularly attractive for developing countries where it is not feasible to use external inputs to boost crop production. Here, agroforestry may be part of the solution, as a "low external input sustainable agriculture" technology. Interesting results have been obtained in the humid tropics, e.g. by alley cropping with leguminous shrubs such as *Leucaena leucocephala* (Kang and Reynolds 1986; Lal 1991; Kang 1993). However, few experiences from semi-arid climates have been thoroughly analysed, and the detailed accounts that do exist most commonly refer to specific cases where synergistic interactions between woody plants and herbs or crops seem to occur. The case of *Faidherbia albida* in West Africa is one of the best known; but evidence is accumulating that in many situations, and particularly in semi-arid regions, competitive relationships between woody plants and crops predominate. It is unrealistic to extrapolate incidental cases as long as the underlying ecological processes and interactions between and trees and crops are imperfectly understood.

Back in 1982, the Club du Sahel requested that data on woody plants in the Sahel region be synthesised, to provide basic information so that the attention given to woody plants in rural development programmes could be improved. This request triggered the studies which, 12 years later, have resulted in this book. Meanwhile, most rural development projects in developing countries have included agroforestry as one of their activities. Rarely has so much been expected of a technology for which, at least for semi-arid climates, so little experience has been analysed, and so few basic principles have been developed.

Not until the 1980s did the PPS (Production Primaire au Sahel) research project elucidate the ecological processes determining primary production in semi-arid climates. The project focussed on the herbage layer in the rangeland vegetation in the West African Sahel zone. The results enabled the factor(s) limiting primary production to be identified for specific sites, and the vegetation dynamics to be predicted (Penning de Vries and Djitèye 1991).

This book builds on the results of the PPS project: the approach is ecophysiological. Little attention is given to specific woody species, because the characteristics of these species, the site (soil and climatic) conditions, and the sociocultural factors vary so enormously that a handbook approach is inappropriate. There are no blueprint solutions. The major objectives of this book are therefore to develop and provide basic ecological principles for evaluating the actual and potential role of woody plants in agro-ecosystems in semi-arid regions, and to identify the ecological conditions that determine whether woody plants may fulfil the claims made about their ability to enhance production and ensure sustainability. Hopefully, its use will replace the trial-and-error approach that so often produces disappointing results for planners, development workers and farmers. The economic and socio-cultural aspects associated with woody plants and agroforestry systems are not dealt with specifically, although it is acknowledged that such aspects are of great importance for a successful agroforestry system. That is one of the reasons why this book indicates how to quantify in detail the plant production of herbs only, and of mixed vegetations of herbs and woody plants. Knowledge of both levels is a prerequisite to account for socio-economic conditions.

Most of the data on which the analysis is based are from the semi-arid Sahel and Sudan zones of West Africa, henceforth referred to as SSWA. The seven countries in this area are Burkina-Faso, Chad, Gambia, Mali, Mauritania, Niger and Senegal. Data from other semi-arid regions have been included whenever considered useful, to supply missing information, or to illustrate deviations. As a result, the general insights are applicable to most semi-arid regions, as is demonstrated in Chapter 5.

In this book the concept of "semi-arid region" has been interpreted broadly in order to cover the transition from water- to nutrient-limited plant production, and the transition from open shrubland savanna to dense savanna woodland. The book therefore covers an area whose dry boundary lies in the arid climatic zone, where pastoral land use is still possible, and whose moist boundary lies in the sub-humid climatic zone, where agriculture is still limited to one cropping cycle because of a distinct dry season. Most of the area lies in the semi-arid climatic zone, where agrosylvopastoral land-use systems predominate.

The synthesis of the past and present importance of woody plants in the rangeland vegetation (Chap. 2) provides ecological insights into vegetation dynamics including woody plants, and provides insight into the impact of natural and human factors. Chapter 3 presents an analysis of production data of woody plants, including the associated nutrient dynamics, in relation to soil and climatic conditions. The focus on the natural vegetation is essential, so that fundamental ecological processes in situations with little human impact can be investigated. In subsequent chapters the results are extrapolated to croplands and pastures with woody plants. Hence, in most cases "herbs" or "herbaceous layer" refers to herbs, grasses and/or annual crops (e.g. fodder crops, cereals).

The first three chapters provide data that allow the influence of woody plants on plant production factors to be analysed quantitatively. Chapter 4 can be considered as the core of the book, dealing with interactions between trees and herbs, the development of fallowland vegetation, and the influences of woody plants on the availability of water, soil organic matter and light, and nutrient dynamics. In this chapter, plant production is determined for situations with and without woody plants, in relation to climatic and soil conditions, based on simulation models predicting primary production limited by light, water and/or nutrients. The models have been modified in order to include the specific role of woody plants.

The analytical approach adopted means that the insights developed in this book can be applied to any location in semi-arid regions for which some basic ecological data are available. Users less interested in vegetation ecology, or with insufficient background or time available, could confine themselves to reading Chapter 5, which summarises the major conclusions and presents some recommendations.

2 The Distribution and Canopy Cover of Woody Species

2.1 Introduction and Methodology

2.1.1 Introduction

This chapter analyses the distribution and importance of woody plants in the Sahel and Sudan zones in West Africa (SSWA), according to climatic zones (Sect. 2.1.2), landscape and geomorphological units (Sect. 2.1.3) and land-use types (Sect. 2.1.4). Canopy cover is utilised as the parameter indicating woody species' importance (Sect. 2.1.5).

Distribution and canopy cover data in the three land-use types are analysed for the pre-drought years before 1970 (Sect. 2.2.2), the years between the two droughts 1975–1982 (Sect. 2.2.1) and the post-drought years, beginning in the mid-1980s (Sect. 2.2.2). Climatic influences (Sect. 2.2.2) and regional ecological influences (Sect. 2.3) are analysed and compared with the human influences of fire, livestock grazing and browsing, agriculture and wood exploitation (Sect. 2.4).

2.1.2 Climatic Zones

The SSWA in West Africa is located at 10–16° N latitude. Pre-1970 average annual rainfall of 100–1200 mm defines the climatic zones (Fig. 2.1; Table 2.1). The driest zone, the Saharan Sahel, lies north of the SSWA. Within the SSWA, from north to south, average annual rainfall progressively increases from the northern Sahel, the southern Sahel, the northern Sudan to the southern Sudan. As these climatic zones are also associated with vegetation and land-use characteristics, they can also be referred to as agro-ecological zones.

The SSWA is characterised by one dry season and one rainy season. As average annual rainfall increases toward the south, rainy seasons lengthen. The rainfall variability, the frequency of below average rainfall years, and the intensity of dry years decrease toward the south.

Average annual rainfall has significantly and persistently decreased in recent years. Two major droughts have occurred, the driest years of

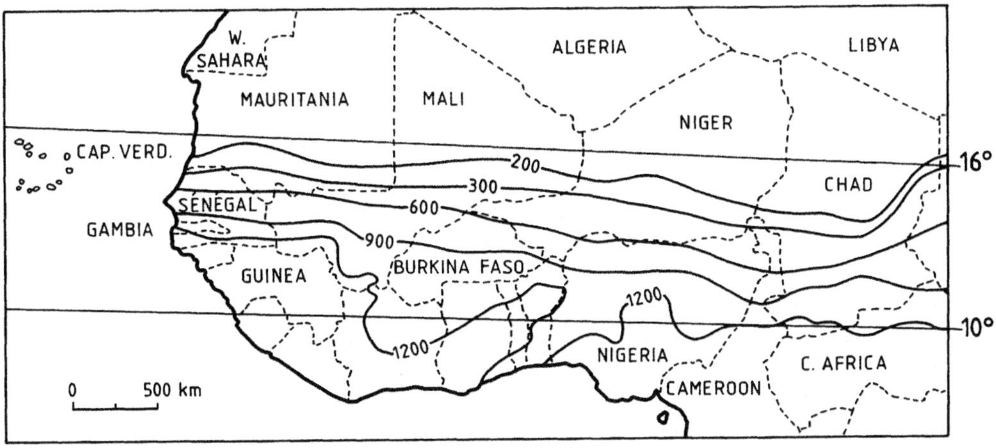

Fig. 2.1. West African average annual rainfall isohyetes, mm year^{-1}. (After Breman and de Ridder 1991)

Table 2.1. Climatic characteristics of the Saharan Sahel, northern Sahel, southern Sahel, northern Sudan and southern Sudan climatic zones. (Kessler and Ohler 1983; PIRL 1988; Le Houérou 1989; Penning de Vries and Djitèye 1991)

Climatic zone	Saharan Sahel	Northern Sahel	Southern Sahel	Northern Sudan	Southern Sudan
Climatic type	Arid	(Semi)-arid	Semi-arid	Semi-arid/ sub-humid	Semi-arid/ sub-humid
Annual rainfall (mm yr^{-1})					
– average year	<150	150–350	350–600	600–900	900–1200
– dry year	0–75	75–200	200–425	425–750	750–1100
– 1970–1990	<100	100–200	200–425	425–750	750–1000
Average temperature (°C)					
– rainy season	30	31	30	29	28
– dry season	<25	26	25	26	27
Potential evapotranspiration (mm yr^{-1})	1800–2500	2000–2500	1800–2300	1600–2000	<1600

which were in 1972–1973 and 1983–1984. In the last 20 years, in every climatic zone, average annual rainfall was 100–200 mm lower than that recorded before the droughts (e.g. PIRL 1988; ICRISAT 1990).

A dry year is defined as one in which the rainfall is exceeded in 9 out of 10 years (Penning de Vries and Djitèye 1991). In the SSWA, dry years tend to occur in clusters. The clustering tendency may occur partly through the influence of earlier dry years on subsequent years (Nicholson

1982). The average annual rainfall in dry years strongly limits the range of possible, sustainable land uses.

SSWA temperatures are consistently high, especially during the rainy season, and are almost nowhere moderated by high elevations or large bodies of water, contrary to some other semi-arid regions. No other semi-arid region with similar average annual rainfall has a dry season so dry or a potential evapotranspiration (PET) so high (Sect. 5.2). Average temperatures and PET in the SSWA, as indicated in Table 2.1, may be influenced by the reduced rainfall of recent years. The northern SSWA climate endures the greatest extremes, with highly variable rainfall in space and time, high day and night temperature fluctuations, strong winds and PET that exceeds rainfall even in the rainy season. Dry season air humidity is very low, as low as 20% in the northern Sahel zone, but humidity rises steadily some months before the rainy season, as the Inter Tropical Convergence Zone approaches.

2.1.3 Landscapes, Geomorphological Units and Groundwater

Three major landscapes occur in the SSWA: the sandy eolian, the skeletal and the fluvial. Landscapes are mainly related to geomorphol-

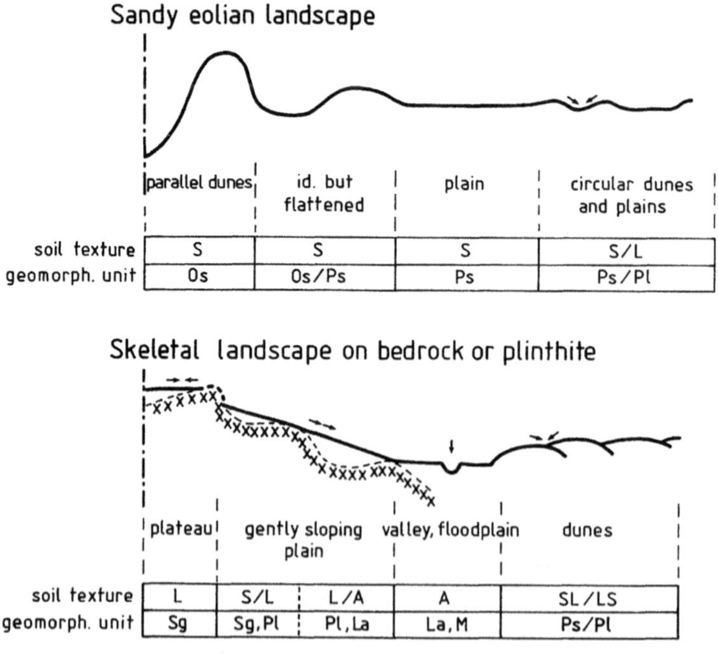

Fig. 2.2. Cross sections of the sandy eolian and skeletal landscapes in the SSWA, with indications of geomorphological units and soil texture. → Run-off; ××× bedrock; *S* sand; *L* loam; *A* clay. (After Penning de Vries and Djitèye 1991)

Table 2.2. Area, population, land-use types and landscapes of the four climatic zones in the SSWA

Climatic zone	Northern Sahel	Southern Sahel	Northern Sudan	Southern Sudan
Area (million ha)[a]	101	52	38	40
(% of total)	44	23	16	17
Sustainable population (persons km^{-2}):[a]				
– based on crop cultivation	3	12	17	25
– based on animal husbandry	2	5	7	10
– total	5	17	24	35
Actual population (1990) (persons km^{-2})[b]	6	25	36	20
Ratio of actual (1990) to sustainable population:	1.2	1.5	1.5	0.6
Present relative importance of land-use types (% area per climatic zone):[c]				
– crop fields	5–10	15–25	20–30	10–15
– young fallowlands	5–10	10–20	15–25	20–30
– old fallowlands, rangelands and forests	~80	~60	~50	~70
Relative importance of landscapes (% area per climatic zone):[d]				
– sandy eolian	65	40	25	15
– skeletal: – shallow soils	25	35	40	40
– plains with deep soils	5	10	20	30
– fluviatile	5	15	15	15

[a] Based on a year of average rainfall (Kessler and Ohler 1983; World Bank 1985; van Keulen and Breman 1990).
[b] Extrapolation to 1990 assumes 3% population growth and migration into the Sudan zones (World Bank 1985).
[c] Estimates based on land required to feed 1990 population in accordance with productivity per zone (van Keulen and Breman 1990).
[d] Based on regional CILSS studies (CILSS 1981, 1982).

ogical factors (Penning de Vries and Djitèye 1991). Typical topographical sequences of the sandy eolian and skeletal landscapes are presented in Fig. 2.2. The importance of the three landscapes varies per climatic zone (Table 2.2).

The classification of geomorphological units is based on the French system where units are defined according to topography, geomorphology and soils. The numerous French vegetation studies in the Sahel zone and northern Sudan zone use the following classification, as presented by Boudet (1978), which has been adopted in this book:

Os: undulating dunes, sandy soils
Ps: gently sloping plains, sandy soils
Pl: plains, loamy soils

La/R: plains, shallow loamy-clay soils
RF: surface bedrock or plinthite, with or without gravel
S/R: bedrock or plinthite beneath sandy, shallow soils
La: valleys and depressions, (loamy-)clay soils
M: temporary lakes.

For the geomorphological classification of the southern Sudan zone, the following units are employed (Geerling 1973, 1976).

Sq: bedrock or plinthite beneath shallow, poorly developed loamy soils (= S/R, La/R or RF)
Q: deep, well-drained, sandy soils (= Ps)
Sg: well-drained, often degraded or rocky, sandy-loam soils (= Pl)
Ts: valleys and depressions, temporarily inundated, hydromorphic and often shallow, clayey soils (= transition from Pl to La)
Sp: deep, well-drained, heavy, clayey soils (= La, M)

Figure 2.2 shows the distribution of the geomorphological units in the sandy eolian and skeletal landscapes. Units Os and Ps dominate in the sandy eolian landscape. Units S/R, La/R and RF are distinguished in some of the vegetation studies, in others they are recognised collectively in the unit Sq: shallow, rocky soils of the skeletal landscape. Unit Pl mainly occurs in the skeletal landscape. Units La and M dominate in the fluvial landscape, and occasionally occur in other landscapes (Table 2.2).

Although groundwater table depths influence woody plant distribution, they could not be used as a determinant of woody plant distribution because they do not correspond to landscape types or geomorphological units. Average groundwater depths increase from 10 m or less in the Sudan zones to more than 50 m in the northern Sahel zone (Penning de Vries and Djitèye 1991). The groundwater table depth is generally shallower in the lower reaches of the topographic sequences. Bedrock permeability determines whether water can percolate. Where the bedrock is sandstone, water can percolate deep into the rock fissures, but water may stagnate on plains with a shallow plinthite layer or in depressions with heavy soils. With recent droughts came widespread land degradation, so that runoff increased on slopes, in spite of the lower rainfall; the groundwater table dropped considerably in the higher reaches of the topographic sequences. As a result, lakes which had not existed before the droughts were formed in depressions in the lower reaches of the topographic sequences. Toward the southern zones, conditions improve for the recharge of the ground water tables (IWACO 1990).

2.1.4 Land-Use Types

Three land-use types reflect the extent and nature of human influences on the woody vegetation:

1. agricultural fields;
2. young fallowlands (of 5 years or less);
3. old fallowlands, rangelands and forests.

In the above sequence human influence decreases, as the distribution and canopy cover of woody species more closely resembles the natural vegetation.

In the Sahel zones rangelands predominate. Most areas are unsuitable for cultivation because of poor soils and inadequate water for cultivation or drinking. In the Sudan zones natural open woodlands predominate. Some areas are unsuitable for cultivation because of flooding, shallow soils or infectious diseases.

Since colonial times, state forests have been located along streams where the black fly occurs which causes river blindness, or along railways for wood supply. Apart from these state forests, most suitable soils in the southern Sahel and northern Sudan zones have been cultivated at some time in the past. The woody vegetation of old fallowlands is different from that of the natural vegetation (Sect. 2.5.3).

The relative importance of land-use types in the four climatic zones (Table 2.2) is based on soil studies indicating land suitability for agriculture and livestock husbandry, and on current population densities. Sustainable population densities for different land-use systems are given for each climatic zone and are based on years with average rainfall.

The northern Sahel zone has the lowest potential for crop cultivation. In a dry year, crops fail completely. Here livestock husbandry prevails, despite large fluctuations in fodder availability related to rainfall variability. The southern Sahel zone has some potential for both crops and livestock husbandry. The southern Sahel and northern Sudan zones are excessively exploited. Population pressure forces people to clear pastures and forests which are only marginally suited for cultivation. Soon after, the soils must be abandoned in a severely degraded state. The importance of the land-use type "old fallowlands, rangelands and forests" is most limited in the southern Sahel and northern Sudan zones, because of population pressure. Only the southern Sudan zone currently remains well within the range of sustainable exploitation (Table 2.2).

2.1.5 Canopy Cover

In semi-arid regions, woody plant canopy cover is considered a useful indicator of the distribution of woody plants and their influence on plant production factors and interactions between woody plants and herbs. Other parameters are considered less suitable. For instance, density does not reflect the relative influence of large and small woody plants; "diameter at breast height" is difficult to apply where woody plants have different relations between this unit and height and canopy cover, and commonly have multiple stems. As the best of several imperfect alternatives, therefore, canopy cover is used as a leading parameter in this book.

Canopy cover provides information about a woody plant or woody plant community's historical ability to produce leaves and branches. Canopy cover also provides information about a woody plant or woody plant community's ability to intercept light for photosynthetic processes. Unjustified conclusions can be avoided by remembering that a plant's historical ability not necessarily determines its potential ability, and by remembering that plants do not live by light alone. Other factors, such as available water and soil nutrients, must also be considered to determine photosynthetic activities of woody plants (Smith 1986).

The canopy cover (CC) of individual woody plants or of woody plants per hectare, can be calculated as follows:

1. CC (individual) = πr^2, representing the area directly under the crown, expressed in m^2 (r = average horizontal distance from trunk to extremities of branches).
2. CC ha^{-1} = the sum of the areas (πr^2) for all woody plants per ha, expressed as a percentage: % total area canopy cover in one ha.

Canopy cover data has been extrapolated from other studies using several methods:

1. Data on woody plant densities and trunk diameter classes were converted into canopy cover by the formula:
   CC ha^{-1} = the sum of all CC (individual) in each diameter class, where CC (individual) is determined by $0.007 \times (2\pi r)^2$ (Sect. 3.4.2).
2. Data on tree and shrub densities per species, where no information on trunk diameter was available, were converted into canopy cover by applying the data from Table 2.3, which indicates average canopy cover per woody species of individuals with a minimal height of 0.5 m. The "average" column indicates trees and shrubs which had been cut or browsed, and the "maximum" column indicates the maximum canopy cover per species of all individuals measured.
3. Most French (IEMVT) studies designate five classes of canopy cover (Boudet 1971). For most classes the average canopy cover within each class was used.
4. The data from Table 2.3 were also used to convert the density of dead trees and shrubs, as determined by vegetation studies after the 1970s drought period, into canopy cover. In this case, the average of the "average" and "maximum" canopy cover values was used, because observations indicate that the drought killed a higher percentage of large trees than small ones.

Table 2.3. Average and maximum canopy cover of woody species in the SSWA, of woody plants exceeding a height of 0.5 m. [Based on unpublished data from Mali (Breman), Burkina Faso (Kessler) and northern Cameroon (Geerling)]

	Canopy cover (m^2)		No. of observations
	Maximum	Average	
Acacia macrostachya	32	14	8
Acacia nilotica	78	24	9
Acacia raddiana	78	17	8
Acacia senegal	8	3	36
Acacia seyal	28	6	36
Afzelia africana		100	?
Anogeissus leiocarpus	113	51	12
Azadirachta indica	84	14	36
Balanites aegyptiaca	50	9	45
Bombax costatum	113	26	10
Boscia angustifolia	7	1	24
Boscia senegalensis	7	1	31
Burkea africana		40	?
Combretum aculeatum	2	1	17
Combretum ghasalense	20	10	23
Combretum glutinosum	80	8	24
Combretum micranthum	28	4	52
Commiphora africana	79	15	48
Crossopterix febrifuga		15	?
Detarium microcarpum	28	5	12
Dichrostachys glomerata	2	1	22
Diospyros mespiliformis	26	15	8
Faidherbia albida	298	46	66
Ficus gnaphalocarpa	309	110	16
Grewia bicolor	10	4	42
Guiera senegalensis	13	2	102
Isoberlinia doka		38	?
Khaya senegalensis	254	97	4
Lannea acida	55	26	14
Lannea microcarpa	152	52	83
Leptadenia pyrotechnica	2	1	4
Maerua crassifolia	7	3	5
Parkia biglobosa	380	165	19
Piliostigma reticulatum	13	3	12
Pterocarpus lucens	79	26	37
Pterocarpus erinaceus	84	37	14
Rophira lanceolata		83	?
Salvadora persica	10	6	2
Sclerocarya birrea	85	49	74
Tamarindus indica	161	82	8
Terminalia avicennioides	28	13	7
Terminalia laxiflora		11	?
Terminalia macroptera		19	?
Vitellaria paradoxa	154	54	158
Ziziphus mauritiana	25	4	75

2.2 Distribution and Canopy Cover Data

2.2.1 1975–1982

2.2.1.1 Land-Use Type "Old Fallowlands, Rangelands and Forests"

Tables 2.4 and 2.5 present a summary of measured and estimated canopy cover for the most important tree and shrub species within "old fallowlands, rangelands and forests." Of the three land-use types, this one is least influenced by human activities; here, woody species distribution and canopy cover mostly reflect variations in natural influences. The data, gathered from 1975–1982, refer to the period between the droughts of the 1970s and 1980s. Most available vegetation studies are from this period. The studies were conducted in Burkina Faso, Mali, Mauritania, Niger, Senegal, Chad, Cameroon, Ivory Coast and Nigeria (Geerling 1973, 1976; Gaston 1974, 1975; Bille 1977; Cornet and Poupon 1977; Toutain and de Wispelaere 1977; Afolayan 1978; Boudet 1978, 1979; Toutain 1978; Breman and Cissé 1979; Grouzis 1979; Lamotte 1979; Dekker 1980; Poupon 1980; Toutain and Piot 1980; de Wispelaere 1980; Aweto 1981; Breman 1981; Penning de Vries and Djitèye 1991).

The available data are clustered per geomorphological unit and per climatic zone; data are not available for every combination of geomorphological unit and climatic zone.

Some woody species, known for their high forage value or other qualities, do not appear in this analysis, either because their canopy cover is too low, or their distribution is too localised. These species, clustered under the heading "others", include *Bauhinia rufescens*, *Cadaba* spp., *Capparis* spp., *Feretia apodanthera*, *Faidherbia albida*, *Hyphaene thebaica* and *Mitragyna inermis*. Some species have a distribution restricted to the eastern or western part of the SSWA. The corresponding canopy cover values are indicated between brackets in Tables 2.4–2.7. Some dominant woody species have indicative value for specific geomorphological units. For example, *Afzelia africana* indicates sandy soils in the Sudan zones, *Isoberlinia doka* indicates heavy soils, and *Terminalia macroptera* indicates hydromorphic soils. Other species occur on various soil types, e.g. *Acacia seyal* in Niono, Mali, occurs both on humid valley vertisols and on coarse sandy soils.

Figure 2.4 shows the total canopy cover per climatic zone and geomorphological unit for the period 1975–1982, and averages per climatic zone based on the relative importance of the landscapes (Table 2.2).

Table 2.4. Canopy cover of common woody species in the northern and southern Sahel zones on land-use type "old fallowlands, rangelands and forests", *after* the 1970's drought period, per geomorphological unit

Climatic zone	Northern Sahel						Southern Sahel						
Pre-drought rainfall (mm)	150–350						350–600						
Geomorphological unit	Os	Ps	PL	RF	S/R	M	Os	Ps	Pl	La	RF	S/R	M
Acacia ehrenbergiana			1			[a]				[a]			
Acacia nilotica						[c]				(1.5)			[c]
Acacia raddiana	2	1	0.5		[b]	[c]	[a]	(0.5)	[a]				
Acacia senegal	[b]	[b]	[a]		[a]		[b]	[b]	[b]		[a]	[a]	
Acacia seyal						[c]		[a]	(3)	3.5			[c]
Anogeissus leiocarpus						[a]				[b]			[c]
Balanites aegyptiaca	0.5	0.5	[b]	[a]	[a]	[c]	[b]	[b]	2	[a]	[b]	(0.5)	
Boscia senegalensis		[a]	[a]	1	[b]	[a]		[a]	[b]	[b]	[b]	[b]	
Combretum aculeatum							[a]	[a]	[a]	[a]			
Combretum glutinosum							(2)	1.5	0.5	1		3	
Combretum micranthum				[a]						(1.5)	2.5	[a]	
Commiphora africana	1	0.5		1	2		[b]	[b]	(1.5)		[b]	0.5	
Euphorbia balsamifera				(1)	5							(2)	
Grewia bicolor				[b]	[b]				[a]	[b]	[b]	[b]	
Guiera senegalensis				[a]			0.5	0.5	0.5	2.5	1.5	2	
Leptadenia pyrotechnica	[b]	[b]			[a]		[b]						
Maerua crassifolia	[a]	[b]		[a]	[b]				[a]				
Pterocarpens lucens				4						(2.5)	13	[b]	
Sclerocarya birrea							[a]	[a]					
Ziziphus mauritiana			[a]			[c]			1.5	0.5	[a]	[a]	[c]
Others							(1.5)			(1)		(0.5)	[c]
Total	3.5	2.5	2	7	8	0–50	4	3	10	14	17	9	0–75
of which leguminous	2	1	1.5	4	[b]	[c]	[b]	0.5	4	7.5	13	[b]	[c]

[a] Low canopy cover, local presence.
[b] Low canopy cover, general presence.
[c] High canopy cover, local presence.

Example of the numeric indicators:
1: 1% canopy cover, present throughout zone; (1): 1% canopy cover, variable within zone.
As leguminous plants are considered only those that from nodules (Sect. 4.5.4.8).

Table 2.5. Canopy cover of common woody species in the northern and southern Sudan zones on land-use type "old fallowlands, rangelands and forests", *after* the 1970s drought period, per geomorphological unit

Climatic zone Pre-droupht rainfall (mm)	Northern Sudan 600–900					Southern Sudan 900–1200				
Geomorphological unit	Os	Ps	Pl	La	RF	Ps	Pl	La/R	La	RF
Acacia dudgeoni				(1)		b	1	b	(2)	2
Acacia macrostachya				(3)			b		1	1
Acacia nilotica				b				b	b	
Acacia seyal		(1.5)	(2.5)	2.5			b	b	(2)	
Afzelia africana						(8)	1	1		
Anogeissus leiocarpus			(1.5)	2	a	1	2	b	(2)	
Balanites aegyptiaca	a	b	a	b			b			b
Bombax costatum		b	2	a	b	b	(1)	b	b	
Burkea africana						2	2	3		3
Combretum aculeatum		a	a				b	b	b	
Combretum glutinosum	8	4.5	5	2	10	2	2	b	b	4
Combretum micranthum		0.5	5	5			b			
Combertum nigricans	(0.5)		(1)	2	(2.5)		2	2		2
Commiphora africana	a	b	(0.5)							
Crossopterix febrifuga						1	1	1		1
Daniellia oliveri						1	2	3		

Detarium microcarpum						1	1	1		1
Gardenia spec.						b	(1)	b		1
Grewia bicolor	a	a	(0.5)	0.5	0.5					
Guiera senegalensis	2	2	2.5	2	1	b	(2)	b	(3)	(2)
Isoberlinia doka						b	1	(5)		
Khaya senegalensis							b	b	(2)	
Lannea spec.			b		a	1	2	1	(1)	2
Parkia bigobosa							b	(1)	a	
Piliostigma spec.		(1)	(1)	b	a	b	1	(2)	1	a
Pterocarpus erinaceus						2	1	1		b
Pterocarpus lucens			(3)	(10)	(5)					
Sclerocarya birrea	a	1.5	3		b		(1)	b	b	a
Terminalia avicennoides	0.5					(3)	(1)	b	b	
Terminalia laxiflora						(2)	(3)	(2)	(2)	b
Terminalia macroptera						b	(1)	(1)	(2)	
Vitellaria paradoxa			b	b		(3)	(2)	(3)	(2)	
Ziziphus mauritiana		a	0.5	a	(0.5)	a	(1)	(2)		b
Others			(2.5)	(1.5)	(1.5)	6	3	2	4	6
Total	11	10	24	32	23	33	35	31	24	25
of which leguminous	b	1.5	5.5	16	1.5	2	2.5	2.5	5	3

See legend Table 2.4.

Table 2.6. Canopy cover of common woody species in the northern and southern Sahel zones on land-use type "old fallowlands, rangelands and forests", *before* the 1970s drought period, per geomorphological unit

Climatic zone Pre-drought rainfall (mm)	Saharan-Sahel <150 mm			Northern Sahel 150–350								Southern Sahel 350–600						
Geomorphological unit	Os	Ps	Pl	Os	Ps	Pl	La	La/R	RF	S/R	M	Os	Ps	Pl	La	RF	S/R	M
Acacia ehrenbergiana			b			10	b	(2)			a				a			
Acacia nilotica											c				(1.5)			c
Acacia raddiana	2.5	5	b	3	1.5	1	(3)	(1)		b	c	(0.5)	(1)	(0.5)				
Acacia senegal				0.5	b	b		b		a		b	b	b		a	a	
Acacia seyal									a		c		(0.5)	(4)	4			c
Anogeissus leiocarpus											a				b			c
Balanites aegyptiaca	b		b	0.5	1	1	3	0.5	a	a	c	b	0.5	3	a	b	(0.5)	
Boscia senegalensis					a	b	b	a	1.5	1	a		a	a	b	b	b	
Combretum aculeatum				a	a	a	a	b				a	a	a	a			
Combretum glutinosum												4	3.5	4.5	1		4	
Combretum micranthum									(2.5)						(1.5)	3	0.5	
Commiphora africana	b	b	b	1.5	0.5		a	0.5	1	2		b	b	(1.5)		b	0.5	
Euphorbia balsamifera									(1)	6							(2)	
Grewia bicolor							b		1	(3)				b	b	b	b	
Guiera senegalensis									a			0.5	1	2.5	2.5	(1.5)	3	
Leptadenia pyrotechnica	b	b		b	b					a		a						
Maerua crassifolia	b	b		a	b	b		a	a	b				a				
Pterocarpus lucens									5						(2.5)	25	0.5	
Sclerocarya birrea												a	a					
Ziziphus mauritiana						1		0.5			c			2.5	1	a	a	c
Others											c	(1)			(1)	(1)	(1)	c
Total	3	5	0.5	5.5	3	13	6	4.5	12	12	0–50	5	6.5	19	15	30	12	0–75
of which leguminous	2.5	5	b	3.5	1.5	11	3	3	5	b	c	0.5	1.5	4.5	8	25	0.5	c

See legend Table 2.4.

Table 2.7. Canopy cover of common woody species in the northern Sudan zone on landuse type "old fallowlands, rangelands and forests", *before* the 1970s drought period, per geomorphological unit

Climatic zone Pre-drought rainfall (mm)	Northern Sudan 600–900				
Geomorphological unit	Os	Ps	Pl	La	RF
Acacia dudgeoni				(1)	
Acacia macrostachya				(3)	
Acacia nilotica				b	
Acacia seyal		(1.5)	(3)	(2.5)	
Afzelia africana					
Anogeissus leiocarpus			(2)	2.5	a
Balanites aegyptiaca	a	a	a		
Bombax costatum			3	b	b
Burkea africana					
Combretum aculeatum		a	a		
Combretum glutinosum	12	10.5	6.5	2	12
Combretum micranthum			0.5	5	5
Combretum nigricans	1.5		(1)	2	(2.5)
Commiphora africana	a	a	(0.5)		
Crossopterix febrifuga					
Daniellia oliveri					
Detarium microcarpum					
Gardenia spec.					
Grewia bicolor	a	a	(0.5)	0.5	0.5
Guiera senegalensis	2	3	2.5	2	2
Isoberlinia doka					
Khaya senegalensis					
Lannea spec.			b		a
Parkia biglobosa					
Pilliostigma spec.		b	(0.5)	b	a
Pterocarpus erinaceus			b		b
Pterocarpus lucens			5	(10)	2
Sclerocarya birrea	a	1.5	3.5		b
Terminalia avicennoides	5.5				
Terminalia laxiflora					
Terminalia macroptera					
Vitellaria paradoxa			b	b	
Ziziphus mauritiana		a	(0.5)	a	(0.5)
Others			(2)	(1.5)	(1.5)
Total	21	17	31	32	26
of which leguminous	b	1.5	8	16	2

See legend Table 2.4.

2.2.1.2 Land-Use Types "Agricultural Fields" and "Young Fallowlands"

On agricultural fields and young fallowlands, variations in canopy cover mainly reflect variations in the nature and intensity of human activities.

Agricultural fields in the SSWA are characterised by trees spaced in a regular, dispersed configuration. As the cultivation period increases and as crop cultivation becomes more permanent, so-called tree parklands develop. On parklands, woody species are planted and protected because of their valuable products. On temporarily cultivated fields, further from the village, woody plant distribution is less regular and species diversity is more variable.

Parklands dominated by *Faidherbia albida* have received considerable attention because of the generally assumed beneficial effects of the trees on soil fertility and crop productivity. *Vitellaria paradoxa* and *Parkia biglobosa*-dominated parklands, however, are more common. These trees occur rarely in the natural vegetation, but are common on old fallowlands, and therefore appear in the land-use classification as constituents of rangeland vegetation (Tables 2.4 and 2.5). Other trees commonly found in the parklands, especially on deep soils, are *Adansonia digitata*, *Hyphaene thebaica*, *Lannea acida*, *Bombax costatum* and *Tamarindus indica*. Parkland trees are well developed compared to those in the natural vegetation.

Tree density and canopy cover of parklands varies considerably. Few systematic studies have been carried out on parklands. Most studies are from relatively favourable sites with conspicuously well developed dense parklands. Canopy cover in parklands varies from 0–50% and averages from 5–10%. Canopy cover in *Faidherbia albida* parklands is at most 40%; highest canopy cover has been recorded in stands of *Borassus aethiopium*. On irrigated fields, most trees have been removed. The large variations in canopy cover are attributed to farmer attitudes toward trees on cultivated fields, to the tree distribution in the natural vegetation, and to the period of cultivation (Pullan 1974). Parklands are discussed further in Section 2.4.3, with tree densities presented in Table 2.14.

When agricultural fields are fallowed, a bushy vegetation develops. Examples of the vegetation structure of 4-year-old fallowlands are given in Fig. 2.3. The woody vegetation of young fallowlands in the Sahel zones is highly variable, largely because of varying intensities of livestock husbandry and wood collection. Young fallowlands are generally characterised by a low species diversity and a predominance of shrubs. Species of the genus Combretaceae, such as *Combretum glutinosum*, *Guiera senegalensis* and *Combretum nigricans*, are often dominant shrubs of young fallowlands. In the Sahel zones and on degraded soils, regeneration of the woody species on young fallowlands may be poor or even absent, due to low rainfall, low water infiltration, or both. Heavy browsing prevents the development of woody seedlings. Regeneration of woody

Fig. 2.3. Height class frequency distribution of shrubs in two 4-year old fallowlands varying in grazing intensity. (After Ohler 1983)

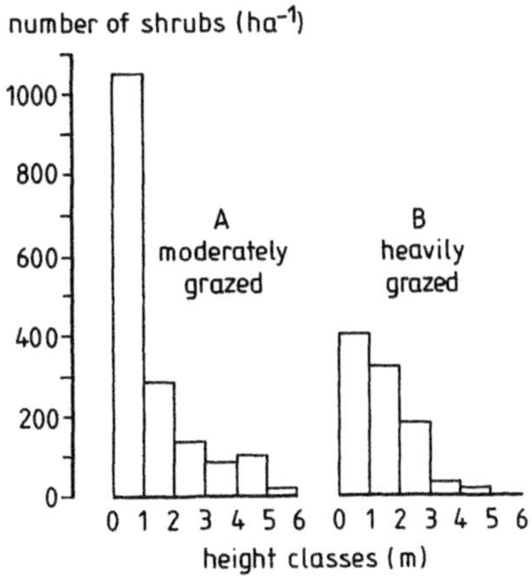

plants in fallowlands originates from seed germination, underground stumps and rootstocks (Aweto 1981; Stromgaard 1985). Under the present conditions of land degradation and high exploitation level, generative regeneration is rarely successful (Gijsbers et al. 1994).

In the Sudan zones, *Piliostigma* spp. and *Detarium microcarpum* are also common on young fallowlands. These shrubs are relatively unpalatable and resistant to defoliation, cutting and fire. On young fallowlands with favourable water availability and without heavy exploitation, shrub canopy cover may exceed 75%. The dense shrublands development in the Sudan zones is related to bush encroachment (Sect. 2.4.3.2).

If the fallow period proceeds, the vegetation gradually becomes more similar to the "natural" vegetation. The colonizing shrub species are characterised by shallow, extended root systems and rapid growth (r reproduction strategy). Trees that establish later in the succession often have a taproot and are characterised by relatively slow growth (K reproduction strategy) (Pianka 1970). The development of the fallowland vegetation will be hampered if exploitation levels are high or if soil degradation is severe (Sect. 2.4.3).

2.2.2 Climatic Influences

2.2.2.1 Introduction

The depth of soil wetting, the period of soil wetting and the depth of the ground water table are important factors determing the establishment,

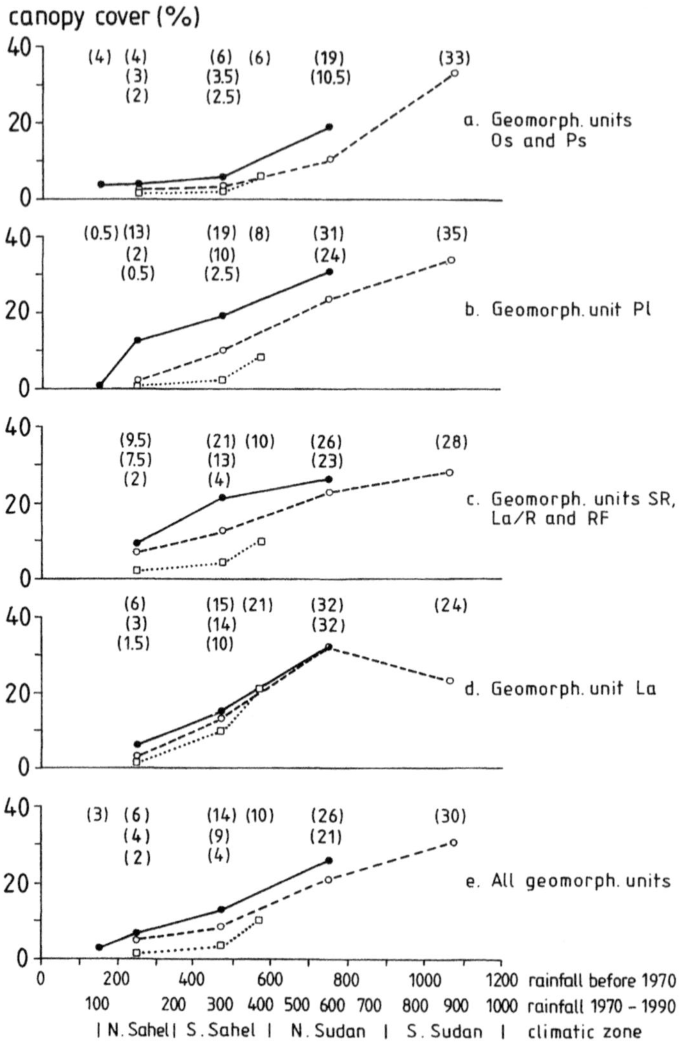

Fig. 2.4. Average canopy cover per climatic zone and per geomorphological unit(s) (a–e) in the SSWA, for different periods in relation to drought. ● Prior to 1970; ○ about 1980; □ end 1980s

survival and productivity of woody species. The importance of these factors and their interrelationships vary by geomorphological unit and by climatic zone.

The influence of both drought periods on woody plant communities and on woody species are reflected in a comparison of post-1970s drought canopy covers (Tables 2.4 and 2.5) with pre-1970s drought canopy covers (Tables 2.6 and 2.7) (Boudet 1961, 1971, 1972; Mosnier 1961; Audry and

Rossetti 1962; Gaston 1967; Breman, pers. observ.). The information is summarised in Fig. 2.4.

A number of studies and data from the post-1980s drought period are analysed to determine recent development (Mali: Benjaminsen 1988; Hiernaux et al. 1990; and Breman and Diarra, in prep.; Burkina Faso: Lamers and Roorda 1991).

During the last 20 years not only has rainfall decreased, but livestock densities, agriculture and wood cutting have increased. The effects of human influences are dealt with in Section 2.4.

2.2.2.2 Influence of Recent Drought Periods on Woody Plant Communities

The distribution and canopy cover of woody plants in the pre-drought period before 1970 is considered as a maximum, because the 1920–1970 rainfalls were relatively high (Nicholson 1982).

In sandy soils of undulating dunes (unit Os) and gently sloping plains (unit Ps), water infiltrates quickly. Canopy cover in these units is typically low, especially in the northern Sahel zone. Observations suggest that after the drought periods, woody plants almost completely disappeared in the Saharan Sahel zone. In the Os and Ps units of the northern Sahel zone, the pre-1970 canopy cover of 4% dropped to 3% by 1980, and to 2% by the end of the 1980s. In the Os and Ps units of the southern Sahel zone, the pre-1970 canopy cover of 6% dropped to 3.5% by 1980, and to 2.5% by the end of the 1980s. Sandy soils are less prevalent in the Sudan zones. On the Os and Ps units of the northern Sudan zone, the canopy cover before 1970 of 20% dropped to 10% by 1980.

In the Sahel zones on the Os and Ps units, rainfall in most years is inadequate to establish the dry season subsoil water reserves essential to survival of woody seedlings (Sect. 2.4.3). Woody seedlings only survive years of sufficient and well distributed rainfall (Toutain and Piot 1980; Barry et al. 1983; Le Houérou 1989). Woody plant regeneration is encouraged by the high water infiltration rate on sandy soils. In the northern Sahel zone of Mali, for instance, more than the usual amount of rain fell in 1978, and it was followed by a flush of *Euphorbia balsamifera* germination (Breman, pers. observ.). The early growth of woody plants, however, is limited by competition for the limited water resources with the herbaceous layer.

An absolute decrease of canopy cover was most dramatic on the sandy loamy plains (unit Pl) and shallow, rocky soils (unit Sq) of the skeletal landscape. In the northern Sahel zone, pre-1970 canopy covers of 10–13% dropped to 2–7.5% by 1980, and to below 2% by the late 1980s. In the southern Sahel zone, pre-1970 canopy covers of 19–21% dropped to 10–13% by 1980 and to 2.5–10% by the late 1980s (Fig. 2.4). In the southern Sahel zone, canopy cover varies considerably: for instance, remnants of natural vegetation survive between heavily degraded sites. These remnants become less and less common toward the north. In the skeletal landscape of the Sahel zones canopy cover was highest on the sandy-loam plains (unit Pl) before 1970, but by the end of the 1980s, canopy cover was higher on the shallow, rocky soils (unit Sq). By the end of the 1980s, in the Sahel zones, only 4–13% of the pre-1970 canopy cover on the Pl unit remained. On the Pl unit in the northern Sudan zone, canopy cover decreased from 31 to 24% between 1970 and 1980. Observations indicate that by the end of the 1980s drought period canopy cover dropped further. Canopy cover in the southern Sudan zone probably is still about 30% on unit Pl and about 25% on unit Sq.

Although loamy soils provide better growth conditions for woody plants than sandy soils (Sect. 2.4), once degraded, on loamy soil surfaces crusts develop. The sealed surfaces prevent water infiltration and thus prevent the establishment of woody plants. Land degradation, therefore, is more irreversible on loamy soils than on sandy soils. Once a threshold is passed, regeneration is difficult. Runoff of the sealed surfaces is high, and the remaining vegetation becomes concentrated in the (micro-) depressions where the water collects. The result is a "tiger bush", a shrub-dominated vegetation pattern characterised by scarce vegetation on runoff surfaces, and concentrated vegetation in depressions (Sect. 2.4). Tiger bush is spreading south into the northern Sudan zone (Le Houérou 1989). In the Sudan zones an increasing density of woody plants, in spite of a stable or even decreasing canopy cover, may be related to a development from tree savanna to shrub savanna (Benjaminsen 1988), or to the regeneration of once degraded areas. In 1990, in central Burkina Faso, shrub canopy cover was 1.5 times higher than tree canopy cover (Kessler, unpubl. data).

On rocky soils, the decrease of canopy cover is less dramatic, because water flows off rocky surfaces into pockets between rocks and into rock fissures, and the woody plants can profit from this runoff.

Canopy cover in most fluvial landscapes (units La and M) was scarcely influenced by the drought periods. However, in the northern Sahel zone, the valleys did not capture sufficient rainfall for woody plants survival, and recent data show that the canopy cover on floodplains varies from 1–20%. South of the northern Sahel zone, canopy cover in the fluvial landscape increased from 1970–1980 (Bille 1977; Boudet 1979; Barry et al. 1983). The increase was probably due to increased runoff captured from the surrounding land.

Canopy cover can be highly variable within a fluvial landscape. Dense woody stands may occur alongside floodplains devoid of vegetation. The amount of water flowing into a fluvial landscape depends on the amount of rainfall, and the size and infiltration capacity of the water catchment. Many woody plants die if not adapted to the increased rates of flooding.

Canopy cover in the fluvial landscape is lower in the southern Sudan zone than in the northern Sudan zone, because of the greater run-on from loamy sealed soil surfaces and higher rainfall. Few woody plants survive in and around shallow lakes (unit M). Those that do include *Mitragyna inermis*, *Mimosa pigra* and *Acacia seyal* in the Sahel zones, and palm trees such as *Borassus aethiopum*, *Hyphaene thebaica*, and *Phoenix reclinata* in the Sudan zones (Menaut 1983). *Acacia nilotica*, dominating moist valleys, dies when a site is flooded for too long. Dense *Acacia raddiana* stands, however, appeared in the Niger delta in Mali, where permanent lakes dried up.

A number of studies analyse the dynamics of canopy cover in relation to geomorphologic units and topographic sequence. In the southern Sahel zone, 90% of *Guiera senegalensis* died on the upper parts of the topography, while a 10% increase in canopy cover was recorded in the valleys (Bille 1977). In the sandy plains of the northern Sudan zone, 60% of *Bombax costatum* died, while only 30% died in the valleys (Togola et al. 1975). After the drought periods in the southern Sahel zone, the density of woody plants (mainly *Acacia nilotica* and *Acacia seyal*) nearly doubled in valleys, yet beyond the valleys density decreased (Benjaminsen 1988). Mortality of *Acacia senegal* in the southern Sahel zone was high everywhere, but in valleys nearly 50% of trees thought to be dead resprouted (Poupon 1977a). In the southern Sahel zone in Senegal, tree mortality between 1981 and 1987 increased from 20% in the valleys, to 40% on lower slopes, to 70% on plateaus and sand dunes. In following years (1987–1990) regeneration was highest in the valleys: there was an increase of up to 300% of young trees (Miehe 1990). In the valleys of the Sahel zones in the country of Sudan, canopy cover of *Acacia raddiana* stands was estimated at 16% in 1985 and 14% in 1987. In the same period the canopy cover of *Acacia raddiana* on the sandy plains dropped from 2.5 to 0.5% (Kennenni and van der Maarel 1990).

In the Sahel zones, valleys may serve as refuge areas for woody species susceptible to drought. The distribution of these species may be limited to the valleys during droughts, and may again colonise sandy plains during years when rainfall is adequate (Barry et al. 1983). Where loamy soils prevail on surrounding plains, however, surface sealing limits the return of woody plants. Human influences, such as livestock grazing and dead wood collection from the loamy soils probably contribute to the irreversibility of the degradation process (Sect. 2.4).

Prior to 1970, canopy covers per climatic zone of 3–26%, averaged for all geomorphological units, are linearly related to average annual rainfalls of 100–750 mm (Fig. 2.4e). These canopy cover estimates are comparable to those given by Hiernaux (1983). Following the 1970s drought period, total canopy cover in the land-use type "old fallowlands, rangelands and forests" dropped most dramatically in the Sahel zones. In the northern Sahel zone, average canopy cover decreased from pre-

1970 levels of 6 to 4% in 1980, and to 2% in the late 1980s. In the southern Sahel zone, canopy cover decreased from pre-1970 levels of 14 to 9% in 1980, and to 7% in the late 1980s. When all land-use types are considered, canopy cover decreased most dramatically in the southern Sahel zone, because a great deal of vegetation there was cleared for crop cultivation.

Observing the post-drought canopy cover data (Fig. 2.4), natural conditions for the establishment and survival of woody plants seem to improve at a threshold of about 500 mm average annual rainfall, in the southern Sahel zone (Fig. 2.4). This phenomenon should be related to drought influences, since human activities are equally intensive throughout the entire southern Sahel zone (Sect. 2.4).

In the northern Sudan zone, canopy cover dropped from 26 to 21% while variation increased: sites with a high canopy cover occur adjacent to areas of abundant woody regeneration and to bare areas. The canopy cover of the southern Sudan zone was hardly influenced by the droughts, but shifts in species composition occurred.

2.2.2.3 Influence of Recent Drought Periods on Woody Species

Woody species have adaptations to survive under arid and semi-arid conditions. Xeromorphism is seen in the many morphological and physiological adaptations to prevent water loss and to enhance water storage. These adaptations include sclerophyllous leaves, thickened epidermis, sunken and protected stomata and small leaves, sometimes reduced to scales. Even in the southern Sudan zone, many plants have xeromorphic adaptations. Other than fan palms, large-leafed woody plants are rare in the SSWA. Hard leaves with a thick cuticle are common among evergreen and semi-evergreen species. Soft leaves are generally shed at the beginning of the dry season (Menaut 1983).

During recent years, a greater proportion of woody species with xerophytic adaptationa have been observed (Le Houérou 1989). There is an increasing proportion of deciduous spiny microphyllous species (*Acacia* spp., *Dichrostachys glomerata*), of ephedroid species (*Leptadenia pyrotechnica*, *Capparis decidua*), and of evergreen mesophyllous, sclerophyllous species (*Balanites aegyptiaca*, *Boscia senegalensis*), although within populations of these species large scale mortality occurred. Species with xerophytic adaptations tend to replace broadleaved species such as *Terminalia* spp., *Pterocarpus lucens*, *Anogeissus leiocarpus* and the Combretaceae.

The importance of some woody species has considerably decreased because of the droughts. In the sandy landscape of the northern Sahel zone, many *Acacia raddiana* died. In the southern Sahel zone, *Acacia senegal* and *Commiphora africana* showed high mortality. On rocky soils, many *Pterocarpus lucens* died in the western SSWA, as well as many *Combretum aculeatum* and *Grewia bicolor*. On all soil types,

Combretum glutinosum, *Guiera senegalensis*, and *Terminalia* spp. were greatly influenced by the drought periods, even in the northern Sudan zone. In the northern Sudan zone, *Pterocarpus erinaceus*, *Burkea africana*, *Prosopis africana* and *Khaya senegalensis* were particularly affected during the last few decades. Most of the above species are intensively exploited for their products (Sect. 2.4). In the Sudan zones, the possible effects of the drop of the groundwater table on deep rooting woody plants should be considered as well.

In general, the highest mortality rates due to the drought periods occur at the northern limit of the geographic range of many woody species. Of all woody species occurring in the southern Sahel zone in Senegal, *Sclerocarya birrea* showed by far highest mortality, virtually disappearing from upland areas (Miehe 1990). Togola et al. (1975) found mortality of *Bombax costatum* to increase proportionally towards the north, reaching 100% mortality where annual rainfall had been less than 350 mm during 2 successive years, which is considered the minimum for survival of this species. Site-specific differences in mortality are the consequence of differences in moisture availability mainly. The distribution of *Acacia senegal* is well documented: in the 1930s, *Acacia senegal* was found south of the 350-mm isohyet only (corresponding to the Senegal river) (Trochain 1940). During the relatively humid period of 1920–1970, its distribution extended towards the north. In the early 1970s, important populations of the species could be found in Mauretania, north of the Senegal river. Now *Acacia senegal* has almost completely disappeared north of the river (Breman, pers. observ.; RAMS 1980). Likewise Ohler (1982) notes that *Vitellaria paradoxa* used to occur up to 500 mm rainfall (pre-drought rainfall data), whereas now it occurs south of the 700-mm isohyet mainly, corresponding more or less to the 500-mm isohyet of the last 20 years.

The summarised data do not show an increased canopy cover for any woody species. Several studies, however, mention an important increase of woody species such as *Balanites aegyptiaca*, *Acacia senegal*, *Acacia raddiana* (Benjaminsen 1988; Miehe 1990). Such observations can be confirmed by unpublished data of the authors. The increase is most important in valleys and is basically an encroachment of spiny, drought resistant Sahelian shrub or small tree species into the Sudan zones.

2.3 An Ecological Interpretation of Woody Vegetation in the SSWA

2.3.1 Introduction: Water Availability and Soil Texture

In semi-arid regions, water availability, varying in time and space, primarily determines vegetation structure and floristic composition. Seasonality in the SSWA, especially in the Sahel zones, is characterised by an extremely long dry season and a hot and short rainy season (Sect. 2.1.2). The pattern is unfavourable for the establishment and survival of woody plants. In areas with similar but bimodal annual rainfall, such as East Africa, canopy cover is greater (e.g. van Wijngaarden 1985). Even in northern Australia, where the climate shows most similarity with the SSWA, canopy cover is greater (Williams 1981).

Woody plants depend on the amount of water that infiltrates to soil layers beyond the reach of herbaceous roots, enabling woody plants to survive the dry season (Fig. 2.7). The ratio of topsoil to subsoil moisture determines the balance between woody and herbaceous vegetation components in savannas (Walter 1973). In the Sahel zones, water infiltration until 1.5–2 m was found to result in an increased nutrient uptake by the herbaceous layer (Breman and de Ridder 1991). Deeper infiltration did not result in greater nutrient uptake by the herbaceous layer, and sometimes even reduced total nutrient uptake. Presumably, water infiltration deeper than 2 m is not available to herbs.

Rainfall and soil type determine infiltration depth and the duration of water availability in soil. Soil type determines water infiltration rate, run-off and redistribution patterns, water holding capacity and soil depth. The relationship between water infiltration and the average depth of soil wetting depends on soil texture (Fig. 2.5). On sandy soils water infiltration is homogeneous, so soil wetting is proportional to absolute rainfall.

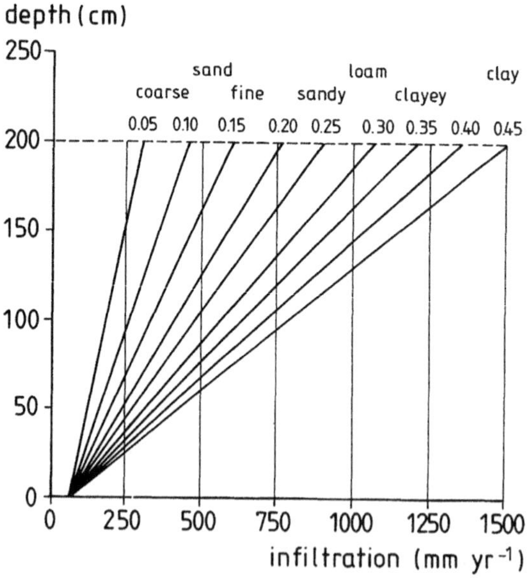

Fig. 2.5. Influence of soil texture on the relationship between water infiltration and depth of soil wetting, characterised by soil water content at field capacity (cm^3 water cm^{-3} soil). (Breman and de Ridder 1991)

But sandy soils can store little water, so water leaches easily. On fine sandy soils, 2 m of soil wetting requires 450–550 mm of infiltrated water (Fig. 2.5). Assuming an average runoff coefficient of 0.20 (Breman and de Ridder 1991) this corresponds to an annual average rainfall of 600–700 mm. On loamy, shallow or rocky soils, water runs off crusted and compacted soils, rocks or rock outcrops, and collects in large and small catchments, such as large valleys or between rocks. Even at low absolute rainfalls, water can infiltrate to the subsoil in these catchments, and provide woody plants with a competitive advantage against the herbaceous layer. Water catchment sites often have finely textured soils which can hold a great deal of water, but much water is required to wet soil to sufficient depth. For loam and clay soils, 850–1200 mm and 1200–1400 mm of infiltrated water, respectively, are required (Fig. 2.5). Most areas in the southern Sahel zone and the Sudan zones collect this amount of water in valleys. In the southern Sudan zone, however, excessive water collection can cause flooding. If flooding persists, woody plants die and a swamp savanna develops.

According to variation in annual rainfall, woody plants can obtain adequate water by migrating to different soils where their water requirements are met. Thus, woody species "move along the catena" to meet specific water requirements, such as *Acacia seyal* "moving to higher positions" with increasing rainfall.

Other conditions being equal, woody plants grow better on fertile soils. For a given average rainfall, canopy cover is higher on a soil with high fertility than on a poor soil. As a woody plant develops a larger canopy and a more extended root system, a competitive advantage in relation to herbs is built up, capturing more light, nutrients and water.

2.3.2 African Savannas

The SSWA is entirely in the savanna zone. The savanna biome type refers to wooded grasslands of the tropics and subtropics (Bourlière 1983; Menaut 1983; White 1983; Tothill and Mott 1985; Breman and de Ridder 1991). The savanna biome type includes grasslands with a sparse woody cover (Menaut 1983), steppes with at least a 10% canopy cover of dwarfed shrubs (Shmida and Burgess 1988) and woodlands with a sporadic grass cover. Thus, savannas are the transition between closed forests and grasslands or deserts. The major determinant for classification of savanna vegetation types is rainfall.

Woodlands have 20% or more canopy cover and little or no woody understory. Tall grasses can dominate the herbaceous layer, which may be sparse (Menaut 1983). A woodland savanna or open woodland is dominated by trees with up to 20% canopy cover. A shrub savanna or open shrub land is dominated by shrubs with up to 20% canopy cover.

Of northern Sahel woody species 45% are evergreen or semi-evergreen. In Australia, evergreens and semi-evergreens are dominant, but in other African savannas, drought deciduous woody species are dominant (Shmida and Burgess 1988). The similarities probably reflect the severity of the Sahelian and north Australian climates. In the Sudan zones, drought deciduous woody species are more common (Sect. 3.5.2.1; Table 3.12).

Trees and shrubs of a wooded savanna are often gnarled and stunted, while those of woodlands have straight boles. Trees of at least 7 m dominate tree savannas, shrubs up to 7 m dominate shrub savannas (Aubréville 1963). Shrubs have also been defined as only 0.7–2.0 m tall (Shmida and Burgess 1988). Woody species which do not fall into one of the above categories can be identified as trees or shrubs by criteria such as the position of the reserve buds.

Within climatic zones, vegetation structure and floristic composition are related to geomorphological units, because the units reflect soil texture, fertility, depth and water-holding capacity, influencing nutrient and water availability. In comparison to herbs, the major floristic elements of the woody layer usually relate to a more detailed level of environmental variation (e.g. Geerling 1987). This can be expected since the dominant herbs are functionally related to different, usually broader, environmental influences than the dominant woody plants (Tothill and Mott 1985).

Recently developed theories on rangeland dynamics in arid and semi-arid regions expressed doubts on the ecological model describing a succession development toward a climax (Westoby et al. 1989; Behnke and Scoones 1991; Friedel 1991). A state-and-transition model has been proposed, describing vegetation development through series of alternative states, and series of transitions between states. Vegetation may enter a transition as a result of natural or human-related influences. Once a threshold is crossed to a more degraded state, characterised by a lower productivity, the state of higher productivity is difficult to recapture. The state-and-transition model is suitable where rainfall is low and erratic, and primary production is variable. Such regions may be characterised as being governed by non-equilibrium dynamics.

2.3.3 The Sahel Zones

From north to south in the Sahel zones, rainfall increases and open shrub savanna gives way to open tree savanna (Figs. 2.6, 2.7). The groundwater table usually lies below the reach of tree roots. In the northern Sahel zone, on sandy soils with homogeneous water infiltration, the soil is not wetted deep enough for roots of woody plants to success-

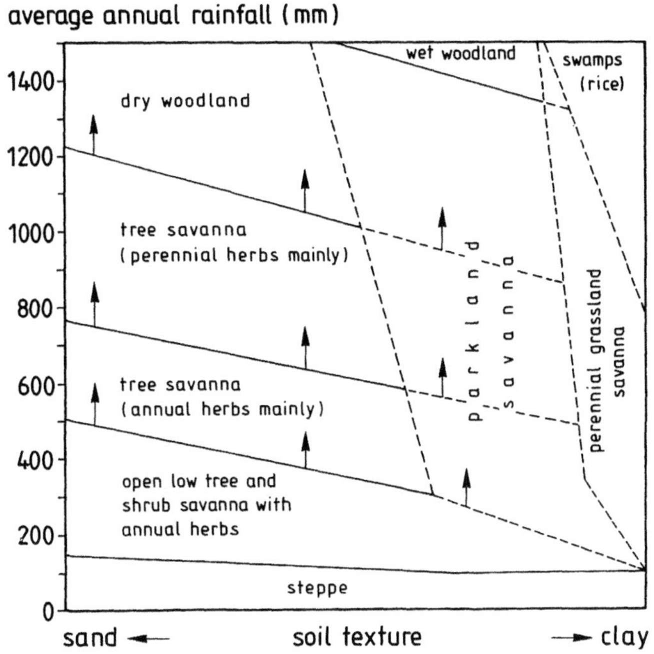

Fig. 2.6. Vegetation types in relation to average annual rainfall and soil texture in the SSWA. → Change in boundaries of vegetation types 1970–1990. (Based on Tothill and Mott 1985)

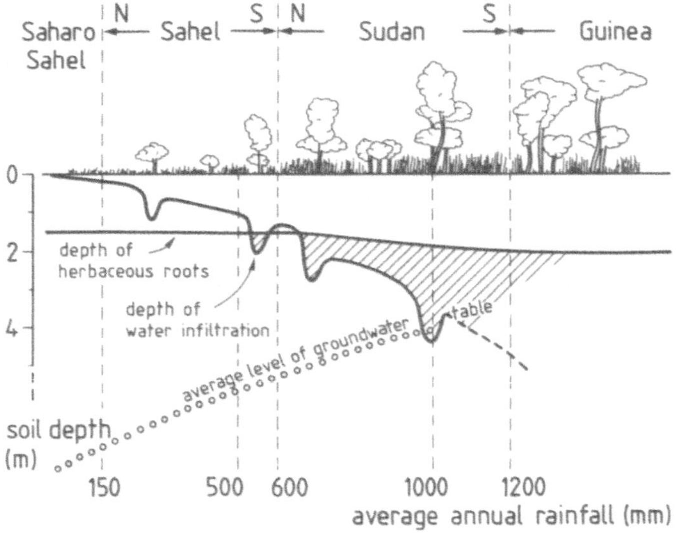

Fig. 2.7. Canopy cover per climatic zone in the SSWA in relation to average maximum depth of soil wetting during the rainy season

fully compete with annual species. Here the fibrous root system of herbs is most efficient in extracting water (Shmida and Burgess 1988). Since photoperiodism determines the growth period of annual species, some water remains after the annual species have died, which can assist woody plant survival (Penning de Vries and Djitèye 1991). In Australia and southern Africa, the regular woody plant distribution pattern on sandy soils appears to be caused by intra-specific competition for soil moisture and nutrients (Smith and Goodman 1986; Wotton and Sparrow 1991). In southern Africa, the distance between neighbouring trees is proportional to the canopy cover of the tree, suggesting that total canopy cover is related to availability of, and competition for resources.

In the Sahel zones, particularly in the driest part, woody plants establish in years with good rainfall only. Such suitable conditions are relatively rare events in semi-arid regions (Westoby et al. 1989; Friedel 1991). Where long-term average rainfall is 250 mm, about 1 out of 10 years' rainfall is at least 400 mm, which is considered sufficient for establishment of woody plants on sandy soils (Sect. 2.3.1). Concentrated rainstorms are beneficial for woody plant establishment because they create subsoil water reserves, whereas well-distributed rainfall stimulates the herbaceous layer (Shmida and Burgess 1988). After a period of wet years in the Sahel zones, woody species and annual species with a long growth cycle enlarge their area of distribution northwards. For example, *Acacia senegal* became more common beyond its usual northern limits in 1920–1970, years of good rainfall (Sect. 2.2.2.3). In the same way, species typical to the Sudan zones may extend their limits north into the Sahel zones. In the dry years which inevitably follow, competition for water will be severe. This is particularly so for woody plants, since annual species with a long growth cycle also enlarged their area of distribution northwards during preceding moist years, and more easily use limited available water resources. As a result, the woody plants die when their water reserves are exhausted (Togola et al. 1975), unless groundwater can be reached by taproots. In water catchment areas, woody plants are unlikely to develop taproots because of aeration problems in heavily textured soils.

Annual rainfall in the Sahel zones has decreased during recent years with 150–200 mm (Table 2.1), explaining the high tree mortality as rainwater no longer infiltrates sufficiently deep. Also, typical Sahelian species extended their limits south into the Sudan zones.

Fire is not an important influence in the Sahel zones (Sect. 2.4.2).

In the southern Sahel zone, sites with water beyond the reach of herbaceous roots are more numerous, as rainfall and the proportion of non-sandy soils increase. Conditions for woody plant establishment improve considerably at 500–600 mm of rainfall (Fig. 2.4). Nevertheless, the southern Sahel zone landscape remains open, because conditions suitable for woody plants are limited.

Microhabitats are important to Sahelian vegetation. Water beyond the reach of herbaceous roots creates a niche differentiation between woody plants and herbs in run-on sites on loamy soils, on rocky soils and in valleys. Niche differentiation also occurs when wind or water erosion removes the sandy topsoil layer, baring a loamy subsoil. Because of the processes of water redistribution, canopy cover in the Sahel zones reached 13–19% on loamy soils, 10–21% on rocky soils and 6–15% in valleys and depressions in the good rainfall years before 1970 (Fig. 2.4). Tiger bush, often associated with the skeletal landscape, is found on shallow, gravel soils in the Sahel zones. The proportion of vegetated groves to bare intergrove soils varies. The regular tiger bush pattern probably occurs where excessive runoff flows over short expanses of loamy, crusted soils, towards catchments, where both woody plants and herbs establish (White 1970). Similar patterns are found in Australia, on gentle slopes where overland sheet flow is an important phenomenon (Tongway and Ludwig 1990).

Canopy cover varies most in valleys because of varying soil textures and varying amounts of water collected. Flooding can occur as a result of heavy rainfall. Even in the northern Sahel zone, large lakes develop every year. Land degradation, leading to increased runoff from the uplands, has greatly increased the incidence of flooding. Flooding harms most woody plants. In flooded areas, a perennial grassland savanna develops. Poor drainage may also occur on shallow, fine soils and in depressions.

On sandy soils in valleys, soil aeration and drainage are adequate to support an abundance of woody plants. However, canopy cover in north Sahelian valleys dropped from 6 to 1.5% in the past 20 years, while canopy cover in south Sahelian valleys only dropped from 15 to 10%. On temporary lakes, woody species adapted to regular flooding survive along the edges, where the water is shallow.

Termite mounds have specific characteristics: fine soil particles, accumulated nutrients and organic matter (Lavelle et al. 1990). In the Sahel zones, termite mounds are usually bare because water infiltration is low. Water runs off the termite mounds, so water and nutrients tend to be available near them, creating favourable growth conditions for woody plants (Menaut et al. 1985).

The absolute availability of nutrients in the Sahel zones is too limited to enable woody plants to develop an enlarged tree-litter-soil nutrient cycle, which would enable the woody plant canopy to become dense enough to profit from light competition, nor to develop a root system taking advantage by water and nutrient interception. This explains why in the northern Sahel zone, woody plants cannot successfully compete with herbs. With increasing rainfall the absolute availability of nutrients gradually increases and as a consequence also the potential advantage of woody plants (Kessler and Breman 1991).

2.3.4 The Sudan Zones

In the Sudan zones, absolute water availability is higher, dry years are less extreme and less frequent and the groundwater table is more shallow. Under these conditions, woody plants have better possibilities of surviving the dry season. The canopy cover in the Sudan zones reach average values of 20–30% on sandy soils and 30% on other soil types. Locally canopy cover may reach as high as 70% on sandy depressions (Fig. 2.8), or drop to zero on shallow soils on bedrock, or on floodplains.

Soils in the Sudan zones are poor, and often shallow and rocky. Strings of woody plants occur in the grassy landscape, following cracks in the shallow bedrock or plinthite layer where roots can penetrate. The landscape remains open where conditions favour the herbaceous layer, especially on light and shallow soils. The deeper rooting perennials, replacing annuals within the herbaceous layer, compete more strongly with woody plants. Drought periods have less influence on woody plants in the Sudan zones, because more water is available (Fig. 2.4).

In the southern Sudan zone, woodland vegetation develops except on shallow or hardpan soils and floodplains. The average canopy cover in the fluvial landscape actually decreases as annual rainfall exceeds 1000 mm. The average canopy cover of the southern Sudan zone does not exceed 30%, because floodplains, shallow soils and fires are common.

South of the southern Sudan zone, in the Guinean climatic zone, average annual rainfall exceeds 1200 mm, and the seasons are less pronounced. To develop a closed forest, dry seasons must not persist

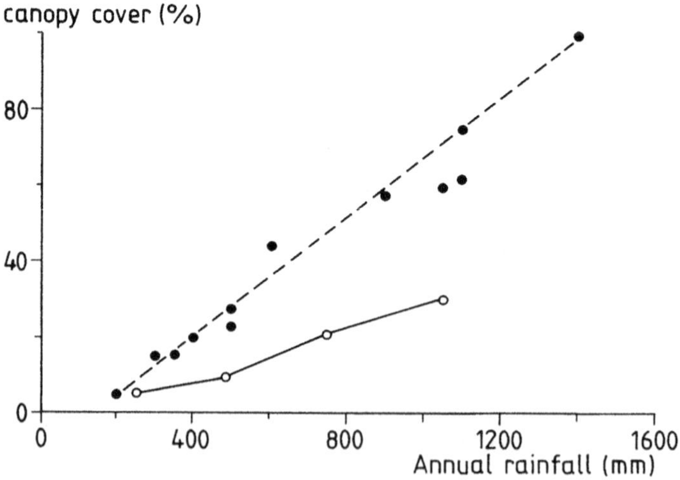

Fig. 2.8. Relation between average annual rainfall and average or maximum canopy cover values in the SSWA. ● Maximum canopy cover recorded; ○ average canopy cover

longer than 3 months (Hopkins 1965). With average rainfall of 1200 mm in the SSWA, the dry season may, however, still extend over 6 months! In West Africa, therefore, an average annual rainfall of 1700 mm on typical soils is necessary to develop a closed forest. A closed forest develops more easily on the more fertile and medium to heavy textured soils, and where rainfall distribution is less seasonal. Thus the absolute minimum for a closed forest in West Africa is 1400 mm annual rainfall on fertile soils (Hopkins 1965). With sufficient rainfall, closed forest communities could develop on all geomorphological units except for impoverished, well-drained sandy soils and soils subject to prolonged flooding.

In wooded savannas, fire strongly influences vegetation structure. In most wooded savannas, woody plants grow in clusters, surrounded by grasslands. The origin of such a distribution may be the occurrene of fires where grasses dominate, limiting the establishment of woody plants (Menaut 1983). The influence of fire lessens as the rainy and dry seasons become less pronounced. The influence of fire on woody plants is discussed in Section 2.4.

In the Sudan zones, adequate rain falls on termite mounds to wet them thoroughly, so vegetation can benefit from the nutrients available on the mounds. Certain woody species preferentially establish on the termite mounds. Thus a pattern develops which is opposite to that of the Sahel zones (Menaut et al. 1985; Lavelle et al. 1990).

With increasing rainfall, absolute nutrient availability also increases and tree roots have more potential to capture leached nutrients. Without the woody vegetation, the absolute availability of nutrients would decrease. On sandy soils, this phenomenon becomes apparent with rainfall exceeding 450 mm; on loamy and clay soils, the phenomenon becomes apparent with rainfall exceeding 850 and 1200 mm respectively, assuming little runoff. Here woody plants favoured by greater nutrient availability and light interception compete better with annuals. The increased plant biomass leads to increased soil organic matter and a higher cation exchange capacity (Sect. 4.6).

On infertile soils perennials compete better than annuals, because perennials store nutrients in seasons of high availability and translocate nutrients in seasons of need (Stuart Chapin 1980). The presence of woody plants provokes a horizontal heterogeneity of nutrient availability from which herbs may benefit (Sect. 4.5). For instance, the grass *Panicum maximum* reaches its highest cover under trees (Sect. 4.2).

2.3.5 Conclusions

In the SSWA, and increasingly so in the northern zones, conditions for woody plant establishment and survival are unfavourable. Canopy cover is lower than in East Africa, where total rainfall is similar, but better distributed. In the SSWA, there is a linear relation between average rainfall and maximum observed canopy covers (Fig. 2.8). Average canopy cover is much lower: during recent years, woody species distribution and canopy cover have dramatically changed in the land-use type "old fallow-lands, rangelands and forests" (Fig. 2.4). In the Sahel zones, canopy cover decreased up to a factor of 10, with important variations by geomorphological unit.

The recent drought periods led to a reduced average annual rainfall of 100–200 mm in the last 20 years. A comparison of pre-drought and post-drought data shows that north of about 500 mm rainfall in the southern Sahel zone, the reduction of canopy cover due to the drought periods is most dramatic.

It is questionable whether one can speak of a desertification of the Sahel zones, and a corresponding sahelinisation of the Sudan zones. Observations of woody species mortality may reflect natural, recurring phenomena. In the Sahel zones, woody vegetation is opportunistic. Species extend their limits in good rainfall years, and shrink or move their limits during droughts. A maximal canopy cover for the Sahel zones can be based on observations following the good rainfall years of 1920–1970. The vegetation dynamics in the Sahel zones are well described by non-equilibrium dynamics. Some observers noted that the transitions following droughts apparently do not involve crossing the thresholds which prevent restoration to the previous situation (Friedel 1991). However, human activities in the Sahel zones may impede the regeneration process after drought periods and thus can cause large scale transitions to the lands' current stable states of low productivity.

In the Sudan zones, drought influences several woody species, but overall canopy cover remains relatively constant because of shifts in species composition. In the Sudan zones, human influences have an increasingly important influence on woody plants, compared to natural influences (Sect. 2.4).

2.4 Human Influences on Woody Plants

2.4.1 Introduction

Four exploitative human influences on woody plants are considered: burning, livestock grazing and browsing, agricultural clearing and wood cutting.

The combined influences of fire, grazing and browsing created the African savannas long before humans lit fires or herded livestock (Geerling and Diakité 1988). Humans are now responsible for most fires, and domesticated herbivores damage more vegetation than wild herbivores ever did, because undependable forage and predators control wild herbivore populations (Taylor and Walker 1978; Werger 1983). Land clearing and wood cutting have no natural counterparts.

The following sections analyse short and long term, direct and indirect, influences of exploitative human land uses on woody vegetation. Indirect effects are mainly focussed on influences on long term nutrient and water availability in the ecosystem. Water and nutrient cycling are discussed in more detail in Chapter 4.

2.4.2 Fire

2.4.2.1 Direct Effects

The most important direct effects of burning depend on timing: whether the fire occurs early or late in the dry season. Burning is less destructive at the beginning of the dry season, when vegetation is moist. Some woody plants shed their leaves in a fire, then grow new leaves, others do not. Fires that burn early in the dry season kill small plants mainly. Burning is most destructive at the end of the dry season because vegetation is driest and it destroys trees and shrubs which translocated mineral reserves to growing buds and young leaves. Trees are likely to die in fires of the late dry season if herders lop off branches and leave them near the trees to serve as browse (Cissé 1984; Egging 1990). Late fires destroy most woody plants shorter than 2 m, and 20–50% of those from 2–5 m.

Perennial grasses grow nutritious leaves after early burning, but if burnt too frequently or too late in the dry season, perennials gradually produce less and are more likely to die. This gradual mortality of perennial herbs is related to exhaustion of nutrient reserves (Penning de Vries and Djitèye 1991). Woody plants respond in much the same way.

Meteorological conditions and fuel availability largely determine fire frequency and intensity. In the SSWA, long dry seasons and low air humidity cause rapid drying of litter and topsoil; these conditions favour fires. Fires are not important where rainfall is low: if less rain falls, less vegetation grows, so less fuel is available.

Soil type determines the pattern of the herbaceous vegetation. Fires are rare or absent when annual herbaceous production is less than $1.0\,t\,ha^{-1}$, even when the herbaceous layer is evenly distributed. In sandy eolian landscapes with predominantly sandy soils, fires therefore occur with 300 mm of annual rainfall only, if the herbaceous biomass is evenly spread. On loamy and rocky soils which dominate the skeletal landscape, irregular fuel distribution limits fires to zones with more than 500 mm annual rainfall. Fires may be expected annually when annual herbaceous layer production exceeds $2.5\,t\,ha^{-1}$ (Penning de Vries and

Djitèye 1991), but this figure depends on the extent of herbaceous layer exploitation. If the herbs are grazed, fire is less likely, but if it is not exploited, litter accumulates, and fire is more likely. Although abundant vegetation grows in depressions and valleys of fluvial landscapes, biomass remains moist until late in the dry season. At the early dry season, the distribution of dry biomass is still too patchy or scarce to sustain a big fire. Floodplains are often deliberately burnt at the end of the dry season.

The influence of fires on past and present distribution and canopy cover of woody plants are considered by climatic zone and geomorphological unit (Tables 2.4–2.7). In the Sahel zones, fire-related influences, and differences between fire regimes, are less important. Where (over-) grazing severely reduces herbaceous biomass, fires can occur in years of good rainfall only. Fires are rare in the northern Sahel zones, but 15% of the southern Sahel zone is burnt every year (Penning de Vries and Djitèye 1991). Here, woody plant scarcity on floodplains is partly due to deliberate and frequent burning late in the dry season. In the northern Sudan zone, 60% of natural vegetation burns each year, in the southern Sudan zone 90% (Fig. 2.9). Land clearing and grazing increasingly limit fuel availability, creating fuel distribution patterns less vulnerable to fire.

In Burkina Faso, nearly 80% of burning takes place in November and December, and only 6% of fires occur at the dry season's end (Parnot

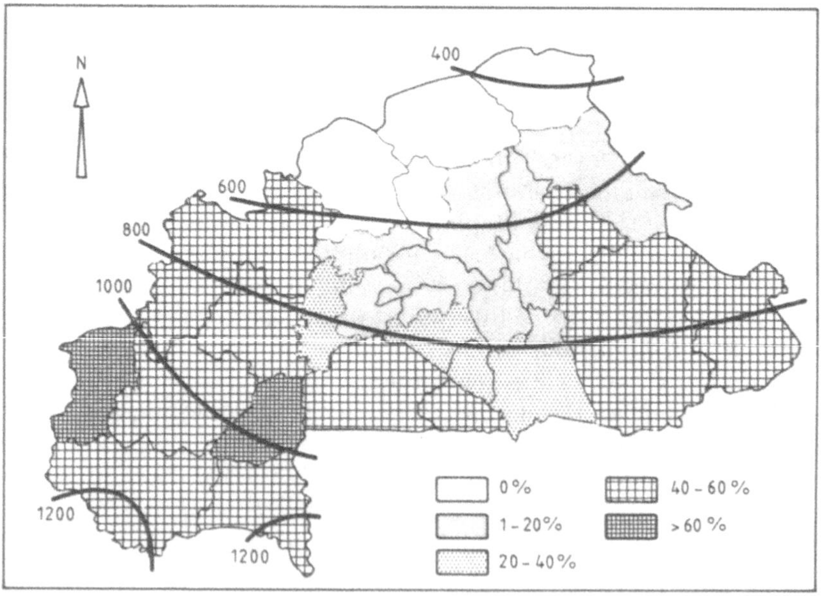

Fig. 2.9. Average annual rainfall isohyetes in Burkina Faso (in mm), and percentage of land burnt after 1986 dry season, per Province. (After Parnot 1988)

1988). However, Egging (1990) found that in the southern Sudan zone of Burkina Faso fires occur between October and April, with about an equal area burnt each month. At the end of the dry season, 90% of the rangeland area is burnt.

In nutrient poor ecosystems, woody plants have mechanisms for rapid nutrient recovery and reduction of nutrient losses (Boerner 1982; Adams and Attiwill 1984). Adaptations to fire include thick bark (*Vitellaria paradoxa*, *Cussonia barteri*), and regeneration after burning, because the growing tip lies below the soil surface (*Pterocarpus* spp., *Combretum* spp. and *Gardenia* spp.). Some woody species have deep taproots, lignotubers or extensive underground organs (Werger 1983; Frost 1985a; Stromgaard 1986) and some have a long period of seed and fruit dispersal (Tolsma 1989). Fruits and seeds of many woody plants are sensitive to fires, such as the Combretaceae, Caesalpiniaceae and Mimosaceae. Other species need fire to break seed dormancy (Tybirk 1991).

Effects on woody plants of three fire regimes, no burning, early dry season annual burning and late dry season annual burning, are quantified for the southern Sudan zone, summarizing data from three long-term studies (Tables 2.8 and 2.12). No burning is clearly most beneficial for woody plants productiviy and diversity, and the prevalence of annual burning in the Sudan zones is probably the most important factor explaining why average canopy cover does not exceed 30%, although most woody plants have adaptations to fire.

2.4.2.2 *Indirect Effects*

Burning leads to volatile nutrients being lost from organic material and soil (nitrogen, sulphur and carbon mainly). Non-volatile nutrients are partly lost in smoke, or are leached or eroded. In the Sudan zones, frequent burning late in the dry season results in more soil erosion ($0.4\,t\,ha^{-1}$), than burning early in the dry season ($0.1\,t\,ha^{-1}$) or no burning ($0.05\,t\,ha^{-1}$) (Roose 1981). The short-term effects of burning are positive for topsoil nutrient availability: burning brings about mineralisation of organic matter, which increases nutrient availability to plants; but wind and water can rapidly disperse mineralised nutrients in the ash.

Burning stimulates regeneration and growth of many plants, partly because burning makes nutrients more available, and partly because burning removes shading litter (Attiwill 1981; Boerner 1982). Perennial grasses, for instance, grow well after burning, unless previous exploitation has depleted their reserves (Breman and de Ridder 1991).

Much carbon is lost in fires. Nevertheless, minor fires may increase topsoil carbon availability, largely as a result of partly combusted material. Additionally, newly available nutrients may stimulate plant productivity and the resultant growth increases carbon in the topsoil (Raison 1979).

In semi-arid regions, less than half of the nitrogen is lost in a fire (Boerner 1982; Mott et al. 1982; Szott et al. 1991). Even during extremely hot fires, only the most shallow layer of soil becomes hot enough to volatilise nitrogen (Trapnell et al. 1975; Raison 1979; Ram and Ramakrishnan 1988). A minor fire may increase topsoil N availability. This is probably due to the improved P availability, stimulating nitrogen fixation and plant productivity (Raison 1979). Increased microbial activities may promote nitrification, stimulating nitrogen recovery (Ram and Ramakrishnan 1988). Burning may destroy or neutralise soil and litter toxins that inhibit nitrification in unburned vegetation (Boerner 1982).

In temperate, semi-arid regions 10% of total phosphorus is lost by fire (Boerner 1982). On tropical forests, burns improve the availability of mineralised phosphorus up to 500% (Nye and Greenland 1960; Nair 1984). In savanna ecosystems, up to 25% of mineralised phosphorus is lost in smoke, while some is immobilised by cation exchange capacity adsorption and P fixation. Burned plots show an increase in calcium and magnesium in the topsoil (Kellman et al. 1985).

Long-term experiments in semi-arid Zambia show that soil organic matter content and N concentrations do not differ under continuous early or late burning regimes from those in protected soil (Trapnell et al. 1975), because termites remove unburnt litter from the protected sites. Concentrations of phosphorus, calcium and magnesium are higher shortly after burning.

Fires influence the competitive relationship between the herbaceous and woody layers. In the Sudan zones, absence of fire leads to increased woody plant diversity and canopy cover (Table 2.8), and probably a reduced herbaceous production. In general, fire favours the herbaceous layer, which rapidly benefits from the increased nutrients available in the topsoil, and from the increased light available when shading vegetation and litter is burnt (Boerner 1982; Werger 1983; Penning de Vries and Djitèye 1991).

Absence of fire stimulates development of the woody vegetation but reduces mineral cycling. Burning reduces canopy cover and woody plant productivity in the Sudan zones (Tables 2.8 and 2.12), leading to a

Table 2.8. Influence of three fire regimes on woody plants in the southern Sudan zone. (Ramsay and Innes 1963; Afolayan 1979; Brookman-Ammissah et al. 1980)

	Without fire	Early burning	Late burning
No. of species	32–43	14–33	13–16
Large shrub and tree density ha^{-1}	400–800	160–400	80–140
Seedlings and small shrub density ha^{-1}	1000–2500	500–1500	0–500

greater proportion of the nutrient pool stored in the soil rather than in the biomass. Although immediate effects are unclear, frequent burning and burning late in the dry season reduces vegetation cover and soil biological activities and gradually leads to soil erosion and degradation. Nutrient losses outweigh the gains of the herbaceous layer resulting from an improved ability to compete with woody plants. The vegetation in the Sahel zones is so degraded that burning should be avoided whenever possible.

2.4.3 Livestock Grazing and Browsing

2.4.3.1 Direct Effects

To study the direct effects of grazing on woody plant production and survival, researchers throughout the world have attempted to simulate herbivore browsing by cutting or lopping leaves, twigs and small branches. These studies do not simulate all aspects of browsing, such as the selective browse of certain plant organs, trampled soil, dung, and herbivore saliva, which is thought to stimulate leaf growth (McNaugton 1983).

The effects of exploitation on woody plants depend on the amount of water and nutrients removed, the season and frequency of removal, and plant ability to endure stress, which varies by species and individual, but which improves with favourable growth conditions (Garrison 1953; Lay 1965; Leigh and Mulham 1970; Willard and McKell 1978; Thalen 1979; Benjamin 1980; Menke and Trlica 1981; Tothill and Mott 1985; Roy 1991). From these studies it can be concluded that light exploitation of 25–50% foliage removal, late in the growing season, may double foliage production for a limited time; 25–50% foliage removal, early in the growing season, lengthens the duration of foliage production over that of untreated plants. The plants which have some foliage removed, allocate carbohydrates and nutrients to meristems, possibly at the expense of other tissues. Intensive exploitation early in the growing season kills woody plants after several years, even if short-term production seems promising. Woody plants are not harmed by a combination of heavy dormant season exploitation followed by an absence of growing season exploitation, probably because root growth and the restoration of carbohydrate reserves proceed unhindered.

Cutting experiments conducted in Niono, Mali from 1978–1984, at 570 mm annual rainfall, studied the influence of season and frequency of cutting and mode of exploitation on the survival and productivity of three fodder trees (*Pterocarpus lucens*, *Faidherbia albida* and *Acacia seyal*) and three fodder shrubs (*Combretum aculeatum*, *Cadaba farinosa* and *Feretia apodanthera*) (Tables 2.9 and 2.10). Regular cutting of

Table 2.9. Influence of the frequency and seasonal timing of cutting all secondary branches of three fodder species on mortality (†) and foliage production (FP) (including production prior to mortality), expressed as a percentage of the mortality and production of untreated specimens. (Cissé 1980, 1984)

Cutting frequency per year	Season: before or after rains	*Pterocarpus lucens*		*Faidherbia albida*		*Acacia seyal*	
		† (%)	FP (%)	† (%)	FP (%)	† (%)	FP (%)
2	Before + after	100	−75	30	+103	100	−33
1	Before	100	−89	0	+ 67	100	− 7
1	After	0	−68	0	+184	100	−45
1/2	Before	10	−76	0	+ 10	100	−91
1/2	After	10	−64	0	+115	40	−43

Table 2.10. Influence of defoliation regimes on foliage production (FP), expressed as percentage of foliage production of untreated shrubs, and on plant recovery in the year after treatment (RE), number of shrubs expressed as percentage of the number in the year of treatment. (Cissé 1980, 1984)

Defoliation regime	*Combretum aculeatum*		*Cadaba farinosa*		*Feretia apodanthera*
	FP (%)	RE (%)	FP (%)	RE (%)	FP (%)
PD^{30} 0	51	8	67	34	174
TD^{30} 0	42	46	77	44	133
TD^{15} 0	29	0	44	23	84
TD^{15} 1	50	26	70	40	147
TD^{15} 2	92	68	35	43	119
TD^{15} 3	105	122	22	14	105
TD^{15} 4	67	146	11	27	78
TD^{15} 5	43	199			77

PD = partial defoliation.
TD = total defoliation.
15–30 = defoliation once every half month or complete month.
0, 1, 2, . . . = first defoliation at start of rainy season, 1 month later, 2 months later etc.

secondary branches of *Pterocarpus lucens* and *Acacia seyal* reduce foliage production, irrespective of the period and frequency of cutting. *Pterocarpus lucens* production increases slightly, but only after the first cut. Cumulative foliage production of untreated plants surpasses that of plants which had foliage cut. Most *Acacia* die after 4 years of cutting, and most *Pterocarpus* die by the end of the trial. Cumulative foliage production of *Faidherbia* was stimulated by regular cutting, although two cuttings per year result in 30% mortality after six years (Table 2.9). Fruit production of cut *Faidherbia* is surpassed by that of untreated trees. All three species are harmed by being cut twice a year. Low-frequency cutting after the growing season is less harmful, or even beneficial to *Faidherbia*. *Pterocarpus* shows a prolonged foliage production into the dry season due to cutting, an advantage from the point of view of the pastoralist (Cissé and Wilson 1984). If continued, the 6-year experiment would probably have led to increased mortality.

Acacia seyal does not die when 30–50% of its branches were cut each year for 15 years. After 3 years, however, the treated trees produce less wood and foliage than untreated trees. When all branches were lopped, *Acacia seyal* die after several years (Toutain and Piot 1980). When trunks were cut through, *Acacia seyal* promptly die.

The three fodder shrubs were also partially defoliated and cut once a month. This regime supports productivity and survival better than the other cutting regimes (Table 2.10). More foliage grows back when defoliation or cutting takes place during the growing season, as compared to the dormant season. *Cadaba farinosa* does not show increased regrowth after any period of defoliation, probably because it was first defoliated at an advanced stage of its plant cycle. Shrub recovery in the year after defoliation is best if defoliation occurred late (Combretaceae), or recovery is poor anyhow (*Cadaba*).

Fifty percent removal of peak foliage production in a typical southern Sahel woody vegetation is estimated to lead to a 30% reduction in productivity in the following year, when lost phosphorus is not balanced with additional P uptake. Fruits are particularly nutrient-rich. Exploitation of the annual fruit production of *Faidherbia albida* trees equals an annual loss of phosphorus equal to 3.5 times the total aboveground P store fixed in wood and branches (Tolsma 1989).

Woody plants vary considerably regarding their susceptibility to exploitation. Some woody species, such as *Maerua* spp., *Boscia* spp., *Cadaba* spp. and *Grewia* spp., are easily harmed by exploitation (Toutain and Piot 1980). Other species are more hardy: *Acacia seyal*, *Acacia tortilis*, *Combretum aculeatum* and *Balanites aegyptiaca* quickly resprout after defoliation. Several *Acacia* species in Tanzania have chlorophyll in their twig and branch epidermis; after defoliation, photosynthesis can proceed (Pellew 1980). The same phenomenon can be observed for *Balanites aegyptiaca* in the SSWA. Spines and thorns provide some protection against browsing, but do not protect fully against camels and goats. *Balanites aegyptiaca* and several *Acacia* species develop stunted branches and stiff thorns, and assume compact and impenetrable habits (e.g. Seif el Din and Obeid 1971). Chemical deterrents to browsing include polyphenols, which reduce foliage palatability and digestibility (Sect. 3.6). The protein content of foliage from *Cadaba farinosa* and *Combretaceae aculeatum* shows an inverse relationship with frequency of defoliation, reducing their palatability, and suggesting that the internal nitrogen reserves are being exhausted (Piot 1980).

2.4.3.2 *Indirect Effects*

The long-term influence of intensive grazing on the survival and productivity of woody vegetation depends on both direct and indirect effects. Indirect effects include influences on water availability, vegetation structure, competition with the herbaceous layer, the survival of seeds and seedlings and vegetation composition.

On the arid Egyptian coast, shrub browsing results in reduced leaf transpiration in the rainy season, so water percolates into deep subsoil storage. Shrubs use the stored water to extend foliage production through the dry season and into the next growing season. Apparently, soil fertility is not a limiting factor. Less subsoil water is stored under shrubs which were not browsed. Although browsed and not browsed shrubs produce similar amounts of foliage, the foliage of the not browsed shrubs dries out and is useless as forage, contrary to the browsed shrubs. The positive response to browsing depends on the shrubs having adequate defenses against browsing (van Duivenbooden 1985).

On sandy soils, intensive grazing reduces the herbaceous cover, so more water percolates to the subsoil, and is available to the roots of woody plants. On loamy, crust-forming soils, the reduced herbaceous cover permits increased runoff and concentration of water and nutrients in depressions. Although on a regional scale water availability is reduced, water and nutrient availability in depressions may double that of areas which are not grazed (Sect. 4.5.4; Table 4.16) and a greater share of the water is stored in the subsoil, accessible to deep rooting woody plants.

Bush encroachment refers to the replacement of a savanna or woodland vegetation with trees towards a dense, impenetrable, often spiny bush vegetation with a low species diversity of unpalatable shrubs mainly. This shrub vegetation constitutes a natural protection against grazing. Bush encroachment processes have been described for northern America since long ago and were ascribed to the influence of grazing mainly (Buffington and Herbel 1965). It is also well known from East and South Africa (Walker et al. 1981; Tolsma 1989). Bush encroachment can be a rapid process: in Botswana woody biomass tripled within 25 years due to the increase of shrub canopy cover from 4 to 17% (van Vegten 1983). Though the process of bush encroachment is not completely understood, it is related to a reduction of water competition between woody and herbaceous plants due to intensive grazing of the latter. Bush encroachment is likely to occur on sandy soils where water reserves remain available in the subsoil. Sandy soils should be at least 2 m deep, so that water can be stored below the reach of herbaceous roots, to obtain an equilibrium between herbs and woody plants in a grazed situation (Mott et al. 1982). On loamy, crust forming soils, clusters of woody vegetation also develop in the run-on depressions.

These phenomena are influenced by rainfall distribution, droughts, shrub water use efficiency and vulnerability to grazing, as well as grazing frequency and intensity. Bush encroachment is more likely in conditions of higher, more evenly distributed annual rainfalls (Walker and Noy-Meir 1980; Walker et al. 1981; Werger 1983; Tolsma 1989). Some trampling and moderate grazing can break up soil crusts (Walker et al. 1981) but too intensive grazing can have an opposite effect and would stimulate erosion. Subsequent water concentration in run-on sites can stimulate woody plants.

Grazing stimulates woody plant seed germination and seedling establishment. Herbivores disseminate seeds and pods of many species. Woody legumes depend on herbivores and birds as the primary or secondary means of seed dissemination (Tybirk 1991). Many seeds disseminated by herbivores can only germinate well after ruminal acids have eliminated dormancy. *Acacia raddiana* germination also depends on whether seeds have passed through a herbivore rumen because if they have not, insects infest them (Lamprey et al. 1974). Additionally, favourable microhabitats for woody seedling growth are provided by animal dung and reduction of competition by the herbaceous cover caused by grazing (Walker et al. 1981; Archer 1989; Tolsma 1989; Tybirk 1991).

Herbivores may favour certain woody species. The rapid spread of *Prosopis glandulosa* in the United States was associated with the introduction of livestock (Archer et al. 1988). In good rainfall years in the southern Sahel zone of Senegal, grazed spiny shrubs regenerated better than shrubs protected against grazing. Other woody plants, however, do not regenerate well when grazed, and prolonged intensive grazing eliminates seedlings of all species (Miehe 1990). Some researchers argue that woody vegetation recovery after droughts requires total protection against grazing (Cornet and Poupon 1977; Bille 1980; de Wispelaere 1983). However, factors such as grazing intensity, frequency and seedling composition must be considered.

To study long-term effects of grazing on woody plants, single regions may be submitted to different grazing regimes, and comparative analyses can be made of regions with different grazing intensities.

In the SSWA, studies of the first type include monitoring of protected enclosures established on formerly heavily grazed land. Regardless of soil type, protection leads to an increased density of woody plants without shifts in woody plant composition (Table 2.11). However, neither Cissé (1983b) nor Granier (1977) included a comparison with vegetation dynamics outside the exclosures. Toutain and Piot (1980) found that on loamy-clay soils in the southern Sahel zone regeneration inside the exclosure was better than on control plots, but on sandy soils the opposite was true. As these studies were conducted in years after the first drought period, regeneration of woody plants in the exclosures is

Table 2.11. Influence of soil texture on regeneration of woody plants protected from grazing, in the southern Sahel zone: initial woody plant density ($n\,ha^{-1}$) and percentage density increase (%) in relation to period of protection

Sand		Loam		Loamy-clay		Clay		Source	Years of protection
$n\,ha^{-1}$	(%)	$n\,ha^{-1}$	(%)	$n\,ha^{-1}$	(%)	$n\,ha^{-1}$	(%)		
149	+94	309	+18	561	+110	241	+54	Cissé (1983b)	5
31	+77[a]			217	+53[b]			Toutain and Piot (1980)	3
3	+100	198	+7			703	+29	Granier (1977)	1

[a] in control plot 116% increase.
[b] in control pott 10% increase.

Table 2.12. Influence of two grazing and three burning regimes, in the northern Sudan zone in Cameroon, on wood production (WP) and percentage wood volume increase (VI) of the natural vegetation. (Peltier and Eyog-Matig 1989)

	No grazing		Grazing	
	WP ($m^3\,ha^{-1}\,yr^{-1}$)	VI (%)	WP ($m^3\,ha^{-1}\,yr^{-1}$)	VI (%)
No fire	0.50	55	0.39	68
Early burning	0.15	12	0.20	28
Late burning	0.18	18	0.41	40

probably determined by better rainfall in the first place, especially in the Sahel zones.

Woody plant regeneration is better on sandy soils than on loamy soils. On loamy, crust forming soils, recovery is often hampered by reduced water infiltration, even in well protected enclosures. Such a situation can be interpreted as a threshold having been crossed towards a less productive state (Westoby et al. 1989). To recover the previous vegetation, the soil crust must be broken up. Regeneration on sandy soils is easier, provided sufficient seed is available.

In the northern Sudan zone of Cameroon, a 3-year study (Peltier and Eyog-Matig 1989) showed that grazing stimulates wood production, possibly because fire intensity and competition with the herbaceous layer are reduced (Table 2.12). In the southern Sudan zone of Mali, the response to grazing protection depends on the initial vegetation cover (van der Poel and Kaya 1990). After 4 years of protection, initially bare sites remain bare, but on sites with at least 10% herbaceous or woody cover, both increase proportionately with annual rainfall.

Table 2.13. Influence of grazing intensity on canopy cover (% CC), per climatic zone and geomorphological unit, with indication of those woody species showing increased canopy cover due to grazing

Climatic zone	Geomorphological unit[a]	Grazing intensity		Increasing species
		Moderate (% CC)	High (% CC)	
Northern Sahel	Os, Ps	0–5	0–20	*A. raddiana*, *Balanites*
	Pl	0–10	0	
	La	0–5	0–5	
Southern Sahal	Os, Ps	2–6	0–40	*A. raddiana*, *Balanites*, *Combretaceae*
	Pl	5–15	0–5	
	La	5–25	0–40	*A. nilotica*, *A. seyal*
Northern Sudan	Ps	10–15	5–30	*Acacia* spp., *Combretaceae*
	Pl	20–30	5–40	*Piliostigma reticulatum*
	La	30–50	10–75	*Acacia* spp., *Combretaceae*

[a] Section 2.1.3.

Studies of the second type, comparing vegetation characteristics of regions which differ in grazing intensities, are difficult because environmental conditions vary between regions. Table 2.13 is based on an interpretation of data used for Tables 2.4–2.7, for sites of known grazing intensities. The influence of grazing intensity on woody plants is evaluated per climatic zone and geomorphological unit. Two categories of grazing intensity are distinguished: moderate grazing intensity refers to a typical grazing intensity far from villages, and high grazing intensity refers to a grazing intensity near villages or water points. Woody plants vary in their response to intensive grazing (Table 2.13). On sandy soils or fluvial landscapes, intensive grazing may lead to an increase in canopy cover, but a strong reduction is also possible. On loamy soils, especially in dry zones, canopy cover is reduced by intensive grazing because infiltration is reduced. However, highest canopy covers occur on fallowlands and lands near natural or artificial water points (i.e. where grazing pressure is high). Higher rainfalls favour the more positive influences of grazing on woody plants.

Table 2.13 lists woody species that increase when grazed. The increase is probably related to bush encroachment. *Acacia raddiana* and *Balanites aegyptiaca* increase when grazed in the Sahel zones. The *Combretaceae* and *Piliostigma* spp. dominate intensively grazed areas in the Sudan zones. *Detarium microcarpum*, *Parinari curatellifolia*, *Hymenocordia acida*, *Acacia* spp. and Combretaceae increase by grazing in the southern Sudan zone (Hoffmann 1985).

Many people believe that grazing and browsing harms woody plants. On the contrary, both direct and indirect effects may be positive. Grazing may stimulate woody species in a process of bush encroachment. Bush encroachment is however much less common in the SSWA than in East African regions of similar annual rainfall. The factors which contribute to the difference are several.

1. In East African regions of similar annual rainfall, rains generally fall twice a year: droughts are less severe and conditions for woody plants are more favourable (Sect. 3.5). To stimulate in the SSWA, particularly in the Sahel zones, the woody cover against the odds of the natural environment, a higher grazing pressure is required to sufficiently reduce the (competition from) herbage layer. With higher rainfalls in the Sudan zones, natural conditions become increasingly favourable for woody plant establishment (Sect. 3.4).
2. In the SSWA, rangelands and water resources are available for one limited period per year. Whereas in the Sahel zones low forage limits livestock densities and leads to pastoralists taking their livestock south in the dry season, in the Sudan zones low pasture quality limits livestock densities. Therefore, the SSWA has overall lower densities of wild and domestic herbivores than East Africa, where rains fall twice a year and fertile soils are more common (Breman and de Ridder 1991). Grazing pressure in the SSWA is lower because herbivore densities are lower, and because grazing occurs for only a limited period per year. In the Sahel zones, bush encroachment occurs mainly near villages and water points where livestock is concentrated. Bush encroachment is more common in the Sudan zones.
3. Following a period of gradual build up of livestock densities, livestock soon overgrazed the land when droughts occurred. The rapid change of exploitation pressure has not allowed a change in woody plant composition, towards woody species better adapted to drought and grazing pressure. Bush encroachment can only occur where such unpalatable and/or grazing-resistant species were already there. In East Africa, wild herbivores have grazed the land intensively for millennia. Drought and grazing-resistant species are more common there, presumably because they had time to adapt to the natural conditions.
4. Due to the single annual peak of rainfall in the SSWA, rain-fed agriculture extends further into the arid zone than it does in East Africa. Land clearing and population densities are therefore more pronounced in arid regions of the SSWA. Agricultural activities are concentrated on the most productive sites, where bush encroachment would otherwise be most likely to occur.

5. On loam and clay, where conditions for woody plants are more favourable than on sand, the soil is easily degraded by overgrazing. Water infiltration and vegetation cover easily decline, especially in the Sahel zones.
6. Increased production related to bush encroachment near settlements and water supplies is often masked by firewood and fodder harvesting. The harvesting opens up the vegetation so that livestock can penetrate. Bush encroachment can be considered a natural mechanism to prevent land degradation, but the combination of bush encroachment and intensive wood harvesting leads to severe degradation. Bush encroachment, and the related reduction of the herbaceous layer, leads to lower vegetation resilience, i.e. an increased sensitivity to errors in management and exogenous events (Walker et al. 1981). Woody species in the SSWA, even those relatively resilient to browsing, are threatened by the increasing proportion of goats in livestock herds.

2.4.4 Agriculture

In the SSWA farmers used to leave some useful trees on croplands. In the Sahel zones, woody species producing useful products and providing shade are preferred; other trees are cut to knee height and burnt with their branches. Some of the burnt trees die, most resprout every year. With the increasing intensification of crop cultivation, including permanency of cropland use and oxen- or mechanised ploughing, shrubs and trees are increasingly uprooted or rootstocks gradually die.

On the parklands, average canopy covers range from 0–10% in the Sahel zones to 5–20% in the Sudan zones (Table 2.14; Sect. 2.1.2). Average densities range from 5–20 trees per ha. In the Sahel zones, particularly on sandy soils, it is not uncommon to find only one or few trees per cropland. The highest parkland tree densities on record are 300 *Borassus aethiopium* per ha and 50 *Faidherbia albida* per ha. Both trees show little light competition with crops but require specific site conditions. *Borassus* grows only in relatively humid climates. As a result of its erect structure, it casts little shade. *Faidherbia* depends on groundwater. It has leaves in the dry season and loses them in the rainy season. Cereal yields under *Faidherbia* canopy equal or exceed that of the open field (CTFT 1988).

Most common parkland trees shade crops below their canopies. Below *Vitellaria paradoxa*, the most common parkland species, and *Parkia biglobosa*, competition for light, and possibly water and nutrients, leads to yields lower than those of the open field (Kessler 1992). In Burkina Faso *Parkia biglobosa* branches are sometimes cut to reduce shading and

Table 2.14. Tree densities in parklands dominated by *Faidherbia albida*, *Vitellaria paradoxa* or *Parkia biglobosa*

Region		Specification	Density (n ha^{-1})	Source
Faidherbia albida				
Burkina Faso	– Yatenga	Village territory	4	Marchal (1983)
Senegal	– Serer	Oldest occupations	50	Pullan (1977)
		Recent settlements	26	Pullan (1977)
Senegal		Specific region	16	Pullan (1977)
Senegal	– Wolof	Specific region	1–20	Pullan (1977)
		Large region	2–10	Pullan (1977)
Niger	– Interior delta	Oldest occupation	20–15 (17% cover)	Pullan (1977)
		Recent settlements	5–10 (5% cover)	Pullan (1977)
Sudan		Jebel Mara Plateau	12–90	Pullan (1977)
Mali		Bandiagara Plateau	40–50	Pullan (1977)
Vitellaria paradoxa				
Ghana	– Northern region	Large region	1–7	Pullan (1977)
Nigeria	– Kano	Large region	3–6	Pullan (1977)
Mali		Very large region	1–40 (mean = 15)	Ruyssen (1957)
Burkina Faso	– Central Plateau	Village territory	2–20 (mean = 7)	(pers. observ.)
			1–10% cover (mean = 5%)	(pers. ovserv.)
Parkia biglobosa				
Nigeria	– Zaria	Large region	1–14 (mean = 6)	Pullan (1977)
	– Baskararea	Large region	1–3 (mean = 2)	Pullan (1977)

enhance fruit production. The regular dispersion of parkland trees allows some lateral light under the canopies (Sect. 4.7).

Expansion of cropland in the Sahel region may be expected to follow the trend of population growth as most inhabitants have to be self-sufficient and cereal grain yields per unit of area have not increased during recent years. Following population growth, at least 2.5% additional land must be cleared annually. Population is growing fastest in the southern Sahel and the northern Sudan zones, so here most additional land is cleared on marginal land (shallow, rocky, sandy soils and erodible slopes). Agriculture is also spreading north, especially in good rainfall years. Since the 1950s, the limit of crop cultivation has moved 150 km northward, from the 400 mm to the 250 mm isohyet in Niger (Breman and Traoré 1986; Le Houérou 1989). As a result of migration from densely populated regions, much land is cleared in the southern Sudan zone. During the last 30 years, annual cropland expansion in the southern Sudan zone of south-eastern Burkina Faso was 3.3% (CRTO 1988).

Before recent droughts, fields were cultivated for 2 to 10 years, with fallows of 5 to 20 years. Restoration of soil fertility requires about 4 years of fallow for each year of cultivation (Kessler and Ohler 1983). In the Sudan zones, woody plants regenerate more quickly during fallows than they do in the Sahel zones, mainly from underground rootstocks. In the northern Sudan zone, 5% canopy cover can increase to 25% during the first years of fallow, and further increase to 30–40% after 15 years (Aweto 1981; Ohler 1985). In the Sudan zones, shrub canopy cover on young fallowlands can exceed 75%. Such canopy covers exceed those of rangelands, because croplands are selected for good soils and heavy grazing triggers the process of bush encroachment (Sect. 2.4.3.2). Undisturbed fallowland vegetation gradually develops into tree or shrub savanna.

Old fallowland vegetation is different from natural vegetation: useful trees are more common and woody plants vulnerable to grazing are absent (Aweto 1981; PIRL 1988). In the Sudan zones, dominant woody plants in old fallowlands are *Vitellaria paradoxa*, *Daniellia oliveri*, *Adansonia digitata* and the Combretaceae. Woody plants common in natural vegetation are *Isoberlinia doka*, *Afzelia africana* and *Burkea africana*.

Since the drought periods, fallows long enough to favour sustainable cropping are rarely observed. The regeneration and survival of woody plants are reduced by short fallows and mechanisation. Wind erosion on sandy soils (particularly in the Sahel zones) and physical degradation of over-exploited croplands also retard woody plant regeneration. Tree densities on croplands may seem to have remained constant during recent years (e.g. Marchal 1983). However, the tree communities show signs of gradual degradation, having mature trees with too few seedlings, saplings and young trees coming up (Fig. 2.10). In croplands of the northern Sudan zone in Burkina Faso, hardly any (natural) regeneration of *Tamarindus indica*, *Parkia biglobosa* and *Bombax costatum* was

Fig. 2.10. Frequency distribution of karité (*Vitellaria paradoxa*) trees per trunk diameter class, in a parkland region in Burkina Faso. (Gijsbers et al. 1994)

observed (Gijsbers et al. 1994). These species may require specific site conditions, or may be most susceptible to drought and browsing. Most young trees are however deliberately cut. In the Niger delta in Mali, high *Faidherbia albida* densities in the 70–80 cm diameter class corresponded to a period of human depopulation, during which presumably less trees were cut (Gallais 1967).

2.4.5 Wood Exploitation

Firewood supplies the majority of the energy needs in developing countries, 90% in the SSWA. Any woody plant is suitable as firewood, although certain species are preferred for qualities such as desired burning temperatures for different foods and smoke that is less harmful to eyes and lungs. There are, however, only a few woody species providing good construction wood. Examples are *Mitragyna inermis*, *Anogeissus leiocarpus* and *Terminalia* spp. Threats are therefore most acute for construction wood species in densely populated areas (Kessler and Boni 1991). People are particularly willing to plant trees that provide construction wood. Exotics commonly planted for this purpose include *Azadirachta indica*, *Eucalyptus camaldulensis* and *Cassia siamea*, which grow rapidly under various site conditions. The species compete well against the herb layer because of their shallow, extensive root systems (Sect. 4.3).

Most wood is obtained from the natural vegetation. Rural inhabitants use about 1 m^3, or 850 kg, per year (Keita 1982). When wood reserves are scarce, this drops to 300 kg per year, with more than 75% used as firewood. Apparently, reductions to this level do not significantly influence feeding patterns (Kessler and Boni 1991; Wilson 1980). It is assumed that, at average, farmers use about 0.5 m^3 (425 kg) of wood per year, while herders use about 0.2 m^3. Urban inhabitants use more firewood than the rural inhabitants, even though kerosene and gas supply some of the demand for urban cooking fuel.

The majority of wood originates from clearing for croplands. In Burkina Faso, in 1988, the wood cleared from forests and old fallowlands could have supplied the entire national firewood demand. Only around urban centres wood is cut for the purpose of energy needs only. At present rates, wood demand will soon be greater than the amount of wood cleared from forests and old fallowlands.

Annual wood production estimates of the natural vegetation per climatic zone (Sect. 3.2) are used to calculate areas required to supply annual wood requirements. For the present situation, these estimates of wood production may be 1.5–2.0 times too large, as they are based on canopy covers measured after the first drought period on well managed fallowlands (Tables 2.6 and 2.7). Table 2.15 shows the population per

Table 2.15. Wood production and wood consumption per climatic zone; and radii of areas required to cover wood requirements of cities

	Northern Sahel	Southern Sahel	Northern Sudan	Southern Sudan
Annual wood production ($m^3 ha^{-1}$)				
- range	0.0–0.2	0.0–0.4	0.2–1.0	0.7–2.5
- average	0.1	0.2	0.6	1.6
Average wood consumption (m^3 $person^{-1}$ yr^{-1})	0.2	0.5	0.5	0.5
Area needed to cover wood requirements (ha $person^{-1}$)	2	2.5	0.8	0.3
Radius around a city needed to cover wood requirements (km)				
- 1000 inhabitants	4	3	1.5	1
- 10 000 inhabitants	12	9	5	3
- 100 000 inhabitants	40	28	16	10
- 1 000 000 inhabitants	125	90	50	30

climatic zone, based on sustainable wood supply. With respect to wood production and wood requirements per person, the 1990 population densities in all zones are lower than sustainable population densities (Table 2.2). However, wood requirements are not evenly spread; unsustainable harvests of wood occur in densely populated rural areas in the southern Sahel and northern Sudan zones mainly. To supply a Sahelian city of 10 000 persons with a sustainable wood harvest, a 10-km radius, 314 km^2, is required (Table 2.15). As wood is harvested further from cities, resource depletion reaches deeper into rural areas. For instance, wood is gathered up to a 100 km radius from Ouagadougou, Burkina Faso. On the other hand, in and around cities, farmers plant trees to produce marketable construction wood. For example, planting has created a sort of green belt around the city of Kano, Nigeria (Cline-Cole et al. 1988).

2.5 Conclusions

In the SSWA recent changes in canopy cover have been dramatic (Fig. 2.4). Variations in the reduction of canopy cover can be related to climatic zones and geomorphological units. Rainfall and geomorphological characteristics determine the distribution of available water within the soil profile and the redistribution that leads to a heterogeneous

Table 2.16. Influences of drought, grazing, fire, crop cultivation, and wood cutting on canopy cover, per climatic zone

	Northern Sahel	Southern Sahel	Northern Sudan	Southern Sudan
Drought	--	--	-	0
Livestock grazing	-	-/+[a]	+	+
Fire	0[b]	0[b]	-	-
Crop cultivation				
- traditional (extensive)	0	0/-	0/-	0/-
- modern (intensive)	0	--	--	--
Wood cutting	-	--	--	--

+: positive influence.
0: no influence, or does not occur fires.
-: negative influence.
--: very negative influence.

[a] Depending on soil type.
[b] Lack of combustible material prevents.

pattern of water availability. The presence of woody plants is associated with water availability beyond the reach of herbaceous roots (Fig. 2.7).

Drought periods are a natural phenomenon in semi-arid regions. Tree and shrub species expand their ranges in good rainfall years and contract their ranges during droughts. Models of successional development, proceeding toward a vegetative climax, are not useful to describe these vegetation dynamics. In the SSWA, especially in the Sahel zones, models of nonequilibrium dynamics are more applicable.

In the land-use type "old fallowlands, rangelands and forests" canopy cover decreased considerably during recent years. The drought effected some woody species more than others. In the Sahel zones, climatic influences predominate, whereas in the Sudan zones, human influences such as fire, grazing, land clearing and wood cutting predominate (Table 2.16). Canopy cover reduction in this land-use type is most pronounced in the northern Sahel zone. Considering all land-use types, canopy cover reduction is most pronounced in the southern Sahel and northern Sudan zones, due to rapidly increasing populations and rates of land clearing.

Droughts and human activities influence woody plant communities in direct and indirect ways. For instance, in the Sahel zones the direct effects of grazing prevail. Although moderate browsing stimulates productivity, intensive browsing eventually kills trees and shrubs. In the Sudan zones indirect, positive effects prevail: grazing triggers a shift in the competitive relationship between the herbaceous and woody layers. This may lead to the phenomenon of bush encroachment, particularly on sandy soils near villages, if water availability is sufficient (Table 2.13). The extreme climate in the SSWA, the recent droughts and the rapid population growth combine to limit the occurrence of bush encroachment in the SSWA. The combination of bush encroachment, the related reduction of the herbaceous layer and wood cutting leads to particularly severe land degradation.

3 Production of Woody Plants

3.1 Introduction

One of the objectives of management of woody plant communities is to ensure sustainable exploitation of useful products of trees and shrubs. To develop sustainable exploitation requires production estimates of wood, foliage and fruit, their quality and availability in time, under different conditions. Such production estimates provide the basis for an evaluation of primary production of woody plant communities, e.g. in comparison to grasslands without trees (Sect. 4.8).

In Section 3.2, the analysis of annual productivity of woody plant communities in semi-arid regions is focussed the relationship between woody plant productivity and water and nutrient availability. The influences of woody plants on plant production factors is analysed in Chapter 4.

In Section 3.3, net productivity as a function of woody plant biomass and canopy cover is discussed. In Section 3.4, a method is described which employs canopy cover data to estimate production. Sections 3.5 and 3.6 deal with phenological patterns and the nutrient value, digestibility and palatability of woody plant components.

3.2 Productivity and Biomass of Woody Plant Communities

3.2.1 Gross and Net Productivity

Net primary production, in dry matter produced per year, is a standard measurement of productivity. Net production is defined as gross assimilation minus respiration. Gross primary production, CO_2 assimilation, reflects total photosynthetic activity. With an optimal supply of water and nutrients, and in the absence of weeds, pests and diseases, gross primary production is determined by solar radiation, temperature, and whether a species has a C_3 or C_4 photosynthetic pathway. Light interception increases with canopy cover, and gross assimilation of a given area increases relative to canopy closure (van Keulen and Wolf 1986). C_4 plants can produce up to a maximum of 300 kg dry matter per hectare per day: twice as much as C_3 plants. All woody plants are C_3 plants.

Plants respire to maintain themselves and grow their tissues. For plant communities of comparable structure, the rate of protein turnover doubles for every rise of 10 °C, with important consequences for maintenance respiration. At comparable temperatures, the respiration rate increases with the mass of the product formed, and varies with the composition of the end product. Fats and lignin are produced at highest costs. For woody plants in hot climates, relatively large amounts of maintenance and growth respiration are required (van Keulen and Wolf 1986). In other words, tropical woody plant communities use solar energy for primary production less efficiently than temperate ones, and woody plants use solar energy for primary production less efficiently than herbs. Older woody plants devote more gross production to respiration and less to net production. As forests mature, total ecosystem respiration approaches 100% of gross production, and net ecosystem production approaches zero (Lieth and Whittaker 1975).

3.2.2 Biomass and Net Productivity

3.2.2.1 Introduction

The high standing biomass of woody plants is the result of the production of many years, and reflects the respiration of many years, supplying the competing demands of maintenance, growth and recovery from exploitation and stress.

Throughout the world, tree height and trunk diameter at breast height are used to predict production (Lieth and Whittaker 1975; Satoo 1982). These parameters have limited usefulness in semi-arid regions (Sect. 2.1.5), so the relationship between canopy cover and net production, or standing biomass is considered here instead.

As woody plants mature, standing biomass grows according to an S-curve: increasing exponentially in the juvenile growth phase and increasing more slowly in maturity, approaching a maximum as less carbohydrate is devoted to net production.

As biomass increases, consumption of photosynthate by leaf respiration increases in proportion to leaf biomass whereas photosynthesis by leaves reaches an asymptotic upper limit dependant upon radiation intensity and temperature. Maturing woody plants must devote carbohydrate to leaves, wood, roots, fruits and flowers. To support all the non-photosynthetic tissues, leaves must occupy a certain fraction of the total biomass. The fraction required depends on the species and the availability of light, nutrients and water, because leaves grow less productively

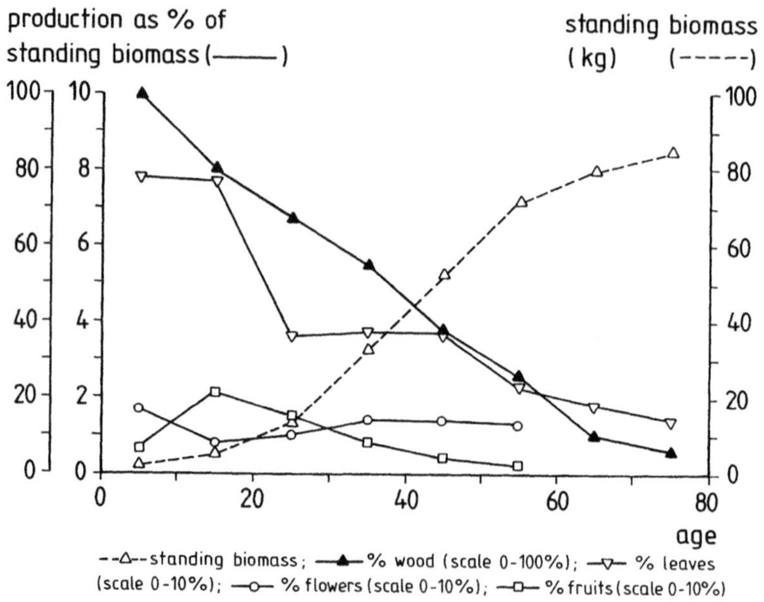

Fig. 3.1. Standing biomass and the annual production of wood, leaves, flowers and fruits of *Acacia senegal* trees, expressed as percentage of standing biomass, in relation to age, Fété Olé, Senegal. (After Poupon 1976)

under poor growing conditions (Stuart Chapin 1980). The ratio of photosynthetic to non-photosynthetic biomass is associated with the limits of woody plant size (Lieth and Whittaker 1975).

The proportions of net production devoted to leaves, wood, roots, flowers and fruits vary with plant age (Fig. 3.1). As the woody plant matures, the proportion of net production devoted to wood decreases at an earlier stage than that devoted to foliage. Absolute foliage production reaches a maximum value at a certain age, more or less corresponding to the time of maximum leaf area.

In semi-arid regions, site quality is related to water and nutrient availability. In water-limited ecosystems in the United States, primary production is related to leaf area index, which in turn is related to site conditions of precipitation, evaporation, and soil water storage (Lieth and Whittaker 1975; Grier and Running 1977). In the Sahel zones, water availability below 250 mm limits primary production, whereas soil fertility rather than water availability tends to limit primary production where water availability exceeds 250 mm (Sect. 4.8.1).

Site productivity indices are generally based on stand height, or height relative to a standard growth curve, for trees of the same age (Lieth and Whittaker 1975; Satoo 1982). On fertile soils, woody plants grow taller and faster in order to intercept as much light as possible (Tilman 1985). As stand densities increase, woody plants devote more resources to grow taller, but accrue less biomass per individual (Cameron et al. 1989). Once crown closure has been reached, stand production remains relatively constant, regardless of age and density (Gill et al. 1987). Where woody plants compete for limiting factors more important than light, stands may reach maximal production before crowns close, while total woody plant production remains relatively constant (Smith and Goodman 1986). Under these conditions, crown size is restricted (Pereira and Pallardy 1989).

On resource-poor sites, woody plants which allocate a greater share of carbohydrates to roots may be favoured over woody plants which allocate a greater share of carbohydrates to leaves, because the former are more likely to acquire adequate nutrients and water to support reproduction at an early age (Lieth and Whittaker 1975; Satoo 1982; Tilman 1985). Woody plants require more root biomass in seasonal climates to store sufficient energy for early rainy season production. Plants on infertile sites tend to live longer and reproduce at a later age than plants on fertile sites, as plants on infertile sites need more time to accumulate enough carbohydrate to support reproduction (Stuart Chapin 1980). Root to shoot ratios, therefore, tend to be relatively high on infertile, dry, or regularly burnt sites (Whittaker 1970; Vitousek and Sanford 1986), especially for immature woody plants. In semi-arid California, 10-year-old chaparral shrubs have a 1.5 root to shoot ratio; at 25 years, the ratio drops to 0.5 (Kummerow and Mangan 1981). In arid shrub lands, root biomass may be five to nine times greater than aboveground biomass (Kummerow et al. 1977).

Woody plant root production is difficult to measure. Most root production estimates assume that the ratio of production to biomass is similar for roots and shoots. In a Venezuelan tropical forest, annual replacement of roots less than 2 mm in diameter is estimated at 25% (Vitousek and Sanford 1986). In semi-arid California and Chile, annual fine root turnover is estimated at 50–75% of annual net production (Kummerow et al. 1977).

3.2.2.2 Biomass and Net Productivity in the SSWA

Average standing biomass of natural woody plant communities is about $2\,m^3\,ha^{-1}$ in the northern Sahel zone and $10\,m^3\,ha^{-1}$ in the southern Sahel zone, increasing to more than $50\,m^3\,ha^{-1}$ with 1200 mm annual rainfall in the southern Sudan zone. Biomass increases more than proportionally with annual rainfall, although this relationship varies with soil site condition. For example, in the northern Sudan zone, forests in fertile valleys may represent $70\,m^3\,ha^{-1}$, while shrub savannas on poor, shallow soils represent only 6 $m^3 ha^{-1}$ (Keita 1982; Catinot 1983; PIRL 1988; FAO 1989a).

Human activities, including fire, influence standing biomass of woody plant communities. Woody plants on lands fallowed for four years have a standing biomass of 1.5 $m^3 ha^{-1}$ under moderate exploitation and 0.5 $m^3 ha^{-1}$ under intensive exploitation (Ohler 1985). Woody biomass on an old fallowland may vary from 0–20 $m^3 ha^{-1}$ (FAO 1989a). The woody biomass on agricultural parklands varies from 10–35 $m^3 ha^{-1}$, depending on tree density and site conditions.

In the SSWA, only few studies analyse the relationship between biomass and net production of woody plant communities. Table 3.1 presents data from Fété Olé, Senegal, with 300 mm average annual rainfall (Bille 1977), from Mare d'Oursi, Burkina Faso, with 500 mm rainfall (Menaut 1983) and from Lamto, Ivory Coast, with 1300 mm rainfall (Lamotte 1979). Average annual rainfalls refer to the pre-drought years. The Ivory Coast data include different vegetation types along a topographic sequence of improving site conditions, where biomass increases from 20 to 60 $t\,ha^{-1}$ and the ratio of net aboveground production to standing biomass increases from 6 to 12%.

In conclusion, with increasing rainfall from the Sahel to the Sudan zones, standing biomass and net productivity increase, the proportion of belowground biomass to total biomass decreases from 50% to 30%, and the proportion of net aboveground production to standing biomass decreases from 15% to 6–12%.

How do these data compare with those from other semi-arid regions? South African tree savannas, with annual rainfall of 300–600 mm, have an average standing biomass of 20–45 $t\,ha^{-1}$, and a ratio of production to biomass of about 10%: 3% wood and 7% leaves and twigs (Werger 1978; Rutherford 1979; Walker 1980). The greater standing biomass in other semi-arid regions results from more favourable conditions for woody plants (Sect. 5.2). For a shrub savanna with a standing biomass of 5 $t\,ha^{-1}$, leaves and twigs occupy 34% of the aboveground biomass (Walker 1980). In a savanna in Zimbabwe at 500 mm annual rainfall, the annual production of leaves and twigs of trees was 3–8% of the standing biomass, and of shrubs was 21–31% (Walker 1980). In North America, the ratio of production to biomass in a semi-arid shrub land may be as high as 19%, the aboveground biomass ranging between 1.5 and 3.0 $t\,ha^{-1}$. The ratio of production to biomass in a forest may be as low as 2%, the aboveground biomass reaching 400 $t\,ha^{-1}$, whereas this ratio varies between 2–12% for trees and 20–40% for shrubs (Lieth and Whittaker 1975).

For the SSWA, few data are available on the belowground biomass of woody plant communities. Poupon (1976) reported 4.0 as the root to shoot ratio for immature *Acacia senegal*, decreasing to 1.0 at 20 years and 0.8 at 35 years. The root to shoot ratio of mature woody plant

Table 3.1. Biomass and annual net production of woody plant communities, at three locations in the SSWA

Average annual rainfall (mm) Vegetation type	300 shrubland steppe	550 dense shrubland	1300 savanna woodland
Aboveground biomass ($t\,ha^{-1}$)	1.9	2.4	20–60[a]
Net production (% of biomass)			
wood	6	9	3–5[a]
leaves etc.	9	10	3–7[a]
Belowground biomass ($t\,ha^{-1}$)	1.6	–	10–30[a]
Net production (% of biomass)	6	–	1.5

[a] Depending on vegetation type along a toposequential gradient.

communities (including immature woody plants) is about 0.5 in the southern Sudan zone, 0.8 in the southern Sahel zone (Table 3.1) and probably higher in the northern Sahel zone. Root to shoot ratios tend to be higher in regularly burnt vegetations. For example, in South African savannas, above 1 m soil depth the root to shoot ratio of woody plants was 1.0 in the absence of burning, and 2.0 under annual burning (Werger 1978).

Root production data in Table 3.1 are estimates from a small number of samples and seem low compared to those from other semi-arid regions. The ratio of root production to biomass probably is higher than the ratio of wood production to biomass and lower than the ratio of total aboveground production to biomass, because root turn over is more rapid than wood turn over (Lieth and Whittaker 1975).

3.2.3 Conclusions

The ratio of net aboveground production to standing biomass decreases from an estimated 15% in the Sahel zones to 6–12% in the southern Sudan zone (Table 3.1). Accordingly, standing biomass and absolute production increase more than proportionally. Root production and belowground biomass probably follow the same trend, but data are not available for confirmation. The root to shoot ratio decreases from 1.0 for mature woody plants in the northern Sahel to 0.5 in the southern Sudan zone.

The production characteristics of woody plant communities in the Sahel zones resemble those of immature and stressed vegetation: low standing biomass, high net production to biomass ratios and high root to shoot ratios. These characteristics are related to poor growing conditions, the occurrence of drought periods and high exploitation levels. A relatively high proportion of the biomass is devoted to foliage production, in spite of a high proportion of evergreen species (Sect. 3.5), because woody plants produce more foliage than wood when total net production is low (Grubb 1989).

3.3 Production of Woody Plant Components

3.3.1 Wood Production

In the following, wood includes branch wood and stem wood. Small twigs (less than 3 mm diameter) are considered as browse; part of them are annually lost. From available data, the ratio of stem wood to branch wood is estimated at 0.5 for the northern Sahel, 1.0 for the southern Sahel; 1.5 for the northern Sudan and 2.0 for the southern Sudan zone (Poupon 1976, 1977a,b; Bille 1977; Piot et al. 1980). These data are comparable to those from south African savannas (Rutherford 1979).

Clément (1982) reviewed pre-drought wood production data for the West African region. Annual average wood production (WP, in $\mathrm{t\,ha^{-1}\,yr^{-1}}$) of unprotected woody plant communities is related to annual rainfall by the following quadratic equation (Fig. 3.2):

$$WP = 0.04 + 0.94\,(R/1000)^2 \quad (R = \text{annual average rainfall in mm}).$$

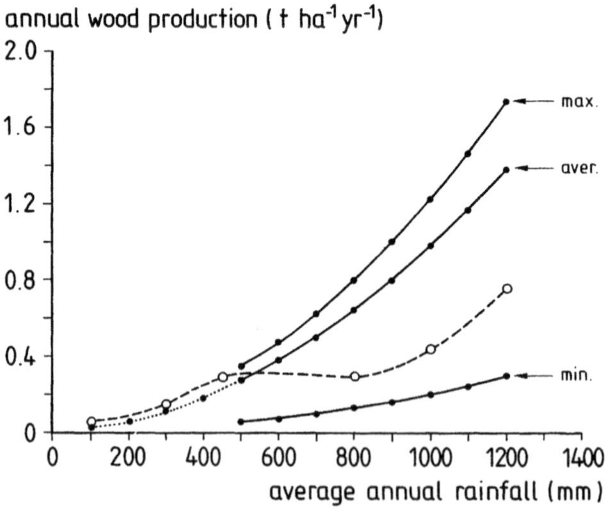

Fig. 3.2. Average, maximum and minimum annual wood production in natural wood communities in the SSWA, in relation to annual rainfall. ● Clément (1982); ○ Keita (1982)

The relationship refers to natural woody vegetations in areas with annual rainfall from 500–1600 mm. In the SSWA, canopy cover is linearly related to average rainfall up to 900 mm (Sect. 2.2.1). Some unpublished field data confirm that average wood production continues to decrease with rainfall less than 500 mm, as indicated in Fig. 3.2. Using the above equation, Sahel zone wood production appears related to canopy cover by a quadratic function.

A reinterpretation of Clément's data shows that wood production on poor sites may be 20% lower, and on rich sites may be 20% higher than indicated by the relationship. "Poor sites" and "rich sites" are subjectively defined, based on soil and vegetation characteristics. With protection against fire and grazing, wood production can be about 1.25 times higher than average wood production. In Mali, however, burning regimes are estimated to reduce the optimum wood production by about 60%, depending on fire frequency and intensity (Jackson et al. 1983).

Degraded lands support a wood production of $0.25 \times R^2$ (Clément 1982), although production seems to approach zero in a severely degraded landscape. As canopy cover has decreased in recent years (Sect. 2.3), current wood production is probably lower than reflected by the above equation.

Comparing wood production estimates from Clément (1982) with those from Keita (1982), both from the same period, the data are different for the Sudan zones (Fig. 3.2). Keita's estimates were based on woody species composition and vegetation structure. These are not the best parameters, because they are, and have been, influenced by humans. For instance, shrub vegetation resulting from human activity on a young fallowland is much more productive than natural shrub vegetation on shallow, rocky soil.

Wood increment depends firstly on site conditions; second, on the development phase of the woody plant community and third, on the species composition of the vegetation.

Two studies illustrate site conditions in the SSWA. At Mare d'Oursi, Burkina Faso, in the southern Sahel zone, site conditions influence *Balanites aegyptiaca* height relative to diameter at breast height (Fig. 3.3). In a valley with favourable site conditions, height relative to diameter at breast height is double that on a slope or sand dune (Piot et al. 1980). In the northern Sahel zone in Senegal, *Acacia raddiana* annual growth rings from 1930–1972 vary from 0–10 mm (Mariaux 1975). No relationship appears between growth ring width and annual rainfall, but growth seems related to rainfall distribution, largely determining the available subsoil water (Sect. 2.3.2). Wood increment of individual woody plants depends on the density of woody plants competing for limited resources. In South Africa, some *Acacia raddiana* were cut from an evenly distributed tree layer, resulting in increased growth of the plants remaining (Smith and Goodman 1986). The same principle probably

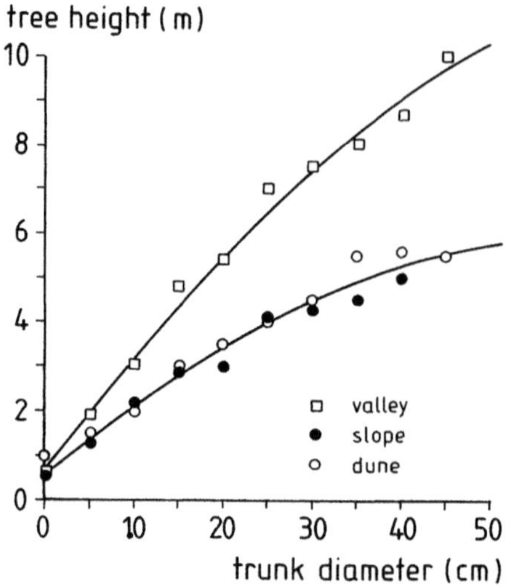

Fig. 3.3. Relationship between trunk diameter and tree height of *Balanites aegyptiaca* at three different sites in Mare d'Oursi, Burkina Faso. (After Piot et al. 1980)

applies to parkland species (Sect. 2.4.3), which grow well in the absence of nearby woody competition, contrary to the same species in dense, natural vegetation.

The ratio of wood production to standing biomass decreases with increasing maturity (Sect. 3.2). In the southern Sahel zone, under moderate exploitation, biomass increases from $0\,t\,ha^{-1}$ in the first year of fallow, to $1.2\,t\,ha^{-1}$ in the fourth year and $5\,t\,ha^{-1}$ after 50 years (Ohler 1985). Accordingly, wood increment as a proportion of standing biomass decreases from almost 100% in the first year, to 50% in the fourth year, to below 5% in 50 years. These data correspond to those recorded in northern Nigeria (Aweto 1981). Under an intensive regime of burning, wood cutting and grazing, biomass increase during fallowing is a factor three lower (Ohler 1985). This corresponds with data from Nigeria (Kemp 1962). In the southern Sahel zone, woody plant regrowth 3–5 years after cutting equals $1.0\,m^3\,ha^{-1}$, almost double the level of wood increment considered average for the zone (Fig. 3.2; Renes and Coulibaly 1988).

In Fété Olé, Senegal, in the northern Sahel zone, the wood increment of *Balanites aegyptiaca* shrubs as a proportion of standing biomass decreases with age: from 15% at 10 years, to 10% at 10–60 years, to 5% at 60–80 years, and 2% for the oldest shrubs (Bille 1977). Maximal standing biomass of *Acacia senegal* trees is reached at an age of 70–100 years (Fig. 3.1) and of *Commiphora africana* trees at a similar age (Poupon 1976, 1977b). *Commiphora africana* wood increment, however, is maximal at an age of 20–50 years.

Table 3.2. Wood production of forestry industrial plantations in the SSWA, non-irrigated and rain-dependent, in relation to annual rainfall. (After Catinot 1983)

Rainfall (mm)	Standing biomass ($m^3 ha^{-1}$)	Production of wood ($m^3 ha^{-1} yr^{-1}$)
<400	No plantations	
400–600	4–16	0.5–2[a]
600–800	8–32	1–4[a]
800–1000	32–64	4–8[a]
1000–1200	48–72	8–12[b]
Irrigated	72–120	18–30[c]

[a] Rotation of 8 years.
[b] Rotation of 6 years.
[c] Depending on amount of water irrigated.

The influence of species composition on growth rates is also illustrated by the comparison between *Commiphora africana* and *Acacia senegal. Commiphora* grows five times as fast as *Acacia* but has extremely light wood. In one year in Fété Olé, Senegal, annual wood increment of six different woody species varied by a factor 4 (Bille 1977). Differences are largest during early growth. In a savanna in Ghana, immature woody plants varied more than mature ones in annual girth increment (Liebermann 1982). Mature trees and shrubs may have more access to production resources.

Human influence may improve site conditions. Table 3.2 shows that wood production of well managed, non-irrigated tree plantations of exotic species may be four times higher than of native vegetation (Fig. 3.2). The exotic species chosen are usually colonisers, with rapid juvenile growth, producing soft wood. The sites chosen have good growing conditions, are prepared by clearing and ploughing, are protected against grazing and fire and are weeded. *Eucalyptus* wood production doubles when weeded once or three times a year.

In reality, such high levels of wood production are not often reached (Sieder 1983). Village plantations produce about half as much wood as values shown in Table 3.2 (Keita 1982). High wood production rates on tree plantations are difficult to sustain, because rapid growth is due to high uptake of limited nutrients.

3.3.2 Foliage Production

Foliage production is usually based on litter fall or peak foliage biomass estimates. Foliage production, however, does not equal annual litter fall or peak foliage biomass, because the average life span of leaves varies (Strojan et al. 1979; Satoo 1982) and because insects consume leaves. In Zimbabwe, for instance, insects consumes 27% of *Faidherbia albida* annual leaf production (Dunham 1989).

In Fété Olé, Senegal, leaves, flowers and fruits constitute about 2/3 of non-wood production, or 6% of the biomass, and twigs constitute the other 1/3. In Mare d'Oursi, Burkina Faso, leaves constitute about 1/2 of the non-wood production, or 5% of the biomass (Table 3.1). The ratio appears related to age: for young *Acacia senegal* trees leaves, flowers and fruits constitute 1/2 of the non-wood production, and twigs constitute the other half. For mature trees, leaves, flowers and fruits constitute 2.2 times as much as twigs (Poupon 1977a). In the southern Sudan zone, twig production is 40–50% of the foliage production (Lamotte 1979).

In Zimbabwe, twigs constitute only 20% of the non-wood production of *Faidherbia albida* (Dunham 1989). In South African savannas, leaf production constitutes 7% of the standing biomass, twig production about 1.5% (Rutherford 1979; Werger 1983).

On the basis of these data, leaves, fruits and flowers are estimated to constitute 60–70% of the total non-wood production in the Sahel zones and 70–80% of the total non-wood production in the Sudan zones. Flowers and fruits constitute a highly variable proportion of the total browse production (Sect. 3.3.3).

Foliage production as a proportion of standing biomass decreases with increasing maturity and standing biomass, from 5–12% for young or small woody plants, to 2% for old or large trees (Tables 3.1, 3.3 and 3.4). Figure 3.1 shows *Acacia senegal* production of wood, leaves, flowers and fruits, as a proportion of standing biomass, as trees mature. *Commiphora africana* allocates production in a similar way during maturation (Poupon 1977b). Although woody species vary in absolute foliage production, the production trends relative to age and size of woody plants are similar. There is no linear relation between biomass and foliage production (Sect. 2.2.1). Small and immature woody plants

Table 3.3. Annual foliage production of five woody species, in relation to age and standing biomass, at Fété Olé, Senegal, 300 mm rainfall. (Bille 1977)

Species	*Guiera senegalensis*	*Grewia bicolor*	*Acacia senegal*	*Balanites aegyptiaca*	*Commiphora africana*
Age (years)	11	26	28	35	63
Standing biomass (kg plant^{-1})	2	21	15	33	55
Annual foliage production					
(% of biomass)	12.0	10.1	7.0	2.5	1.7
(kg plant^{-1})	0.2	2.1	1.1	0.8	0.9

Table 3.4. Annual foliage production of some woody species, expressed as a proportion of the standing biomass, and production per tree

Species	Standing biomass ($kg\,tree^{-1}$)	Foliage production (% of biomass)	Foliage production ($kg\,tree^{-1}$)	Source
Acacia senegal	28	3.7	1.0	Poupon (1976)
Acacia laeta	15	5.3	0.8	Menaut (1983)
Acacia seyal	72[a]	4.6	2.9	Penning de Vries (unpubl. data)
Combretum ghasalense	90[a]	5.0	4.5	Penning de Vries (unpubl. data)
Faidherbia albida	1900	5.1[b]	97.3[b]	Jung (1969)
Pterocarpus lucens	85[a]	4.6	3.9	Penning de Vries (unpubl. data)
	185[a]	3.4	6.3	Penning de Vries (unpubl. data)
Sclerocarya birrea	475[a]	3.0	14.3	Penning de Vries (unpubl. data)

[a] Weight in green matter.
[b] Leaves and flowers.

Table 3.5. Annual browse and foliage production ($kg\,ha^{-1}$) of six woody species, in two years with average rainfall (1971, 1973) a and dry year (1972), in Fété Olé, Senegal. (Poupon and Bille 1974; Bille 1977)

Rainfall	1971 202 mm		1972 33 mm	1973 209 mm	
Woody species	Browse	Foliage	Browse = foliage	Browse	Relation foliage 1971/1972
Acacia senegal	16.1	12.5	6.2	10.3	2.0
Balanites aegyptiaca	22.0	13.9	6.3	15.2	2.2
Boscia senegalensis	?	5.1	3.3	?	1.5
Commiphora africana	8.0	6.6	3.2	9.3	2.1
Grewia bicolor	35.6	27.5	6.6	17.8	4.2
Guiera senegalensis	21.2	16.4	8.3	17.9	2.0

have the highest ratio of foliage production to standing biomass, whereas large and mature woody plants have the highest absolute foliage production.

In Fété Olé, Senegal, foliage production of several woody species is, on average, twice as low in a dry year, 33 mm, than in a year with normal rainfall (200 mm). Species vary in their foliage production response to rainfall (Table 3.5). After a drought period, some species seem to require several years of "normal" rainfall to obtain pre-drought foliage production levels. *Acacia senegal* foliage production varies between depression sites, slightly sloping sites and dune sites (Fig. 3.4).

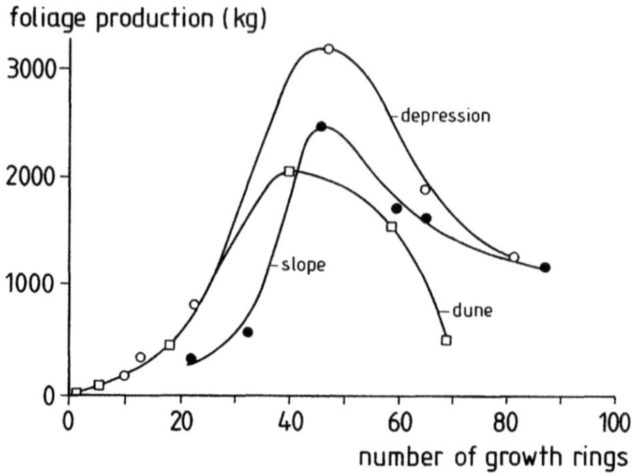

Fig. 3.4. Annual foliage production of *Acacia senegal* trees in relation to age, at different sites at Fété Olé, Senegal. (Poupon 1976)

The highest foliage production on depression sites may have been due to greater availability of either, or both nutrients and water.

3.3.3 Flower and Fruit Production

Flowers and fruits are often considered as a single category. Flowers may be removed by wind or herbivores, may abort or mature. Flower production varies with the species, the individuals within the species and the climate. In the semi-arid Mojave region in the United States, the coefficient of variation of total annual flower production is four times that of total annual foliage production (Strojan et al. 1979). In semi-arid Zimbabwe, *Faidherbia albida* fruit production varies by a factor of 5, both for different trees in one year and for the same tree in different years (Dunham 1989). Fruit production of *Faidherbia albida* may, in exceptional cases, exceed foliage production by 50% (Jung 1969; Bille 1980). However, typical fruit production of *Faidherbia albida*, on average sites, under intensive exploitation, is only 15% of foliage production (CTFT 1988). In Botswana, variable flower production between individuals of a species in a single year are probably partly due to different sites or microhabitats (Tolsma 1989).

From the SSWA, some data are available on fruit production from specific woody species, due to their importance to various consumers. Flower and fruit production of mature *Acacia senegal* trees constitutes about 1.5% of the standing biomass and 1/3 of foliage production. For young fruiting trees, flower and fruit production constitutes less than 1/5 of foliage production (Fig. 3.1). In central Mali, similar patterns appear for *Pterocarpus lucens* and *Boscia senegalensis* (Penning de Vries, unpubl. data). At Fété Olé, Senegal, annual fruit production varies from 0–20% of total non-wood production, in relation to rainfall mainly (Table 3.5). Here, fruit production also shows large variation between woody species: *Balanites aegyptiaca*, *Acacia senegal* and *Grewia bicolor* produce more than *Commiphora africana*, *Guiera senegalensis* and *Adansonia digitata*. Le Houérou (1980a) concludes that fruit production, as a percentage of foliage production, is 30% for *Acacia seyal*, 40% for *Pterocarpus lucens* and 100% for *Faidherbia albida*. Average fruit production tends to constitute less than 10% of non-wood production, and less than 1% of the standing biomass. For comparison, on South African savannas, fruit and flower production represents only 6% of non-wood production (Rutherford 1979).

In Ethiopia, *Acacia raddiana* fruit production is not related to canopy cover (Coppock et al. 1991). In Mali, *Vitellaria paradoxa* (shea butter tree) fruit production varies from 0.5 to 1.5% of the standing biomass (Ruyssen 1957). The annual variability is not directly related to total rainfall, rainfall distribution, rainfall in the preceding year or wind impact. Regular weeding and light burning lead to higher fruit production, but these practices cannot explain all the variability between years.

Woody plant reproduction is often delayed by nutrient stress. In less fertile habitats and where rainfall is low, reproduction is limited to years with better than average conditions (Stuart Chapin 1980).

In Botswana, several woody species were studied in detail (Tolsma 1989). For the flowers to develop into mature fruits, three times as much phosphorus is required as is available in the total phosphorus store in stems and branches. Before being shed, foliage translocates a consider-

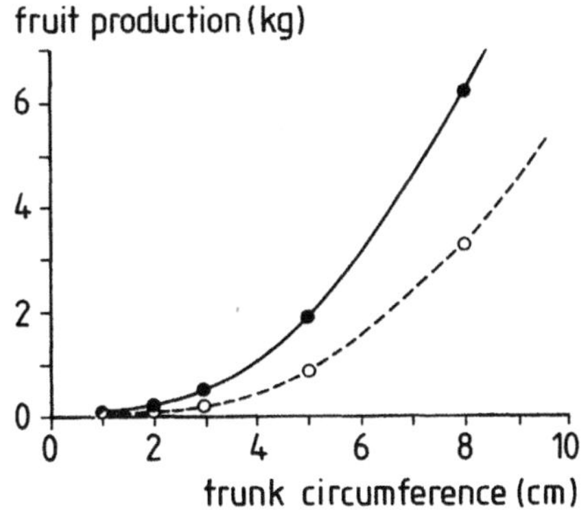

Fig. 3.5. Relationship between *Acacia raddiana* annual fruit production and trunk circumference at two sites differing in soil depth, in Mali. ● shallow soil site; ○ deep soil site. (After Cissé 1983a)

able proportion of nutrients stored (Sect. 4.5.4.6). Most phosphorus is translocated to flowers and fruits. The high P requirements might explain the spontaneous abortion of flowers and the large variation in fruit production between and within tree species. *Acacia raddiana* fruit production was correlated to diameter at breast height for different soil site conditions: on deep soils, three times as much fruit is produced by a tree of given size than on shallow soils (Fig. 3.5). In the northern Sahel zone, fruit production is limited by water availability mainly (e.g. Poupon 1977a).

3.3.4 Conclusions

The ratio of production of leaves, flowers, fruits and wood to standing biomass, decreases with increasing maturity and size. The ratio of production to standing biomass of leaves, flowers and fruits stabilises at an earlier stage than wood (Figs. 3.1 and 3.4). Wood increment decreases to less than 2% of the standing biomass when the woody plant reaches maturity, at 50–100 years of age in the northern Sahel and more than 200 years in the southern Sudan zone. In the natural vegetation, such low values are seldom represented, because few woody plant communities consist exclusively of mature trees.

Wood production is low in the northern Sahel zone, 0–0.2 $t\,ha^{-1}$, and increases more than proportionally with rainfall, to 1.0–1.5 $t\,ha^{-1}$ in the southern Sudan zone (Fig. 3.2). Wood production under local conditions of optimal management and exotic species selection may be seven times higher than these figures (Table 3.2), but this production level is probably not sustainable.

For small or immature woody plants in the Sahel zones, foliage production occupies a higher proportion of standing biomass, up to 10%, compared to the proportion of about 5% in the Sudan zones (Tables 3.1, 3.3 and 3.4; Fig. 3.1). Flower and fruit production vary more than foliage production, between species, between individuals of the same species, and between years.

The relative production of woody plant components varies per climatic zone: in the northern Sahel zone above-ground production is 45% wood, 30% leaves, 17% twigs, and 8% flowers and fruits; in the southern Sudan zone it is 50% wood, 33% leaves, 12% twigs and 5% flowers and fruits. These averages show considerable variation between species and between years.

As soil conditions improve, standing biomass and average plant heights increase. Soil conditions influence wood production (Fig. 3.2), and non-wood production, especially flowers and fruits, because these components are rich in nutrients (Figs. 3.5 and 3.6).

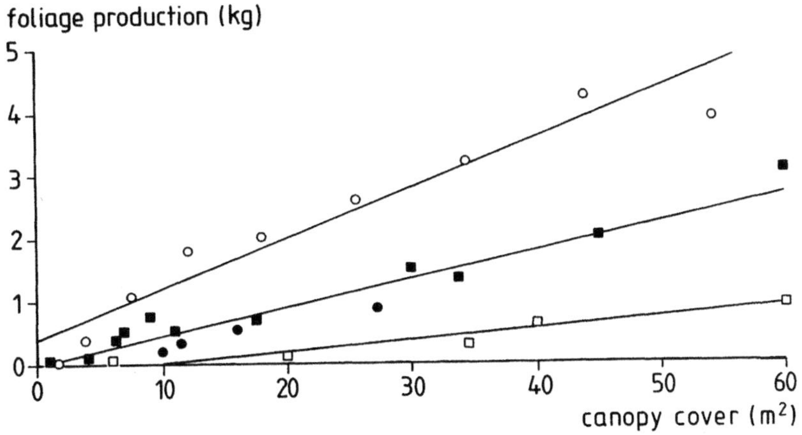

Fig. 3.6. Relationship between annual foliage production and canopy cover of *Commiphora africana* trees at different sites and varying rainfall. ● 200 mm, Fété Olé; □ 33 mm, Fété Olé; ■ 200 mm, Fété Olé; ○ 450 mm, Niono. (After Bille 1977; Poupon 1977b; Cissé 1980)

3.4 Estimating Woody Plant Production

3.4.1 Introduction

In this section, a method is worked out that estimates foliage production from canopy cover data. On the basis of canopy cover in the SSWA and relationships between foliage production and the production of other plant components, total browse and wood production are estimated per climatic zone, geomorphological unit, land-use type and in relation to recent changes. Results obtained are verified with available field data.

3.4.2 Foliage Production and Canopy Cover

The relationship between canopy cover and foliage production is analysed for canopy cover classes in three Sahel locations (Poupon and Bille 1974; Poupon 1976; Bille 1977; Poupon 1977a,b; Cissé 1980; Hiernaux 1980; Piot et al. 1980; Penning de Vries and Djitèye 1991; Menaut 1983). At least five woody species occur at each location (Table 3.6). Tree height correlates to diameter at breast height at all locations. In central Mali, about 600 observations on 26 woody species provides the correlation between canopy cover (in m^2) and trunk circumference (in cm), with a correlation coefficient of 0.97: canopy cover = $0.007 \times (2\pi r)^2$ (Penning

Table 3.6. Annual foliage production of woody plants at three SSWA locations. For seven canopy cover classes (m^2), are indicated foliage production per plant (range and average) and per unit of canopy cover ($g\,m^{-2}$). [Based on a summary of various studies (see text) and Bille 1980]

Location	Fété Olé	Mare d'Oursi	Niono	"Bille (1980)"
Rainfall ($mm\,yr^{-1}$)				
– average	350	550	570	
– in year of observation	200	400	400	
No. of woody species	5	5	8	
Canopy cover class (m^2)				
2.5–7.5 m^2				
$kg\,plant^{-1}$	0.2–1.0 (0.7)	0.3–1.0 (0.8)	0.3–1.2 (0.9)	(0.7)
$g\,m^{-2}$	140	160	180	140
7.5–12.5 m^2				
$kg\,plant^{-1}$	0.5–1.3 (1.1)	0.8–1.5 (1.2)	1.0–2.5 (1.8)	(1.4)
$g\,m^{-2}$	110	120	180	140
12.5–17.5 m^2				
$kg\,plant^{-1}$	0.7–2.5 (1.6)	1.0–2.5 (1.8)	1.3–3.5 (2.4)	(2.2)
$g\,m^{-2}$	110	120	160	140
17.2–25 m^2				
$kg\,plant^{-1}$	1.0–2.0 (1.8)	2.5–4.0 (3.2)	1.5–4.0 (3.4)	(3.0)
$g\,m^{-2}$	80	140	140	140
25–35 m^2				
$kg\,plant^{-1}$	1.3–3.0 (2.0)	3.0–6.0 (4.2)	2.5–7.0 (5.0)	(4.3)
$g\,m^{-2}$	70	140	170	140
35–45 m^2				
$kg\,plant^{-1}$	1.4–4.5 (2.5)	5.0–8.0 (6.0)	3.8–10.0 (5.6)	(5.7)
$g\,m^{-2}$	60	150	140	140
45–55 m^2				
$kg\,plant^{-1}$			4.0–12.5 (8.0)	(7.1)
$g\,m^{-2}$			160	140

de Vries and Djitèye 1991). Similar relationships are available for woody plants in East Africa (UNESCO-IPAL 1982).

This correlation is not applicable for shrubs with canopy covers below 2.5 m^2. Small shrubs also show large variation due to varying growth characteristics and the effects of cutting and browse. On the basis of the available studies, particularly that of Cissé (1980), shrub foliage production is estimated for canopy cover categories of 0.1–1.0 m^2 and 1.0–2.5 m^2 (Table 3.7). Competition with herbs may be responsible for low foliage production of shrubs of 0.1–1.0 m^2 canopy cover. The 1.0–2.5 m^2 shrub category is compatible with the 2.5–7.5 m^2 canopy cover class (Table 3.6).

Table 3.7. Plant parameters and annual foliage production for two shrub size categories

Canopy cover (m^2)	Trunk diameter (cm at basis)	Height (m)	Foliage production		
			($g\,plant^{-1}$)		($g\,m^{-2}$)
			Range	Average	
0.1–1	<3	<1	10–150	50	100
1–2.5	3–10	1–2.5	50–600	300	180

The range of foliage production per canopy class is large, mainly because of interspecies variation, but consistent trends appear for each location. The smallest canopy covers produce proportionately most foliage, with little variation in relation to rainfall. The largest canopy covers produce more than twice as much foliage with 400 mm than with 200 mm rainfall in the year of observation (Table 3.6). As an example, *Commiphora africana*, from which data are available from the three different locations, shows increasing foliage production per unit of canopy cover with increasing rainfall (Fig. 3.6).

On the basis of a summary of studies from West Africa mainly, Bille (1980) proposes the following relation between foliage production and trunk diameter in the Sahel zones:

$$\log(\text{foliage production in kg}) = 2\log(\text{diameter in cm}) + 1.$$

In combination with the relationship between trunk circumference and canopy cover (see above) this gives:

$$\log(\text{foliage production in kg}) = \log(\text{canopy cover in m}^2) + 2.15.$$

The foliage production based on this relationship fits with data from Mare d'Oursi and Niono but overestimates productivity of the highest canopy cover classes at Fété Olé in the northern Sahel zone (Table 3.6). This may be so because drought years particularly influence survival and productivity of the largest woody plants on drier sites, and because on drier sites woody plants are shorter (Fig. 3.3), and so produce less foliage per unit of canopy cover.

3.4.3 Methods to Estimate Woody Plant Productivity

Two methods are tested for their adequacy to estimate woody plant foliage production at specific locations in the SSWA.

Where data are available on woody species density per canopy cover class, foliage production is calculated with the data used for Table 3.6. In the absence of data on the relationship between species canopy cover and foliage production, or on species density per canopy cover class, foliage production is calculated using the average foliage production values from Table 3.6. Only a few studies provide enough data to calculate foliage production by this method (Bille 1977; Hiernaux 1980; Piot et al. 1980; Menaut 1983; Compaoré, unpubl. data). The estimates are usually lower than field measurements (Fig. 3.7), probably because the data used for Table 3.6 was gathered during dry years, and field measurements are either from average rainfall years or are averages over several years.

Where only canopy cover estimates are available, foliage production is estimated by multiplying the estimated production per leaf layer with the number of leaf layers. The number of leaf layers is related to the height of the woody plant, and depends on site conditions. Table 3.9 gives estimates of the number of leaf layers per climatic zone and per geomorphological unit, representing variation in the availability of water and nutrients. The average number of leaf layers per climatic zone is based on the relative importance of the geomorphological units per climatic zone (Table 2.2). At Niono, Mali, woody species are assigned to one of three groups, according to leaf layer production (Table 3.8). Species not listed in Table 3.8 are assigned to the second, average production group.

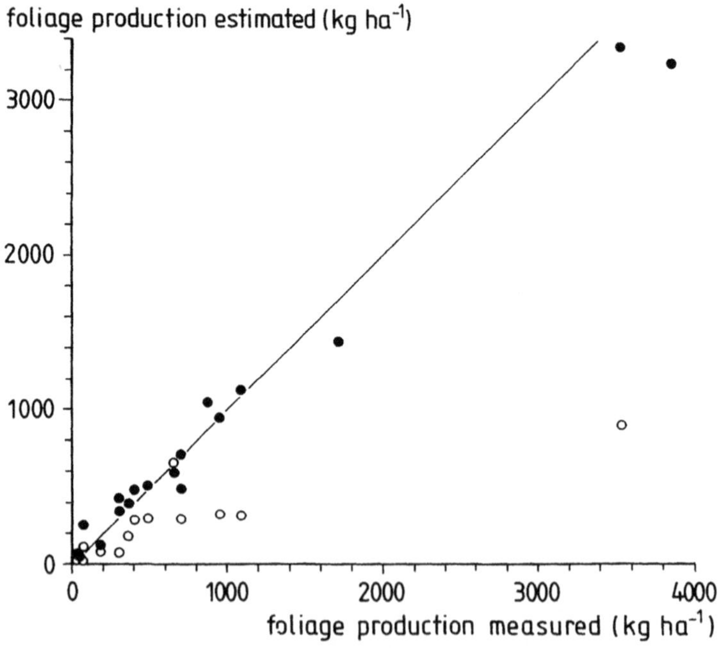

Fig. 3.7. Relationship between measured and estimated annual foliage production by two methods (Sect. 3.4.3). ○ method 1; ● method 2

Table 3.8. Annual foliage production per leaf layer ($g\,m^{-2}$) for three different leaf types, and classification of woody species among the three types

Leaf type Annual foliage production	1 $100\,g\,m^{-2}$	2 $120\,g\,m^{-2}$	3 $140\,g\,m^{-2}$
Acacia raddiana		×	
Acacia senegal		×	
Acacia seyal		×	
Acacia laeta		×	
Boscia senegalensis			×
Balanites aegyptiaca			×
Combretum ghasalenze			×
Combretum glutinosum			×
Combretum aculeatum	×		
Combretum micranthum	×		
Commiphora africana	×		
Euphorbia balsamifera	×		
Grewia bicolor		×	
Guiera senegalensis	×		
Maerua crassifolia		×	
Piliostigma reticulatum			×
Pterocarpus lucens	×		
Sclerocarya birrea	×		
Ziziphus mauritiana	×		

The second method provides a better fit with field measurements (Fig. 3.7). From the northern Sahel and the southern Sudan zone, however, too few field measurements are available for verification. Figure 3.8 shows that foliage production, based on canopy cover estimates in the land-use type "old fallowlands, rangelands and forests" (Fig. 2.4), decreased considerably in recent years. From the relationship between foliage production and production of other woody plant components, varying per climatic zone (Sect. 3.3), total browse and wood production are also determined (Fig. 3.8). For woody plants with known canopy cover and number of leaf layers (determined by site conditions; Table 3.9), this method can be used to estimate production of foliage, fruits, twigs, total browse and wood (Table 3.10).

For 1980, the browse production estimates (Fig. 3.8) in the Sahel zones are lower, but in the northern Sudan zone they are similar to estimates made by Le Houérou (1989). The wood production estimates in the Sahel zones resemble other researchers' estimates, except in the Sudan zones, where estimates are higher than those of other researchers (Fig. 3.3; Clément 1982; Keita 1982; de Winter et al. 1989).

Foliage production per leaf layer or canopy cover has been estimated in other semi-arid regions. In semi-arid East Africa, foliage production is related to canopy cover (van Wijngaarden 1985). However, the relation-

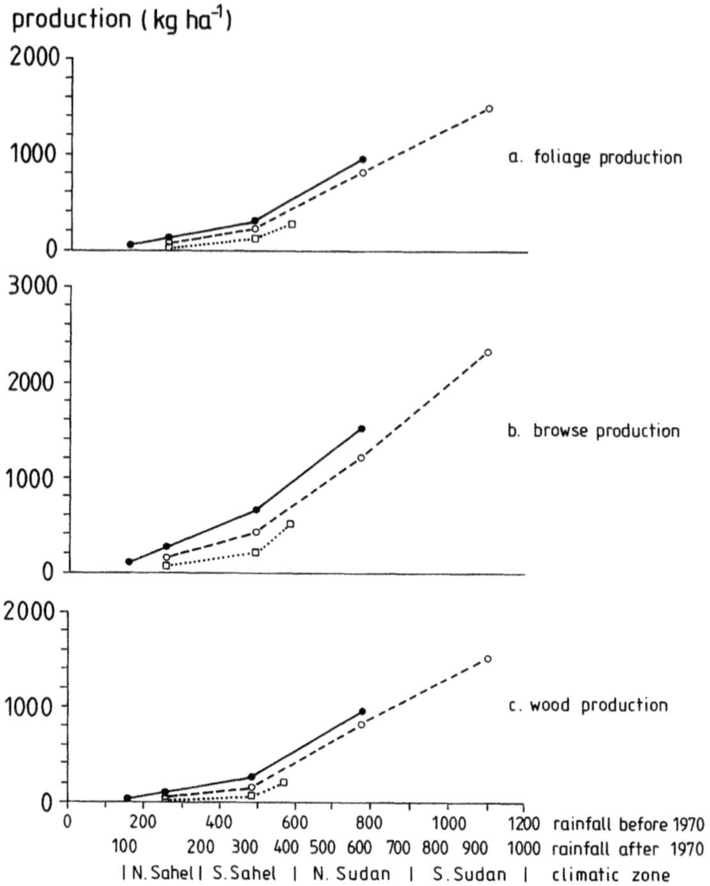

Fig. 3.8. Annual foliage, browse and wood production per climatic zone, based on canopy cover in the land-use type "old fallowlands, rangelands and forests", for different periods in relation to drought. ● prior to 1970; ○ about 1980; □ end 1980s

ship varies between species, relative to grazing intensities: species not grazed have 0.5 $g\,m^{-2}$ canopy cover per mm rainfall; moderately grazed species 2.0 $g\,m^{-2}$ canopy cover per mm rainfall and intensively grazed species 1.0 $g\,m^{-2}$ canopy cover per mm rainfall. Within each grazing intensity, foliage production increases linearly with rainfall. Thus, foliage production estimates within the region, at 100–200 mm rainfall, range from 50 to 250 $g\,m^{-2}$, and average 150 $g\,m^{-2}$. Other estimates in East Africa vary by species, between 40 and 170 $g\,m^{-2}$ foliage production per unit of canopy cover, averaging 100 $g\,m^{-2}$ (UNESCO-IPAL 1982). In Zimbabwe, *Faidherbia albida* foliage production, at 760 mm rainfall, is 340 $g\,m^{-2}$ (Dunham 1989). This is similar to the estimate of 360 $g\,m^{-2}$ for the northern Sudan zone. In India, shrub foliage production, at 370 mm

Table 3.9. Number of leaf layers in natural woody plant communities, per climatic zone and per geomorphological unit

Climatic zone	Geomorphological unit					
	Os	Ps	Pl	Sq	La, M	Average
Northern Sahel	2	2	1	1	3	1.5
Southern Sahel	2	2	2	1	4	2.0
Northern Sudan	3	3	3	2	5	3.0
Southern Sudan	3	4	4	3	6	4.0

Table 3.10. Annual foliage, fruit, twig and total browse production per unit of woody canopy cover ($g m^{-2}$) (a) in the natural vegetation and (b) on agricultural fields

	Foliage		Fruits		Twigs		Browse	
	(a)	(b)	(a)	(b)	(a)	(b)	(a)	(b)
Northern Sahel	180		50		100		330	
Southern Sahel	240	360	65	100	130	195	435	655
Northern Sudan	360	480	75	100	150	200	585	780
Southern Sudan	480	600	80	110	185	220	745	930

rainfall, is $100 g m^{-2}$ (Kaul and Ganguli 1963) which is lower than the estimate for the northern Sahel zone, $180 g m^{-2}$. For chaparral shrubs in semi-arid California, foliage production is $280 g m^{-2}$ (Kummerow and Mangan 1981), corresponding to estimates for the southern Sahel zone.

It can be concluded that, despite the unusual climate and soils of the SSWA, woody plant production in the land-use type "old fallowlands, rangelands and forests" is within the same order of magnitude as that of other semi-arid regions. Woody plant production for the northern Sahel zone may have been overestimated.

Land-use types "young fallowlands" and "agricultural fields" are usually located where site conditions are good. On agricultural fields, woody vegetation is characterised by dispersed trees and canopy covers average 5% in the Sahel zones and 12% in the Sudan zones (Sect. 2.4.3). Trees on agricultural fields are usually taller than those in the natural vegetation because of favourable site conditions and reduced competition. As a result, the number of leaf layers is higher. Foliage production of mature trees on agricultural fields can be estimated by the second method (average canopy cover × production per leaf layer × leaf layers):

northern Sahel zone: $3\% \times 120\,g\,m^{-2} \times 3.5 = 110\,kg\,ha^{-1}$
southern Sahel zone: $6\% \times 120\,g\,m^{-2} \times 4.0 = 290\,kg\,ha^{-1}$
northern Sudan zone: $10\% \times 120\,g\,m^{-2} \times 6.0 = 720\,kg\,ha^{-1}$
southern Sudan zone: $15\% \times 120\,g\,m^{-2} \times 6.0 = 1080\,kg\,ha^{-1}$.

The woody vegetation of young fallowlands is generally dominated by shrubs. Canopy cover may be high because of bush encroachment, particularly in the Sudan zones (Sect. 2.4.3). Thus, on a young fallowland with a 60% canopy cover of 50% shrubs and 10% trees, characteristic of the northern Sudan zone, foliage production is $1.7\,t\,ha^{-1}$, assuming the shrubs have an average of two leaf layers. In the Sudan zones, wood production is equal to total non-wood production, averaging $2.5\,t\,ha^{-1}$.

3.4.4 Conclusions

Two types of methods are introduced and tested to estimate foliage production of woody plants. The first method is unreliable when applied to large areas and different years. The second method, based on an average foliage production of $120\,g\,m^{-2}$ per leaf layer and a varying number of leaf layers according to climatic zones and geomorphological units, reflects varying site conditions for woody plant growth and productivity (Sect. 3.2). By this method, reliable estimates are obtained of foliage, total browse and wood production, per climatic zone (Fig. 3.8) and unit of canopy cover (Table 3.10), using known relationships between wood, leaf, twig, fruit and foliage production (Sect. 3.3.4). The method can also be applied to woody plants on agricultural fields and fallowlands.

The recent decline of woody plant productivity (Fig. 3.8) is mainly due to a decline in canopy cover during recent years. Other factors which may have contributed to depressed production include the following.

1. Within the woody layer trees are replaced by shrubs. Thus, the number of leaf layers decreases, although canopy cover may increase. As a result, wood production declines more than foliage production, whereas the net production of below-ground biomass increases.
2. The leaf layer suffered reduced productivity over the short term following droughts (Sect. 3.3.2), and over the long term, following lowered ground water tables. The relationship between rainfall in a dry year (Table 2.1) and woody plant production is complex. Whereas some woody species seem insensitive to drought, most species produce less for several years after a dry year or a drought period (Table 3.5).
3. Foliage production per leaf layer varies between species, so changes in species composition, especially those leading to the dominance of a few species, influenced foliage and browse production. Thus, woody plant productivity decreases as the proportion of evergreen species increases.

3.5 Phenology

3.5.1 Introduction

Phenological patterns determine the presence of foliage and fruits of woody plants throughout the year. Phenological patterns differ according to species, genetic variability within a species, plant age, site conditions such as soil type and incidence of fire; climatic factors such as rainfall distribution, air humidity, and temperature; and biotic factors such as pollination, browsing and insect attacks. In semi-arid regions climatic factors and site conditions seem most important. The less these factors vary, the more predictable are phenological patterns.

Leaf sprouting and shedding are more regular and predictable than flowering and fruiting. The initial flush of leaves and twigs, is not a manifestation of growth but a reallocation of stored material from the previous season. Flushing, therefore, may not require an uptake of water and nutrients (Walker 1980); it may be triggered by some external stimulus. For some woody species, flowering is predictable: for instance, flowering may occur after a specific amount or distribution of rainfall, or may be related to soil moisture remaining from the year before (Kummerow et al. 1977; Tolsma 1989; Seghieri 1990). Fruiting is less predictable: it requires specific combinations of rainfall, temperature and photoperiod, which are rare for some species in semi-arid zones (Davies 1976). Even if suitable conditions occur, insects and birds attacks can reduce fruiting.

Woody species are generally classified into two phenological categories, evergreen and deciduous, a classification reflecting different dry season survival strategies. Evergreen species have green leaves throughout the year. New leaves replace old ones before older leaves are shed. Semi-evergreen species are almost without green leaves for two months at the most. Drought deciduous species shed all leaves and have none for at least 2 months. Although deciduous species grow more rapidly under optimum conditions, evergreen species are more resistant to stress, and better adapted to low soil fertility (Stuart Chapin 1980; Menaut 1983).

3.5.2 Phenological Data from the SSWA

3.5.2.1 General

Phenological patterns of woody species in the SSWA are summarised in Table 3.11 (Boudet and Duverger 1961; Bille 1977; Poupon 1979; Cissé 1980; Grouzis and Sicot 1980; Hiernaux 1980; Piot et al. 1980; de Leeuw 1981; Liebermann 1982; Cissé and Wilson 1984; Seghieri 1990; de Bie 1991).

Table 3.11. Phenological patterns of woody species occurring in the SSWA, leaves in upper part of diagram, fruits in lower part

Evergreen riparian species

Acacia ataxacantha
Acacia nilotica
Dichrostachys cinerea
Mimosa pigra
Mitragyna inermis

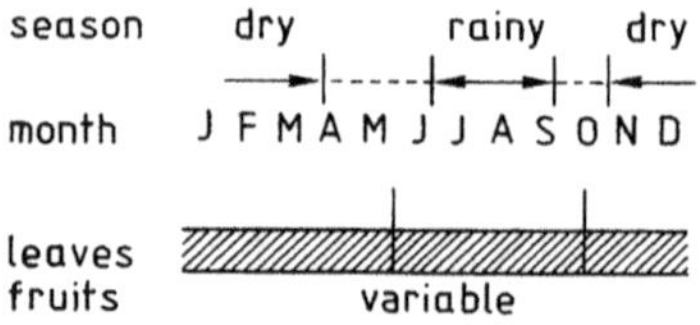

Evergreen upland species

Boscia spp.
Cadaba farinosa
Capparis tomentosa
Diospyros mespiliformis
Leptadenia pyrotechnica
Maerua spp.
Saba senegalensis
Salvadora persica
Tamarindus indica

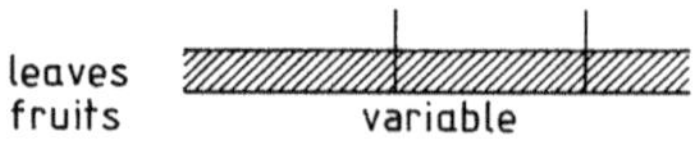

Semi-evergreen species

Acacia ehrenbergiana
Acacia raddiana
Acacia seyal
Annona senegalensis
Balanites aegyptiaca
Cassia sieberiana
Combretum glutinosum
Combretum nigricans
Euphorbia balsamifera
Gardenia spp.
Guiera senegalensis
Isoberlinia doka
Khaya senegalensis
Piliostigma spp.
Prosopis africana
Terminalia laxiflora
Terminalia macroptera
Vitellaria paradoxa
Ximenia americana
Ziziphus spp.

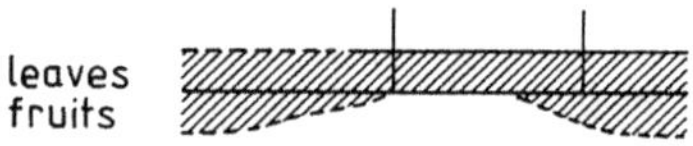

Early flushing deciduous species

Acacia dudgeoni
Acacia senegal
Afzelia africana
Anogeissus leiocarpus
Combretum aculeatum
Combretum micranthum
Combretum nigricans

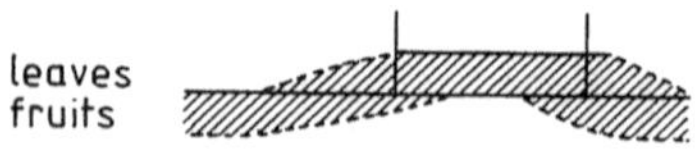

Table 3.11. *(Contd.)*

Commiphora africana
Crossopteryx febrifuga
Daniellia oliveri
Detarium microcarpum
Dichrostachys glomerata
Entada africana
Grewia bicolor
Lannea spp.
Maytenus senegalensis
Parkia biglobosa
Pericopsis laxiflora
Pterocarpus spp.
Sclerocarya birrea
Stereospermum kunthianum
Terminalia avicennioides

Late flushing deciduous species

Acacia macrostachya
Adansonia digitata
Bauhinia rufescens
Bombax costatum
Burkea africana
Feretia apodanthera
Securinega virosa
Sterculia setigera

Faidherbia albida

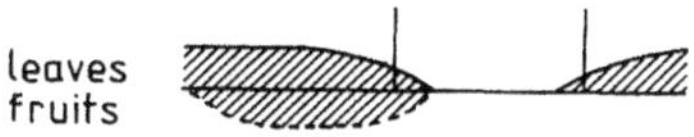

All species with leaves or fruits

Of all species a proportion of plants with leaves or fruits

A proportion of species or of plants with leaves or fruits

Species are classified as evergreen, semi-evergreen and deciduous. Evergreen species are classified into those that occur on riverbanks and floodplains, losing some or most of their leaves during flooding, and those that occur on uplands, with xeric adaptation mechanisms to prevent the leaves from drying. Usually, semi-evergreen species have no leaves from February to March, and flushing occurs either before or after the leaves are shed. Drought deciduous species shed their leaves early in the dry season, and are bare for at least 2 months. Some drought deciduous species flush early, at the end of the dry season, and some flush late, at the beginning of the rainy season. An exceptional adaptation to drought is shown by species associating little leaf biomass with photosynthetic

Table 3.12. Importance of woody species per phenological group, expressed as percentage of total number of species or total canopy cover per climatic zone

	Saharan Sahel	Northern Sahel	Southern Sahel	Northern Sudan	Southern Sudan
Species diversity:					
evergreen	~30	40	30	11	4
semi-evergreen	~50	27	26	29	26
deciduous	~20	33	46	62	72
Canopy cover:					
evergreen	~15	20	2	0	0
semi-evergeen	~80	50	55	54	41
deciduous	~5	30	42	46	59

activity in stems. *Leptadenia pyrotechnica* and *Balanites aegyptiaca* show such characteristics.

The proportion of evergreen woody species, both in terms of contribution to species diversity and canopy cover, decreases from north to south (Table 3.12). For deciduous species the inverse pattern appears. The few evergreen species in the Sudan zones are unimportant as regards their contribution to canopy cover. Semi-evergreen species are important in all climatic zones, particularly as regards their canopy cover.

3.5.2.2 The Sahel Zones

In the Sahel zones, air humidity and water availability may determine woody species flushing (Poupon 1979). *Commiphora africana*, described as a detector of air humidity, flushes when air humidity increases and sheds leaves just after the rainy season (Clanet and Gillet 1980). In northern Cameroon, *Sterculia setigera* and *Sclerocarya birrea* are also described as detectors of air humidity (Seghieri 1990). After flushing, the first rains prompt the first real growth. Incidental showers in the dry season elicite immediate flushing for many species on sandy soils, but for fewer species on clay soils, where plants still have green leaves (pers. observ.). Other woody species do not respond to rainfall, air humidity may be the triggering factor: *Adansonia digitata*, *Bombax costatum* and *Sclerocarya birrea*. On clay soils, *Combretum aculeatum* flushes and flowers in the same period of every year; on sandy soils, flushing and flowering is variable, probably due to greater variability in soil moisture. Variations between years are related to rainfall and insect attacks. Whereas *Acacia raddiana* flushes rapidly after locust attacks, *Balanites aegyptiaca* does not (Grouzis and Sicot 1980).

Several Combretaceae species on fallowlands remain green during the dry season, when the same species on the uplands are bare or have dried up leaves (pers. observ.). On clay soils and on catchments, most woody species keep green leaves longer than on other soils or habitats (Seghieri 1990). In a very dry year, however, *Acacia seyal* shows the opposite pattern (Cissé 1980). The phenology of some woody species varies with water availability of the site: they may be evergreen where soil moisture remains sufficient throughout the year, but become semi-evergreen or deciduous on drier sites (Seghieri 1990). Contrary to observations from Botswana (Skarpe and Bergström 1986), small and young *Combretum glutinosum* shrubs in the southern Sahel zone remain green longer than large or mature shrubs. Variation could be related to different site conditions.

Dry season foliage growth and flowering requires available water in the subsoil, especially for *Faidherbia albida*, which has green leaves during the dry season. Such species must either grow on humid sites or have roots that reach the groundwater. In the dry season, woody plants with green foliage indicate sites with available subsoil water.

For most woody species in the Sahel zones periodicity of flowering and fruiting is quite variable; information in Table 3.11 has indicative value only. Most species show one flowering period per year, but *Acacia senegal* sometimes flowers twice a year (Poupon 1979). *Anogeissus leiocarpus* and *Balanites aegyptiaca* may flower several times per year, although only one flowering results in fruiting (Seghieri 1990). Water availability influences flowering of most species: *Pterocarpus lucens*, *Boscia senegalensis*, *Balanites aegyptiaca*, *Commiphora africana* and *Combretum aculeatum* flowered shortly after a large shower in November. Other species flower around the same time every year: *Bombax costatum* around Christmas and *Sterculia setigera* flowers at the end of the dry season. In a dry year of 100 mm rainfall in Fété Olé, Senegal, most species did not flower at all, whereas for other woody species flowering did not lead to fruit set (Poupon and Bille 1974).

In the Sahel zones, large annual variations in fruit production are common. Attempts failed to explain and predict variations in fruit production of *Vitellaria paradoxa* (Ruyssen 1957).

It can be concluded that in the Sahel zones, water availability is the most important factor determining flushing, green leaf longevity and flowering. Some woody species seem less sensitive to climatic conditions (Grouzis and Sicot 1980).

3.5.2.3 The Sudan Zones

Towards the south, phenological patterns are more uniform and predictable, as rainfall varies less and bush fires are more common. Bush fires tend to occur in the early dry season (Sect. 2.4.1), causing immediate shedding of leaves, and if present flowers and fruits (César and Menaut 1974). Some species do not flush in the rainy season unless a bush fire has occurred in the dry season (Brookman-Amissah et al. 1980; Liebermann 1982). This may be an adaptation to prevent nutrient-rich young leaves from being burnt. Some species immediately produce young foliage after burning in the dry season. In the southern Sudan zone of Burkina Faso, about 75% of woody plants have foliage at the end of the dry season on sites burnt early in the dry season. This dropped to about 55% with no burning, and to about 45% with late burning (de Boer and Kessler 1994).

With increasing distance from a river, an increasing proportion of woody plants shed leaves in the dry season (César and Menaut 1974). Species such as *Cassia sieberiana*, *Guiera senegalensis* and *Sclerocarya birrea* keep green leaves longer in the southern parts of their ranges (de Bie 1991). In the southern Sudan zone, the number of species flushing and flowering relates to rainfall in the three preceding weeks exceeding 35 mm (Liebermann 1982). Individuals of most species show synchronous reproduction, available moisture being the determining factor.

It can be concluded that moisture availability and the incidence of fire determine phenological patterns in the Sudan zones. Photoperiodism seems more common for woody species of humid tropical regions.

Species which fruit in the rainy season produce fleshy fruits with animal-dispersed seeds mainly. Species which fruit in the dry season produce dry fruits with wind-dispersed seeds mainly (Liebermann 1982). Exceptions are *Sclerocarya birrea* and *Ziziphus mauritiana*, both producing fleshy fruits in the dry season.

3.5.2.4 Phenological Patterns and Browse Availability

On the basis of the distribution of woody species (Tables 2.4–2.7) and of the phenological data (Table 3.11) woody species foliage availability can be estimated per climatic zone, for five selected seasons (Table 3.13; Hiernaux 1980; Piot et al. 1980; Cissé 1981; Cissé and Wilson 1984; Penning de Vries and Djitèye 1991). Foliage availability per season is expressed as a proportion of annual foliage production. Variations occur according to geomorphological units, land-use types and microhabitats (Fig. 3.9).

In every climatic zone the largest proportion, 70–80% of foliage biomass, is available in the middle of the rainy season (Table 3.13). The smallest proportion is available late in the dry season, and decreases from 25% in the north, to 10% in the south, corresponding to a decreasing proportion of semi-evergreen species in the woody vegetation. The absolute quantity of foliage in the late dry season (to be determined

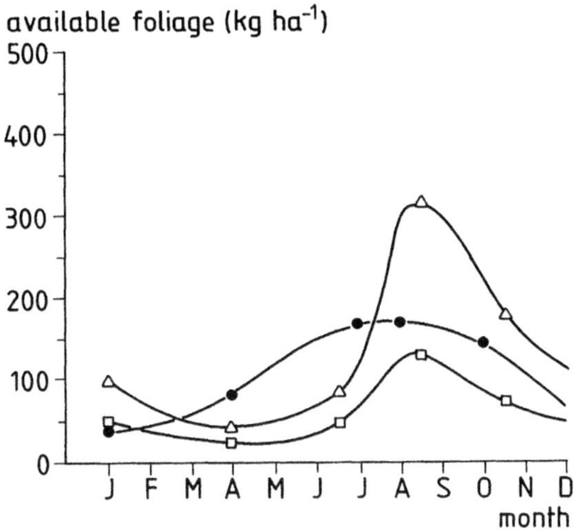

Fig. 3.9. Availability of foliage from woody plants throughout the year, at three sites. △ sandy soil, Niono ranch, Mali; □ loamy soil, Niono ranch, Mali; ● sandy soil, Mare d'Oursi, Burkina Faso. (After Penning de Vries and Djitèye 1991, Mali; Piot et al. 1980, Burkina Faso)

Table 3.13. Foliage availability per selected season and per climatic zone, expressed as percentage of total annual foliage production, based on the distribution of woody species before the drought periods in the SSWA

Season:	Early rainy (May–June)	Middle rainy (July–Sept)	Early dry (Oct–Nov)	Middle dry + cold (Dec–Feb)	Late dry + hot (March–April)
Saharan-Sahel	30	80	40	35	25
Northern Sahel	25	70	45	40	25
Southern Sahel	25	75	50	30[a]	15[a]
Northern Sudan	25	80	45	25[a]	8[a]
Southern Sudan	25	80	50	20[a]	5[a]

[a] Lower in case of bush fires in the early dry season.

from Fig. 3.8 and Table 3.13) decreases from 20 kg ha^{-1} in the northern Sahel zone to 70 kg ha^{-1} in the southern Sudan zone. However, foliage production in the Sudan zones is probably over-estimated, as it is based on the assumption that bush fires occur late in the dry season. On the contrary, most bush fires occur early in the dry season, and leaves are shed at that time (Sect. 2.4.2). On moist sites such as riverbeds, green foliage remains until late in the dry season.

3.5.3 Conclusions

The proportion of evergreen and semi-evergreen species in woody plant communities decreases with increasing rainfall. The proportion of available dry season browse drops from 25–40% of total annual browse in the northern Sahel zone, to 5–25% in the southern Sudan zone. The southern zones, however, have the largest absolute quantities of browse. Water availability is one determinate of periodicity of leaf flushing and shedding, flowering and fruiting; fire is another factor in the Sudan zones. Photoperiodism increases from north to south, as does the predictability of the phenological pattern.

3.6 Nutrient Concentrations, Digestibility and Palatability

3.6.1 Introduction

Nutrient concentrations of woody plant tissues may vary in time, between seasons and years, between sites, e.g. of different fertility, and according to the development stage of the woody plant or woody plant component. Knowledge of nutrient concentrations provides insight into the allocation of nutrients in woody plant tissues, and in nutrient cycling in a vegetation with woody plants. These aspects are considered in detail in Section 4.5.

One indication of the nutritive value of a forage is the digestibility of organic matter (DOM). The apparent DOM is the proportional difference between the quantities consumed and excreted in faeces (Minson 1990). The DOM is generally determined by an in vitro analysis in a laboratory using the ruminal fluid of herbivores, or enzymes (van Soest 1982). In vitro estimates are corrected according to a standard set of in vivo measurements for a given forage and herbivore species.

For temperate grasses and legumes, DOM is closely related to digestible energy. The digestible crude protein fraction is an important component of the organic matter. Inclusion of this variable in predicting digestible energy from DOM values largely reduces the remaining error. DOM is limited by the proportion of crude fibre mainly. In woody plants, crude fibre content is mainly determined by lignin content. Tannins, phenolic compounds in woody plant tissues, also influence DOM. Two major types of tannins occur: hydrolysable and condensed tannins. The two types may have different physiological effects on some herbivores, but have a similar influence on digestion in deer (Hanley et al. 1992). Plant tannins form complexes with proteins and/or with pepsins responsible for protein digestion. The complexes resist microbial activities, so reduce protein digestibility and disrupt the animals' digestion processes (Kumar 1983; Topps 1992). Tannins increase plant resistance against herbivore, insect and fungal attacks (Topps 1992).

Table 3.14. Nitrogen (N) and phosphorus (P) concentrations, digestibility and phenolic concentrations of fruits and foliage of common woody plants in the SSWA

Woody species	No. of samples[a]	N ($g\,kg^{-1}$)		P ($g\,kg^{-1}$)	Phenols ($g\,kg^{-1}$)[c]		Digestibility[d] ($g\,kg^{-1}$)
		Mean	Range[b]		(a)	(b)	
Nodulating woody legumes							
Acacia ataxacantha							
leaves	9	32	19–44			70	160–690
Acacia nilotica							
fruits	5	19	16–21	1.8			510
leaves	10	21	16–28	1.5	284	229	500–690
Acacia raddiana							
fruits	18	25	19–29	2.4			680
leaves	8	27	21–29	2.0			440
Acacia senegal							
fruits	8	25	18–32	2.0			570
leaves	18	32	20–48	2.0	24	11	650
Acacia seyal							
fruits	8	30	28–34				
leaves	28	26	18–38	2.2	170	121	410–850
Dichrostachys cinerea							
fruits	2	25	21–28				
leaves	21	26	14–49	2.0		52	290–760
Faidherbia albida							
fruits	13	19	16–21	1.2	42	30	540–630
leaves	10	25	16–36	2.0	43	28	310–570
Prosopis africana							
leaves	5	22	17–27			160	370–650
Pterocarpus erinaceus							
fruits	3	19	17–24	1.8			
leaves	10	26	22–34	1.8			410–580
Pterocarpus lucens							
fruits	2	26	24–28				630
leaves	29	25	14–40	1.6	16	6	330–670
Non-nodulating woody legumes							
Afzelia africana							
fruits	1	21		1.5			
leaves	3	25	18–29	2.0			560
Bauhinia rufescens							
fruits	8	20	18–28	2.0	62	55	260–500
leaves	9	23	19–35	1.5	34	23	310–650
Danielia oliveri							
leaves	10	22	14–37	2.0			540–650
Isoberlinia doka							
leaves	4	22	16–30				420–690
Piliostigma reticulatum							
fruits	9	12	11–16	1.5			
leaves	20	24	13–37	2.0	42	26	160–650

Table 3.14. *(Contd.)*

Woody species	No. of samples[a]	N ($g\ kg^{-1}$) Mean	N ($g\ kg^{-1}$) Range[b]	P ($g\ kg^{-1}$)	Phenols ($g\ kg^{-1}$)[c] (a)	Phenols ($g\ kg^{-1}$)[c] (b)	Digestibility[d] ($g\ kg^{-1}$)
Piliostigma thonningii							
fruits	2	16					
leaves	5	20	11–24				
Non-legumes							
Anogeissus leiocarpus							
leaves	6	18	13–21	1.2	25	186	210–380
Balanites aegyptiaca							
fruits	6	16	11–18	1.0			
leaves	46	20	11–38	1.0	15	5	590–720
Bombax costatum							
fruits	4	16	6–27	1.3			370–550
Boscia angustifolia							
leaves	16	28	22–35		17	5	310–610
Boscia senegalensis							
fruits	1	27					
leaves	19	34	24–48	1.5	15	6	430–720
Cadaba farinosa							
leaves	19	40	29–50	1.5	8	4	400–760
Combretum aculeatum							
leaves	13	26	16–48	2.2	43	34	550–670
Combretum ghasalense							
leaves	4	21	13–26	2.0		106	430
Combretum glutinosum							
leavels	23	17	7–27	2.0		56	260–650
Combretum micranthum							
leaves	10	22	13–35			54	490–770
Combretum nigricans							
leaves	14	20	14–41			70	200–740
Commiphora africana							
leaves	4	22	18–26	1.7			430
Detarium microcarpum							
leaves	13	14	13–16				580
Gardenia spp.							
fruits	1	9					
leaves	10	12	8–24	2.2			480–650
Grewia bicolor							
fruit	1	8		1.5			400
leaves	13	23	13–32	1.5			550
Guiera senegalensis							
fruits	1	20		2.0			
leaves	18	20	11–32	1.8	53	39	120–610
Khaya senegalensis							
leaves	5	24	12–37	2.0			560
Lannea acida							
leaves	3	16	14–30				

Table 3.14. *(Contd.)*

Woody species	No. of samples[a]	N (g kg^{-1})		P (g kg^{-1})	Phenols (g kg^{-1})[c]		Digestibility[d] (g kg^{-1})
		Mean	Range[b]		(a)	(b)	
Maerua crassifolia							
leaves	19	37	29–45	1.5		9	750
Salvadora persica							
leaves	7	21	16–26	1.0			
Sclerocarya birrea							
fruits	1	11		2.0			
leaves	16	20	10–31	1.5		94	360–870
Terminalia avicennioides							
leaves	6	15	11–18				550–640
Ziziphus mauritiana							
fruits	1	13		1.0			
leaves	24	23	13–30	1.8		31	470–820

[a] Refers to the number of samples monitored for N and P concentrations; the number of samples for digestibility and phenol concentrations are lower.
[b] Lowest value refers to old, senescent leaves; highest value to young, green leaves.
[c] (a) = all phenols, (b) = tannins only.
[d] Digestibility of organic matter in vitro (see text).

This section does not consider other toxic compounds in foliage, such as mimosine in *Leucaena leucocephala*. Comprehensive treatment is given in Lowry (1990).

In Table 3.14, data are summarised on N and P concentrations, organic matter digestibility and concentrations of phenols and tannins, for foliage and fruits of woody species in the SSWA (Fotius and Valenza 1966; Piot 1969; ILCA 1975; Boudet and Toutain 1980; Le Houérou 1980a,b; Piot et al. 1980; Diagayeté 1981; Klein et al. 1981; de Leeuw 1981; de Bie 1991; Fall 1991, and own unpubl. data, mainly from Mali and Burkina Faso). Most available data address N concentrations in leaves. Although N and P concentrations in leaves vary greatly throughout the year, the season and phenological phase of the sample are often not specified. Therefore, averaged data are presented, but only for those woody plants for which at least three values are available. In Table 3.14 a distinction was made between woody legumes with nodules, indicating their capacity for nitrogen fixation (Sect. 4.5.4.8), woody legumes without nodules, and other woody plants. Nitrogen and phosphorus concentrations are summarised for these three groups, per climatic zone, in Table 3.15.

3.6.2 Nutrient Concentrations of Woody Plant Components

3.6.2.1 Browse

Nitrogen

Mean foliage N concentration of all woody plants examined is 24 g kg^{-1}, varying from 16–33 g kg^{-1} (Table 3.15). Highest foliage N concentrations, at average more than 30 g kg^{-1}, are found with *Acacia ataxacantha, Acacia senegal*, *Boscia senegalensis*, *Cadaba farinosa* and *Maerua crassifolia* (Table 3.14). The latter three species are not legumes, but are from the family Capparidaceae. Foliage N concentrations less than 18 g kg^{-1} are found with *Bombax costatum*, *Combretum glutinosum*, *Detarium microcarpum*, *Gardenia* spp., *Lannea acida* and *Terminalia avicennioides*.

Mean foliage N concentrations in nitrogen-fixing woody legumes are about 3 g N kg^{-1} higher than in other woody species (Table 3.15). The difference is small compared to that between leguminous and non-leguminous herbs, at least 10 g N kg^{-1} (Penning de Vries and Djitèye 1991). Mean foliage N concentrations of woody species are higher in the Sahel zones than in the Sudan zones, or are similar (Table 3.15).

For all woody plants examined, foliage N concentrations of leaves decrease with leaf age, from young leaves, to old leaves, to leaf litter (Table 3.14). Figure 3.10 shows average foliage N concentrations for five woody species over the course of one year. In contrast to the two deciduous species, average foliage N concentrations in the evergreen species *Boscia angustifolia* and the two semi-evergreen species *Balanites aegyptiaca* and *Combretum glutinosum* vary little, and do not show any

Table 3.15. N and P concentrations and the P/N ratio of foliage, for nodulating and non-nodulating woody legumes and non-leguminous woody plants, per climatic zone

	Nodulating woody legumes		Non-nodulating woody legumes		Non-leguminous woody plants	
Climatic zone	Sahel	Sudan	Sahel	Sudan	Sahel	Sudan
N concentration (g kg^{-1}):						
mean maximum values	36	32	36	31	33	30
mean of all values	26	23	23	20	24	20
mean minimum values	18	17	15	15	16	13
number of samples	7	7	3	6	18	17
P concentration (g kg^{-1}):						
mean of all values	1.9	1.9	1.8	1.8	1.6	1.8
number of samples	7	5	3	4	14	11
P/N ratio	0.07	0.08	0.08	0.07	0.07	0.09

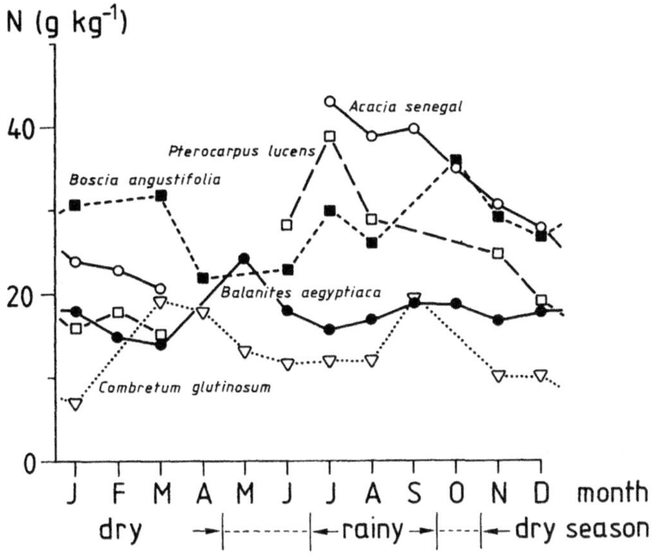

Fig. 3.10. Average nitrogen concentration of leaves throughout the year, for five woody species. ○ *Acacia senegal*, Sahel zones; ● *Balanites aegyptiaca*, Sahel zones; ■ *Boscia angustifolia*, Sudan zones; ▽ *Combretum glutinosum*, Sudan zones; ○ *Pterocarpus lucens*, Sudan zones. (Bernhard-Reversat 1982, Sahel zones; de Bie 1991, Sudan zones)

correlation with season. In evergreen species, different phases of leaf development occur at the same time, on the same tree, but young leaves have the highest nutrient concentrations. In deciduous species, average N concentration varies consistently, corresponding to seasonally determined leaf development (Evans 1979; Bernhard-Reversat 1982; de Bie 1991). There are no significant differences in mean year-round N concentrations between evergreen, semi-evergreen and deciduous woody species.

Mean N concentrations of fruits are generally between those of young and old leaves and vary little with maturity. As a result, in the dry season, fruits have higher N concentrations than leaves. Species that produce new leaves in the dry season are an exception (*Faidherbia albida*).

Phosphorus

There is too little data on phosphorus concentrations to determine significant differences between woody species or different growth stages. Researchers have found that several Combretaceae, *Combretum aculeatum*, *Guiera senegalensis* and *Gardenia erubescens*, have relatively

Table 3.16. N and P concentrations of woody plant foliage in the SSWA, East Africa and in southern Africa

	Legumes		Non-legumes		All	
	N ($g\,kg^{-1}$)	P ($g\,kg^{-1}$)	N ($g\,kg^{-1}$)	P ($g\,kg^{-1}$)	N ($g\,kg^{-1}$)	P ($g\,kg^{-1}$)
This study:						
Sahel zones	29[a]	1.9[a]	24	1.7	18[b]	1.8
Sudan zones	23[a]	1.9[a]	20	1.8	15[a]	1.8
Brinckmann and de Leeuw (1984):						
Sahel zone					24	1.6
Sudan zone					19	1.6
East Africa					23	2.0
Tolsma (1989):						
Botswana	30	1.1				
Le Houérou (1980b):						
West Africa	27	2.5			20	1.5
East Africa	24	2.6			21	2.0

[a] Nodulating woody legumes only.
[b] N concentration of all browse (leaves, twigs and flowers).

high P concentrations in their foliage (Le Houérou 1980b; Toutain 1980) but the summarised data do not confirm this. Average P concentrations of foliage are somewhat lower in the SSWA than in East Africa (Table 3.16), except for nitrogen-fixing woody legumes. For woody species occurring in both West and East Africa, such as *Faidherbia albida*, *Acacia raddiana* and *Acacia senegal*, foliage P concentrations are consistently higher in East Africa (Dunham 1989). This may be related to the greater occurrence of fertile, volcanic soils in East Africa.

In Botswana, P concentrations decrease with increasing maturity. The decrease is less remarkable for evergreen and semi-evergreen species than for deciduous species (Skarpe and Bergström 1986; Tolsma 1989).

General

Nutrient concentrations are also influenced by (1) site conditions, (2) woody plant age, (3) exploitation intensity and (4) location in canopy. It is difficult to draw conclusions from studies which do not consider these parameters.

1. Nutrient concentrations on infertile sites tend to be lower than those on fertile sites (Vitousek and Sanford 1986). For some woody species,

foliage concentrations of nitrogen, phosphorus and potassium on a fertile site near a water point are up to twice as high as those on less fertile sites. In other species, foliage nutrient concentrations do not differ significantly between rich and poor sites (Tolsma 1989; Sect. 4.5.4). A correlation between nutrient concentrations and soil nutrient availability is well documented for herbs in the SSWA (Breman and de Ridder 1991).

2. In India, nutrient concentrations in litter decrease as the woody plant of origin age (Gill et al. 1987). Woody plant ageing, however, is sometimes related to increasing site exploitation, leading to a gradual depletion of the soil. In that case, site condition would be at least as likely to determine the nutrient concentration as woody plants age.
3. Nitrogen concentrations are lower in frequently defoliated shrubs. The reduction may discourage browsing. Site condition may be related to the reduction in N concentration, because overgrazing leads to soil nutrient depletion (Piot 1980).
4. Evans (1979) concludes that in *Gmelina arborea*, N and P concentrations are highest and calcium concentrations are lowest where plant tissues grow most actively in the canopy. Foliage N concentrations vary as much as 70% within canopies.

3.6.2.2 Wood and Roots

The nutrient concentrations in various woody plant components are probably related to each other. In tropical forests, variations in nutrient concentrations in roots and wood are correlated with nutrient concentrations in the foliage of the preceding year (Noij et al. 1990).

Few data are available on nutrient concentrations of twigs, stem wood and branches. Twig nutrient concentrations are higher than those of branches and stem wood, and less than those of foliage. Nutrient concentrations of young *Faidherbia albida* twigs resemble those of foliage, but nutrient concentrations of mature and thick twigs resemble those of stem wood (Sterck et al. 1991). *Acacia senegal* N concentrations are $6.2\,g\,kg^{-1}$ for twigs and $4.8\,g\,kg^{-1}$ for stems and branches (Bernhard-Reversat and Poupon 1980). In Mali, stem wood N concentrations range from $2-20\,g\,kg^{-1}$, with a mean of $5\,g\,kg^{-1}$, and P concentrations range from $50-400\,mg\,kg^{-1}$, with a mean $200\,mg\,kg^{-1}$ (own unpubl. data). These estimates are comparable with data from semi-arid Australia (Beadle and White 1968; Crane and Raison 1980; Turner 1981).

Nitrogen and phosphorus concentrations of wood decrease with increasing woody plant age (Egunjobi and Bada 1979). Nitrogen and phosphorus concentrations in young wood are about twice those of old wood of *Faidherbia albida* (Sterck et al. 1991).

Only one reference on nutrient concentrations of roots was found: a N concentration of 16.6 g kg^{-1} for roots of *Acacia senegal*. The root type was not indicated (Bernhard-Reversat and Poupon 1980).

3.6.3 Digestibility of Forage from Woody Plants

The digestibility of organic matter (DOM) of foliage from woody plants is based on in vitro analyses (Table 3.14). The in vivo standard analyses, used to correct the in vitro results, were done for standard types of herbaceous material. This procedure has limited application to woody plant components, because these components commonly contain substances which influence DOM. More recent studies used woody material as a standard (Diagayeté 1981). The available DOM values are difficult to evaluate, as different in vitro analyses, using different in vivo standards, are involved. Average DOM values for the SSWA range from about 600 g kg^{-1} for young leaves, to 200–500 g kg^{-1} for old leaves. Foliage averages 530 g kg^{-1}.

For a number of forages, degradation profiles of organic matter and crude protein were determined by in sacco analyses (Fall 1991). Comparison of the results with those from Koné (1987; cited by Fall 1991) shows large variation for some species. For example, digestibility of crude proteins in *Faidherbia albida* foliage ranges from 160 g kg^{-1} (ISRA 1983) to 200 and 530 g kg^{-1} (Fall 1991).

For forage of one species, N concentrations often correlate with DOM, both N concentrations and DOM decrease as the leaf matures. Protein digestibility is determined by N and tannin concentrations (Hanley et al. 1992), but for many forages factors other than protein digestibility influence DOM, so that there is little correlation between N concentrations and DOM (Minson 1990). For instance, *Boscia senegalensis* foliage, with high protein concentrations, has a low DOM, whereas *Acacia* spp. foliage commonly has a low concentration of protein and a high DOM. Analyses of the summarised SSWA data show no significant differences in DOM values between nitrogen-fixing woody plants and other woody plants, so other factors than N concentration also influence DOM. Other factors that can be responsible for reduction of DOM are crude fibre, lignin, silica and tannins.

Crude fibre varies from 150–200 g kg^{-1} for young leaves to 150–400 g kg^{-1} for old leaves. Crude fibre in fruits varies from 100–500 g kg^{-1}. Lignin is a component of crude fibre. Lignin concentrations vary from 30–160 g kg^{-1} for leaves and from 80–260 g kg^{-1} for woody legume pods. Lignin concentrations increase with increasing leaf maturity. Lignin concentrations are higher for total browse than for foliage, because of the high lignin concentrations in twigs (Diagayeté 1981). For the Sudan zones, DOM is negatively correlated with lignin concentrations, but N concentrations show no significant correlation with DOM (de Bie 1991).

High silica concentrations probably reduce DOM, but the effect has not been quantified (Le Houérou 1980b). Data on silica concentrations are scarce. *Stereospermum kunthianum* and *Cadaba glandulosum* are reputed to have high silica concentrations. Silica concentrations average $22\,g\,kg^{-1}$ in West Africa and $12\,kg^{-1}$ in East Africa, but the number and species composition of plants studied are not specified (Le Houérou 1980b).

Table 3.14 shows available data on phenolic and tannin concentrations in woody plant foliage. Phenolic concentrations generally range from $10–300\,g\,kg^{-1}$, and tannin concentrations from $5–200\,g\,kg^{-1}$. *Acacia nilotica, Acacia seyal* and *Anogeissus leiocarpus* show highest tannin concentrations. In India, *Acacia nilotica* foliage has higher phenolic concentrations, $325\,g\,kg^{-1}$ (Kumar 1983), than in the SSWA ($64\,g\,kg^{-1}$), although DOM differs little. The differences may be related to differences in tannin compounds.

Phenolic concentrations are highest for young leaves (Fall, pers. comm.), but de Bie (1991) found that in most woody species the total tannin concentration increases with leaf maturity. Highest tannin concentrations in young leaves corresponds to data from Botswana, where phenolic concentrations are highest for young tree leaves and immature fruits and seeds. High phenolic concentrations in young leaves may be a storage of the precursors of cell wall growth (Tolsma 1989).

The relatively low DOM values of woody plant foliage are probably largely due to tannins (McKey et al. 1978; Breman et al. 1979; Diagayeté 1981; Kumar 1983; Waterman et al. 1984). Tannins also probably cause low protein digestibility in woody plant foliage, regardless of N concentrations. Tannins reduce digestible protein by 44% and DOM by 14% (Hanley et al. 1992). Negative, but not significant relationships between DOM and total tannin concentration appear for woody species from the Sahel zones (Diagayeté 1981) and from the Sudan zones (de Bie 1991).

The mean concentration of tannins in vegetation on less fertile soils is $46\,g\,kg^{-1}$, on more fertile soils $25\,g\,kg^{-1}$ (McKey et al. 1978; Cooper and Owen-Smith 1985). The higher tannin concentration may represent an adaptation to reduce nutrient loss by exploitation (Janzen 1974). Foliage tannin concentrations increase after leaf damage, e.g. following browsing (Cooper and Owen-Smith 1985). Tannins particularly reduce forage consumption by herbivores, including many insects, when concentrations exceed a threshold, probably about $50\,g\,kg^{-1}$ (McKey et al. 1978; Panda et al. 1983; Cooper and Owen-Smith 1985). However, pods of woody species with high tannin concentrations, such as *Acacia nilotica*, are readily consumed (Tolsma 1989).

In several studies, a threshold of tannin concentration is associated with low DOM. In Mali, DOM in forage samples decreases from an average of 570 g kg^{-1} in the absence of tannins, to 430 g kg^{-1} for tannin concentrations of at least 50 g kg^{-1} (Diagayeté 1981). Comparison of tannin concentrations and DOM values from Table 3.14 suggests a threshold at about 60 g kg^{-1}; at higher total tannin concentrations DOM is generally lower than 500 g kg^{-1}. In India, DOM in woody plants decreases from 700 g kg^{-1} for foliage without tannins, to less than 100 g kg^{-1} for foliage with total tannin concentrations exceeding 70 g kg^{-1} (Lohan et al. 1980; Panda et al. 1983).

3.6.4 A Comparison with Herbaceous Foliage

Mean N and P concentrations and DOM values are higher throughout the year for most woody plants than for herbaceous foliage. Woody plant foliage may show lower nutrient concentrations in the early rainy season than herbaceous foliage, but in general shows a less dramatic decrease of nutrient concentrations and DOM with increasing maturity. Nutrient concentrations of woody plant components vary less than those of herbs, in relation to annual rainfall, climatic zone and soil characteristics. Nitrogen concentrations of annual plant species may vary by a factor of three according to site conditions (Breman and de Ridder 1991). The lower variability of N concentrations for perennial grasses and woody plants is due to their ability to accumulate, store and translocate nutrients (Sect. 4.5.4).

Tannins are a toxic secondary plant component and a defense mechanism against browsing. Tannins are related to lignosity, and thus are found mainly in woody plants. Alkaloids are more common in herbs. Both alkaloids and tannins reduce browsing of plant components (Lebreton 1982).

3.6.5 Palatability of Forage from Woody Plants

Woody plants forage can be classified according to its palatability. Palatability refers to the rate that animals select and consume forage (Minson 1990). Species palatability values are relative to each other. Where a mixture of forages is available, the selection of one type of forage depends on the availability of alternative types of forage. A relatively unpalatable species, ignored in the midst of more palatable species, is often eaten if nothing else is available.

There is agreement on the palatability of forage of some woody species. There are many species, however, that cannot be classified as palatable or unpalatable: some authors refer to them as palatable whereas

Table 3.17. Palatability of woody plant components for domestic herbivores, per woody species

Woody species	Palatability[a]	Components[b]	Herbivores[c]
Acacia ataxacantha	0	fr, l	c, g, s
Acacia dudgeoni	+	l	all
Acacia nilotica	0	fr !, l	c, g, s
Acacia raddiana	+	fr !, l	c, g, s
Acacia senegal	+	fr !, l	c, g, s
Acacia seyal	+	fr, l	all
Afzelia africana	+	l	all
Anogeissus leiocarpus	0	l	all
Balanites aegyptiaca	+	fr, l	c, g, s
Bauhinia rufescens	+	fr !, l	c, g, s
Bombax costatum	+	fr, l	all
Boscia angustifolia	+	l	all
Boscia senegalensis	+	l	all
Burkea africana	0	l	all
Cadaba farinosa	+	fr, l	all
Capparis tomentosa	0	l	all ?
Combretum aculeatum	+	l	all
Combretum ghasalense	0	l	all
Combretum glutinosum	0	l	all
Combretum micranthum	0	l	g, s
Combretum nigricans	0	l	?
Commiphora africana	+	l	?
Daniellia oliveri	0	l	all
Detarium microcarpum	0	l	?
Dichrostachys cinerea	+	fr, l	all
Dichrostachys glomerata	0	fr, l	all
Euphorbia balsamifera	−		
Faidherbia albida	+	fr !, l	c, g, s
Feretia apodanthera	+	l	all
Gardenia spp.	0	fr !	all
Grewia bicolor	+	fr, l	all
Guiera senegalensis	0	fr, l	all
Isoberlinia doka	0	l	?
Khaya senegalensis	+	fr, l!	all
Lannea spp.	0	l	?
Leptadenia pyrotechnica	0	l	c, g
Maerua crassifolia	+	fr, l	all
Parkia biglobosa	0	l	?
Piliostigma reticulatum	+	fr, l	all
Piliostigma thonningii	+	fr, l	all
Pterocarpus erinaceus	+	fr, l!	all
Pterocarpus lucens	+	fr, l!	all
Salvadora persica	+	fr, l	all
Sclerocarya birrea	+	fr, l	all
Securinega virosa	+	l	all
Terminalia avicennioides	0	fr, l	?
Vitellaria paradoxa	0	fr, l?	?
Ximenia americana	+	fr, l	all
Ziziphus mauritiana	+	fr, l	all

[a] Palatability +: high, 0: moderate, −: low.
[b] Components fr: fruits, l: leaves, !: important.
[c] Herbivores c: camel, g: goat, s: sheep, all: c, g, s and cattle, ?: uncertain.

Table 3.18. Qualitative aspects of highly and moderately palatable woody species in the SSWA

	Highly palatable	Moderately palatable
Average N concentration ($g\,kg^{-1}$)	24	19
Tannin content ($g\,kg^{-1}$)	<60	>60
Average digestibility of organic matter ($g\,kg^{-1}$)	520	350

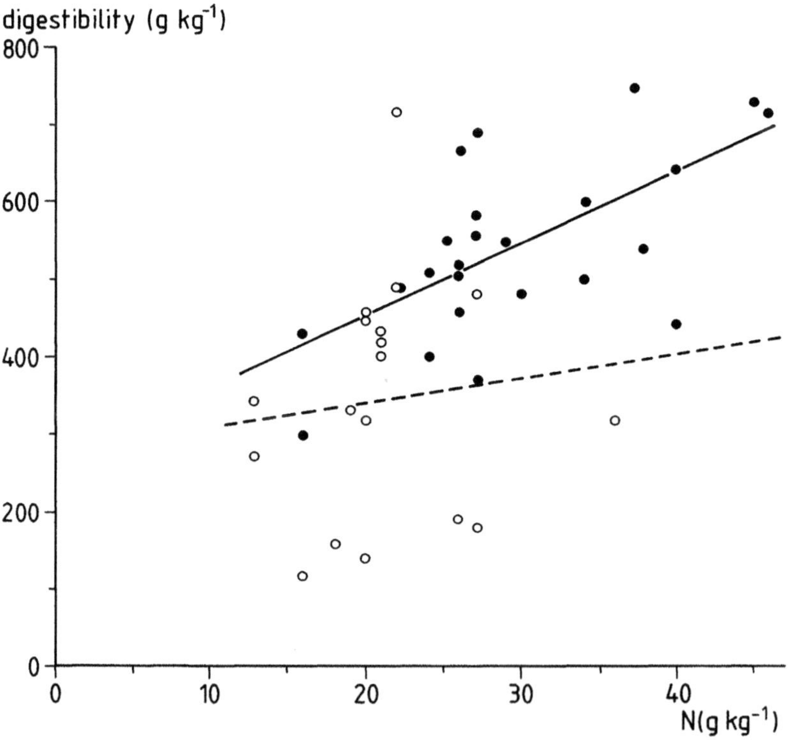

Fig. 3.11. Relationship between nitrogen concentration and digestibility of foliage of palatable and moderately palatable woody plants in the Sahel zones. ● palatable species; ○ moderately palatable species

others have never observed an animal eating them. Table 3:17 classifies woody plant foliage and fruits of the SSWA as highly palatable, moderately palatable and unpalatable for domestic livestock (Boudet and Crue 1966; Piot 1970; Boudet 1971; de Leeuw 1975; Diallo 1978; Boudet and Toutain 1980; Le Houérou 1980a; Piot et al. 1980; Toutain 1980; Klein et al. 1981; Zech and Von Maydell 1981; ISRA 1983; Egging 1990; own unpubl. observ.).

In the Sahel zones, palatable woody plant foliage has significantly higher mean DOM values, 520 g kg^{-1}, than moderately palatable species, 350 g kg^{-1}. Palatable foliage has higher average N concentrations and lower tannin concentrations, less than 60 g kg^{-1}, than moderately palatable species (Table 3.18). Palatable woody species show a linear relationship between foliage N concentrations and DOM (correlation coefficient = 0.64), but moderately palatable species do not show this relationship (correlation coefficient = 0.11) (Fig. 3.11). An exponential relationship between N concentration and DOM describes palatable species even better (correlation coefficient = 0.81), suggesting that DOM reaches a maximum with increasing N concentration.

3.6.6 Conclusions

In semi-arid regions, nutrient concentrations and DOM of woody plant components vary according to woody species, woody plant components and development stage of components. Additional variation is related to site conditions, which may be influenced by previous exploitation intensity. Woody plant foliage nutrient concentrations and DOM vary less by season and site than does herbaceous foliage. Woody plant foliage has relatively high N concentrations during the dry season, so are particularly useful to herbivores as supplementary food. For herbivores, the availability of energy and proteins from woody foliage is limited by low DOM, regardless of high N concentrations. Woody plant foliage DOM values, on average 500 g kg^{-1}, are lower than DOM values of green grasses, 600–800 g kg^{-1}, but higher than DOM values of dried out grasses, 300–500 g kg^{-1}. Tannins reduce DOM of woody plant foliage, particularly when tannin concentrations exceed 60 g kg^{-1}. Tannin concentrations vary according to woody species and plant components. High lignin and silica concentrations may also reduce DOM of woody plant foliage. Woody plant foliage palatability is correlated with DOM, N concentrations and tannin concentrations (Table 3.18). Factors which reduce digestibility may be more important in the SSWA than in East Africa, and may contribute to woody plant protection against herbivore exploitation under resource-poor conditions.

4 The Influence of Woody Plants on Plant Production Factors

4.1 Introduction

In semi-arid regions it can generally be observed that growth of herbs under a woody canopy differs considerably from that in open areas. If growth under the woody canopy is better than elsewhere, it may seem logical to conclude that woody plants are improving the availability of resources required for primary production (light, water and nutrients mainly). Thus one may conclude that "trees are good". Taking the argument one step further, it may be assumed that cutting woody plants will decrease, and planting them will improve herbaceous production.

Because of these far-reaching conclusions, the types of evidence have to be carefully considered. To what extent do such observations allow conclusions on a beneficial role of woody plants for agro-ecosystems in semi-arid regions?

The basic resources needed for growth of any plant are the same. A major question is to what extent woody plants concentrate scarce resources for their own use or whether they also increase total resource availability. The latter may result from a reduction of losses or by niche differentiation. Increased nitrogen fixation and utilisation of water and nutrients from deep soil layers by woody plants are most commonly mentioned. Insight into these interactions can be obtained by analysing resource supply and quantifying processes by which woody plants use and modify availability of resources. On such basis predictions can be made of how best to use the advantages of woody plants: which woody plants are good, under which conditions and for what purpose? What is the order of magnitude of benefits to be expected, in terms of productivity for human use?

In Section 4.2 are summarised available data on correlations between woody- and herb layer productivity from semi-arid regions. The analysis of such data should provide some insight into the conditions determining patterns of woody- and herb layer interactions. These observations do not allow conclusions on causal mechanisms. Correlations between two organisms can always be caused by a common response to environmental factors, with both spatial and temporal variation.

Root systems of woody plants and their dynamics are discussed in Section 4.3. In following sections are analysed the influence of woody plants on the plant production factors water, nutrients and light (Sects. 4.4, 4.5 and 4.7). Specific attention is given to the nutrient cycles of the key nutrients nitrogen and phosphorus. In Section 4.6 the role of soil organic matter is discussed in relation to organic matter inputs from woody plants. In the concluding Section 4.8 the previous information is used to compare productivity of a vegetation without woody plants with a vegetation with woody plants, in relation to the availability of the production resources water, nutrients and light. Obviously, the main focus is on the vegetation with woody plants, on the basis of which conclusions and recommendations will be formulated for the potential role of woody plants in sylvopastoral and agroforestry systems (Chap. 5).

4.2 Interactions Between the Woody and Herbaceous Layer

4.2.1 Introduction

Regarding interactions between woody plants and herbs three types of evidence are considered:

1. Section 4.2.2: relationships between the woody and herbaceous layer. Here are analysed spatial patterns across geomorphological units and microsites within a region and around individual woody plants, as well as temporal variability in each of these situations. The specificity of each woody plant with its prevailing site conditions makes it difficult to extrapolate findings from one particular situation. The historical context of a site should also be taken account of to adequately analyse the actual situation: what was initial soil fertility before establishment of the woody plant, has the location been previously cultivated, did bushfires occur, what is and has been the grazing pressure? Very rarely is such information provided for a research site.
2. Section 4.2.3: documented response to removing or planting woody plants. This type of evidence is more rare, as it takes time, relatively large plots and adequate controls. A major problem is to distinguish between short-term and long-term effects. If woody plants have a positive effect on soil fertility but limit herbaceous production by shading, removing woody plants will lead to a marked increase of grass growth initially, followed by a decline as the positive effect is slowly fading out. Conversely, increasing woody cover would rapidly lead to a negative shading effect whereas the positive soil fertility effect takes more time to develop.

3. Section 4.2.4: herbaceous plant composition in relation to woody plants. Apart from general patterns of interactions between woody and herbage layers, specific herb species show specific response.

4.2.2 Relationships Between the Woody and Herbage Layer

4.2.2.1 Correlations Between Woody and Herbage Layer Productivity

On a regional level different vegetation types occur in association with specific site conditions. Vegetation types differ as regards the cover of the woody and herbage layers. If on a sub-regional level positive correlations between woody and herbage cover occur, it is most probable that both vegetation components commonly respond to environmental factors varying between the two vegetation types.

Both woody cover and herbaceous cover are often highest in places receiving runoff, such as in the *Acacia aneura* groves in Australia (Perry 1970). Here the positive correlation between woody and herbaceous cover is based on both variation in moisture and nutrient availability, the latter being deposited with the runoff water. A positive correlation between the two vegetation components also occurrs on a time scale. In semi-arid Australia, both woody plants and herbs respond to rainfall variations in a similar way: with increasing rainfall between 1973 and 1976 both the density of woody plants and herbaceous production increased up to tenfold (Foran et al. 1982). The increased rainfall leads to increased availability of both moisture and nutrients, the latter becoming more available with increasing time and depth of soil wetting.

Relationships between woody plants and herbs within one vegetation type are investigated in order to eliminate as much as possible the influence of variable environmental conditions. In the *Eucalyptus populnea* vegetation type in Australia most of the variation in herbage production is accounted for by the combination of woody cover and the erosion surface (Wilson et al. 1988). Scanlan and Burrows (1990) found curvilinear regressions between herbage production and *Eucalyptus* tree basal area for sites varying in soil conditions mainly. Even at low tree densities, herbage production is significantly reduced. The rate of decline of herbage production with increasing tree basal area is greatest for those sites with lowest herbage production in the absence of trees, representing lowest site potential. This conclusion was also obtained when comparing overstorey-understorey relationships from different studies. Potential herbage production in the absence of woody plants (X in $t\,ha^{-1}$), indicating site productivity, and basal area of woody plants causing a 50% reduction in that herbage yield (Y in $m^2\,ha^{-1}$), are related as follows: $Y = 1.45 \times X^{1.5}$. As site conditions improve, the curvilinear relationship flattens towards a linear relationship. This finding supports the general observation that woody plants have a greater depressant effect on herbage production in years of low rainfall (Scanlan and Burrows 1990). Thus both water and nutrient availability can influence the competitive relationship between herbs and woody plants.

In semi-arid Kenya van Wijngaarden (1985) found a linear relationship between woody cover and the maximum cover of perennial grasses (Fig. 4.1). Factors such as soil type, fire and grazing determine whether this maximum level is actually reached. Given a certain canopy cover, the maximum cover of perennial grasses level increases with increasing rainfall. The competitive relationship between woody plants and perennial grasses is probably more severe than with annual grasses, in view of the similar root systems and resource requirements of perennial plants. Data from Zimbabwe (Kennard and Walker 1973) show that overall herbage biomass is 4.5 t ha^{-1} in an open canopy situation (low tree densities), as compared to lower values in the closed canopy situation (2.8 t ha^{-1}) or grasslands (3.0 t ha^{-1}).

In the SSWA relationships between woody and herbage cover in vegetation types from different geomorphological units (plateau, slope, valley etc.) are generally positive (Bille 1977; Lamotte 1979). Both vegetation components respond to variations in moisture and/or nutrient availability in a similar way. Thus a pattern occurs with vegetation concentrated in patches. Patchiness is most pronounced in the so-called tiger-bush vegetation type in the Sahel zones (White 1970). The pattern gradually fades out with increasing rainfall and increasing densities of woody plants towards the south.

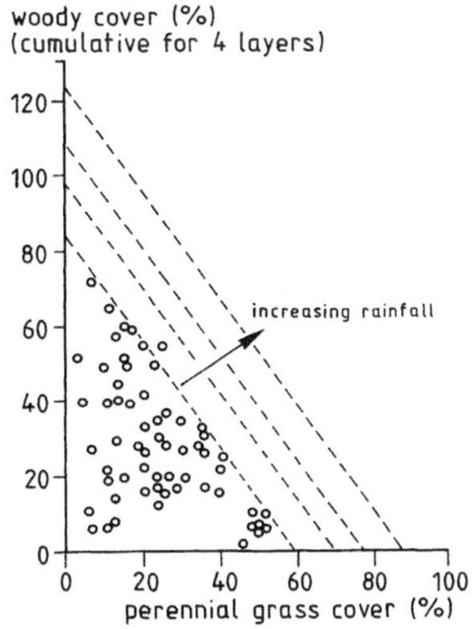

Fig. 4.1. Relationship between cover of perennial grasses and woody canopy cover in relation to annual rainfall in semi-arid Kenya. (After van Wijngaarden 1985; only observations of the most arid zone are plotted)

Plotting woody cover against herbage production of 85 vegetation relevees taken after the first drought period in the Sahel zones in Mali gives a cloud of points: a combination of low woody cover and low herbage production occurs on poor sites (e.g. areas with strong run-off) whereas the combination of high woody cover and high herbage production occurs at moist and fertile sites (e.g. valleys). Both vegetation parameters increase with increasing rainfall towards the south. In order to further investigate possible influence of woody cover on the herbage layer, the analysis was limited to vegetation relevees with distinct water availability and soil fertility. Only the data set limited to the 250–350 mm rainfall zone and sites where no run-off nor run-on occurred, provide a negative, but not significant, correlation between woody cover and

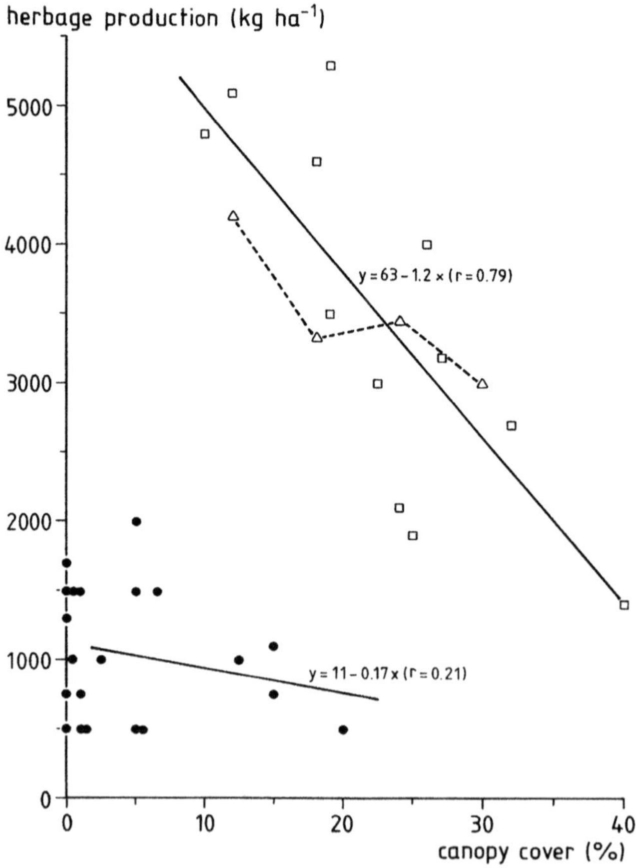

Fig. 4.2. Relationship between annual herbage production and woody canopy cover for three situations in the SSWA. ● natural vegetation in the Sahel zones (Breman, unpubl. data); △ natural vegetation is southern Sudan zone (Ouédraogo 1986); □ natural vegetation in the southern Sudan zone subject to different degrees of clearing. (Sidibé 1976)

herbage production (Fig. 4.2). Up to about 10% canopy cover there is a cloud of points; an increase of canopy cover from 10–20% gives a herbaceous yield reduction of about 50%.

In the Sudan zones the situation is different. Measurements of average herbage biomass in relation to canopy cover in a distinct area (Nazinga reserve) in the southern Sudan zone in Burkina Faso (1100 mm rainfall) provides a significant negative correlation between the two vegetation components (Fig. 4.2). The points representing sites with shallow soil were left out: here both canopy cover and herbaceous biomass are lowest.

In a similar study Sedogo (1988) found average herbage production in a distinct region in the northern Sudan zone of Burkina Faso (800 mm rainfall) to decrease from 1.6 t ha^{-1} to 1.2 t ha^{-1} with increasing density of the woody vegetation from 1500 to 5000 woody plants per hectare.

4.2.2.2 Herbage Production Under Woody Canopies

It is often suggested that a higher herbage biomass under woody plants than in the open field is the common picture. There is evidence from the USA (Tiedemann and Klemmedson 1977; Barth and Klemmedson 1978), Chile (e.g. Ovalle and Avendaño 1987), Australia (e.g. Lowry et al. 1988) and southern Africa (e.g. Hirst 1975). There is, however, also evidence suggesting a reduced herbage biomass under woody plants.

In Texas (750 mm rainfall), in a chaparral vegetation with an average woody shrub cover of 50% cover, there are virtually no herbs under the shrubs (Powell and Box 1967). In California (900 mm rainfall), herbage biomass under blue oak is considerably less than in the open field (Murphy and Crampton 1964). In Queensland in Australia (600 mm rainfall), circles of bare soil, with few herbs only, occur under and around existing trees (Story 1967). Here the apparent suppression of herbage growth exceeds the area of the woody canopy, and is correlated with the presence of lateral roots from the trees. In most other semi-arid regions in Australia, for other woody species, herbage under woody canopies is similar or more luxurious than in the open field.

Regarding the explanation of such patterns, apart from variable water and nutrient availability and shading effects, the following causal mechanisms may be involved as well.

1. Where strong winds prevail, protection of herbs under woody canopies against mechanical wind effects may account for highest herbage biomass under shrubs as compared to open areas (e.g. Halvorson and Patten 1975; Barth and Klemmedson 1978; Montana et al. 1991). But in a desert like vegetation in Chile, highest herbage biomass under shrubs is predominantly due to protection by the low

and spiny shrubs against grazing by wild herbivores. Microclimate factors are relatively unimportant (Jaksic and Fuentes 1980). Effects vary between shrub species. In Arizona (300 mm rainfall) the ratio of herbage biomass under shrub canopies to herbage biomass in the open field varies between 0.33 to 2.12 for different shrub species. The differences can be related to differences in morphology between shrub species: low and dense canopies do not permit herbs to grow underneath, contrary to open, loose canopies (Halvorson and Patten 1975). Here shading may be the governing factor.
2. Allelopathy accounts for lower herbage production under canopies of specific woody species (e.g. Story 1967; Seligman et al. 1991). Allelopathy is the direct or indirect deleterious effect of one plant upon another through the production of chemical inhibitors, which may be volatile substances, root exudates or decomposition residues (Kramer 1983). Allelopathy has been reported for several woody species, such as *Eucalyptus* spp., *Casuarina* spp., *Leucaena leucocephala*, *Prosopis juliflora* and *Tectona grandis*. Slow litter decomposition of such species is responsible for part of the low herbage production (e.g. Jameson 1966). For *Eucalyptus* spp. in India, allelopathic effects predominate over effects related to light, water or nutrient availability (Suresh and Vinaya Rai 1988). Allelopathic effects are less when toxic elements are leached, e.g. in areas with relatively high rainfall (Story 1967; del Moral and Muller 1970). Therefore allelopathic effects are most harmful to herbage production in semi-arid regions (Soni and Vasistha 1991). Under dense stands of *Eucalyptus* in semi-arid regions understorey is suppressed, whereas under humid conditions understorey may even be enhanced (Soni and Vasistha 1991).

There is evidence to assume that the pattern of higher herbage biomass under woody canopies particularly appears for isolated woody plants and for a vegetation with low densities of woody plants. The pattern changes when densities of woody plants or woody cover increases. For instance, under isolated *Albizzia lebbeck* trees, pasture yields are double that in the open field but are comparable when tree stands close (Lowry et al. 1988). In the situation with high tree densities the herbs also dry out more rapidly than with low tree densities. Ahuja et al. (1978) suggest that herbage biomass under woody canopies in semi-arid India depends on tree species, but lowest herbage biomass was found for those woody species occurring at the highest densities.

In savánna vegetation in semi-arid Kenya herbage biomass under woody canopies is two to three times higher than in the open field, irrespective of tree species (Belsky et al. 1989; Weltzin and Coughenor 1990). In semi-arid Zimbabwe (630 mm rainfall) herbage biomass is highest (4.5 t ha^{-1}) under woody canopies in the low tree density situation, as compared to $3\,t\,ha^{-1}$ in the open field and $2.8\,t\,ha^{-1}$ at high densities of woody plants (Kennard and Walker 1973). The common situation in the *Burkea africana* tree savanna in South Africa (600 mm rainfall), where canopy cover is a high 20–60%, is that of $1.4\,t\,ha^{-1}$ under tree canopies, $1.3\,t\,ha^{-1}$ under shrub canopies, and $1.9\,t\,ha^{-1}$ in the open field (Hirst 1975).

High herbage production under woody canopies may be at the expense of the surrounding areas. Stuart-Hill and Tainton (1989) provide evidence for this hypothesis: high herbage biomass under the canopies of dispersed, small *Acacia karroo* trees in semi-arid southern Africa is associated with low herbage biomass at the edge of the woody canopy, gradually increasing at increasing distance from the stem. The influence of a tree could be observed until 9 m from the stem. Such a pattern suggests resource redistribution towards the tree canopy, a process which will become less effective as tree densities increase.

The size of woody plants also influences the pattern of herbage biomass under the woody canopy. Woody plants less than 1 m tall hardly suppress herbage production, contrary to those exceding 2 m in height (Stuart-Hill and Tainton 1989). However, herbage production under trees more than 7 m tall is similar or even higher than in the open field, contrary to woody plants not taller than 3 m (Sanford et al. 1982; Muoghalu and Isichei 1991). Shading may be the determining factor, large trees allowing lateral incoming light when tree densities are not high (Sect. 4.7).

From the SSWA only few data are available on herbage production under woody plants. Herbage biomass under *Acacia senegal* and *Balanites aegyptiaca* trees is twice as high as in the open field in a sandy landscape with deep soils in the southern Sahel zone (Grouzis et al. 1990). In the northern Sudan zone, Sedogo (1988) found highest herbage peak biomass under tree canopies with dense foliage ($2.2\,t\,ha^{-1}$), as compared to $1.6\,t\,ha^{-1}$ for either tree canopies with open-structured foliage or the open field. Highest herbage production under the woody canopy is due to growth continuing for about 1 month longer than in the open field. Under the woody canopies, moisture content of the herbaceous biomass is also highest (Fig. 4.3), suggesting that improved moisture availability is one governing factor (Sedogo 1988). However, differences in growth may also be due to variation in species composition (Sect. 4.2.4). In fallowlands in the northern Sudan zone, herbage production under the canopy of *Parkia biglobosa* trees is $6.9\,t\,ha^{-1}$, compared to $3.3\,t\,ha^{-1}$ in the open field (Sabiiti and Cobbina 1992). This pattern may also result from intensive cultivation of the soil in the open field during preceding years, contrary to the soil under the woody canopy. In the southern Sudan zone in Nigeria highest herbage biomass occurs under a "light" (open) canopy cover, as compared to the open field or a dense woody cover (Sanford et al. 1982).

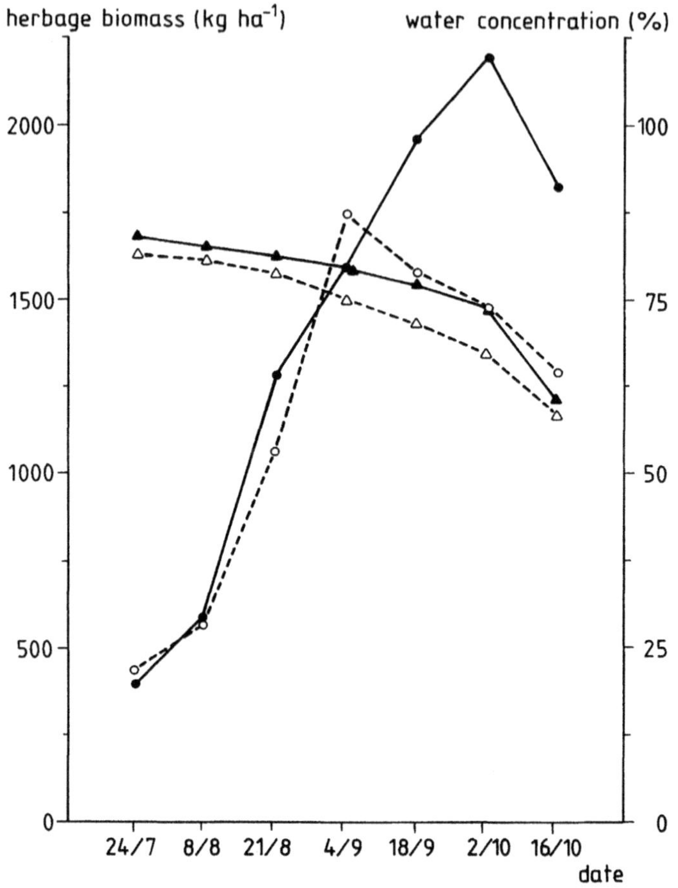

Fig. 4.3. Herbaceous biomass development (○●) and its moisture contents (△▲) during the rainy season, under the woody canopy (●▲) and in the open field (○△). (After Sedogo 1988)

Although in the Sahel zones the pattern of enhanced herbage growth under isolated woody plants (low woody plant densities) is common, it is certainly not a rule. In some years herbage production under woody canopies is less than in open areas (pers. observ.). In Australia, reduced herbage production under woody canopies correlates with the occurrence of distinct dry periods during the rainy season (Mott et al. 1982). This is probably also the case in the Sahel zones. The processes leading to this pattern are discussed in Section 4.2.4. There is evidence to assume that in the SSWA the pattern of enhanced herbage biomass under woody canopies is related to low woody plant densities (Boutrais 1980).

Several factors related to human expoloitation can influence the outcome of the pattern of herbage biomass under woody plants: bushfires (early fires may leave more humid herbs under woody canopies), grazing (herbivores may select relatively palatable herbs under woody canopies), defoliation (browsing) of woody plants (possibly leading to regrowth and increased resource competition), cultivation (poor yields of crops under woody canopies may leave more nutrients for herbs). These factors should be included in any study on relationships between woody plants and herbage production.

4.2.3 Effects on Herbage Production by Bush Clearing or Tree Felling

The influence of woody plants on the herbage layer can also be investigated through the effects of total or partial tree felling or bush clearing.

Results from such studies in temperate zones show an inverse relation between basal cover of trees and herbage production. There are linear, logarithmic and exponential relationships between canopy cover and herbage production (Jameson 1967; Torres 1983). These relationships mostly refer to one woody species only (pine mainly) at different densities, either in natural stands or in plantations.

Similar relationships are available from semi-arid zones in Australia (Walker et al. 1972; Beale 1973; Pressland 1973; Christie 1975; Walker et al. 1986). However, these studies also suggest that herbage production does not increase any more when canopy cover is less than a certain level (Kelly 1977).

In semi-arid Australia, with 500 mm annual rainfall, this critical level was estimated at 2–6 mature *Eucalyptus populnea* trees or 360 shrubs per hectare (Walker et al. 1972). With average rainfall of 750 mm no significant increase in herbage production is recorded with tree densities lower than $80\,ha^{-1}$. However, in a dry year the situation with all trees cleared shows highest herbage production (Walker et al. 1986). In the Australian subtropics (1100 mm rainfall) a similar experiment showed the critical level to depend on the combination of age and density of trees (Fig. 4.4). These two parameters influence canopy cover. The critical level is estimated for dry years at 20–50 mature trees per hectare, corresponding to about 20% canopy cover, and for a year with average rainfall at 50–80 trees, corresponding to about 30% canopy cover.

Bush encroachment (Sect. 2.4.2) causes a reduction of herbage production. Subsequent bush clearing leads to an increase in herbage production, as reported for southern Africa (Donaldson and Kelk 1970; Dye and Spear 1982; Pratchett 1983), Australia (Barnes 1979; Knoop and Walker 1985; Tothill and Mott 1985) and the USA (Murphy and Crampton 1964; Powell and Box 1967; Tiedemann and Klemmedson

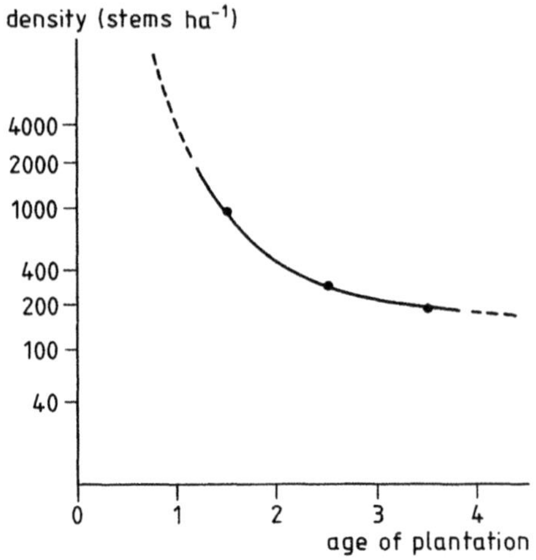

Fig. 4.4. The lowest density of *Eucalyptus populnea* trees at which pasture production is negatively influenced, in relation to age of the plantation. (After Cameron et al. 1989)

1977). In these regions rainfall varies between 300 and 800 mm annually. The increase of herbage production following bush clearing can be up to tenfold, such as in semi-arid South Africa (400 mm rainfall), where bush cover before clearing was about 22% (Donaldson and Kelk 1973). During a dry year in South Africa (200 mm; average rainfall 420 mm), Donaldson (1967) found mortality of the herbage layer to be eight times higher in the situation with a 22% shrub cover than in the situation without shrubs. Earlier conclusions, that even a low shrub cover reduces herbage production, were later found to be incorrect (Aucamp et al. 1983; cited by Stuart-Hill and Tainton 1989). Herbage production in fact remains unaffected by increasing bush densities up to approximately 300 per hectare, and only thereafter starts to decrease.

It seems that, even in a situation of bush encroachment where shallow rooting shrubs predominate, advantages of woody plants in low densities may balance competitive interactions with the herbaceous layer.

Increased herbage production following tree felling or bush clearing may be only a short-term effect. The short-term effect is comparable to high crop production on recently cleared land. It reflects buildup of soil fertility in the phase with woody plants, and liberation of nutrients following clearing of the vegetation (Sect. 4.5.3). After clearing of woody plants, herbage production is generally concentrated where shrubs or trees were standing (Murphy and Crampton 1964; Powell and Box 1967). Farmers in the SSWA know that during several years after cutting a tree, crop production at that site is relatively high (pers. observ.). Tothill and Mott (1985), however, reports of an increased herbage production during 10 years following bush clearing. Maybe production resources that were used by woody plants in the situation before clearing, became permanently available to herbs.

In the SSWA, in the southern Sudan zone in Mali (1100 mm rainfall), herbage production increases with a reduction in canopy cover (Fig. 4.2). It is unknown whether this effect had a permanent character. A linear relationship gives the best fit between canopy cover and herbage production. It seems, however, that herbage production is not influenced until canopy-cover exceeds about 15%. The data from this study correspond to the results from Ouédraogo (1986) (Fig. 4.2).

4.2.4 Herbage Species Composition in Relation to Canopy Cover

In the natural vegetation, herbage species composition under woody canopies generally differs from the open field. A strong dominance under the woody canopy of one grass species commonly occurs: *Cenchrus ciliaris* (Australia: Christie 1975), *Brachiaria leersoides* (Kenya: Weltzin and Coughenor 1990), *Cynodon nlemfuensis* (Kenya: Belsky et al. 1989), *Panicum maximum* (southern Africa: Bosch and Van Wijk 1970; Kennard and Walker 1973; Nigeria: Sabiiti and Cobbina 1992), *Andropogon tectorum* (Nigeria: Sanford et al. 1982), Texas bristlegrass (USA, Wildin 1986), *Lolium multiflorum* (Chile, Ovalle and Avendaño 1987). A strong dominance is often associated with a positive relationship between herbage production and canopy cover.

In the rangelands in Arizona (350 mm rainfall), Yavitt and Smith (1982) found herbage species distribution in a grassland to be random, contrary to mesquite-invaded sites. They found a negative correlation of mesquite with annual grasses but a positive correlation with perennial grasses, the latter occurring under or near the shrub canopies. The scale of pattern roughly corresponds to the distance between shrubs, suggesting that the pattern is caused by the spacing of the shrubs. Similar observations are from a mallee (*Eucalyptus*) community in semi-arid Australia (Whittaker et al. 1979). Such patchiness suggests spatial concentration of production resources by shrubs.

Bush clearing generally leads to changes in herbage species composition. In Zimbabwe, bush clearing leads to the change of dominance from *Panicum maximum* and *Digitaria* to *Heteropogon contortus* and *Urochloa mosambicensis* (Bosch and Van Wijk 1970; Dye and Spear 1982). The first species occurs almost exclusively under tree canopies. In Australia, Walker et al. (1986) found significant changes in herbage composition within 2 years following bush clearing. In South Africa, such a change in herbage species composition was not observed up to 6 years after bush clearing (Donaldson and Kelk 1970). It may be that seed resources of climax grass species were lacking as a result of overgrazing.

Several authors suggest that species showing preference for canopy cover are those adapted to shade conditions and can thus benefit from higher soil fertility and/or water availability under the trees (Bosch and Van Wijk 1970; Kennard and Walker 1973; Veenendaal 1991). However, as Belsky et al. (1989) pointed out, the abrupt change in herbage species composition at the canopy edge is not correlated with changes in shade, soil fertility, temperature or depth of leaf litter, all of which change gradually at increasing distance from the stem. Under woody canopies, perennial grasses are usually more abundant than in the open field, whereas in the open field the proportion of annual and leguminous herbs is usually higher (Tiedemann and Klemmedson 1977; Yavitt and Smith 1982; Joffre and Rambal 1988; Weltzin and Coughenor 1990; Muoghalu and Isichei 1991). In Australia, the Andropogonoid grasses are depressed by woody plants, the Panicoid grasses exhibit the opposite relationship, whereas forbs (non-grasses) are little influenced by woody plants (Scanlan and Burrows 1990). Species diversity in the open field is generally higher than under tree canopies.

Root to shoot ratios of the herbs under woody canopies in Kenya are lower (0.55–1.48) than in the open field (1.56–1.98), corresponding to results from South Africa (Belsky et al. 1989). Lower root to shoot ratios may indicate less (drought or nutrient) stress under woody canopies.

The evidence from the SSWA is scanty. In the Sahel zones 21 herbage species are particularly associated with woody canopies (Grouzis et al. 1990). In the northern Sudan zone herbage biomass under tree canopies is strongly dominated by the annual grass *Pennisetum pedicellatum* (Sedogo 1988; de Bie 1991). In the southern Sudan zone in Nigeria *Andropogon* spp. dominate under woody canopies, whereas *Hyparrhenia* spp. dominate in the open field (Sanford et al. 1982). Species diversity is generally highest in the open areas. Annual species reach highest covers in open areas (e.g. *Loudetia togoensis*) or at sites with low densities of woody plants (e.g. *Andropogon pseudapricus*). Leguminous herbs are rare under woody canopies, contrary to the open field. In the Sahel zones, the proportion of leguminous herbs in the herbage layer decreases from more than 5% where woody cover is 5% to almost zero where woody cover exceeds 30% (Breman and de Ridder 1991).

Detailed monitoring of the herbage layer over several years showed that in the Sahel zones herbage species composition under woody canopies varies between years. Besides, in some years the area under woody canopies remains virtually bare, in other years herbage production is more luxurious under canopies. For an adequate explanation of such variation, seed losses of herbs in the pre-emergence phase should be taken into account (Elberse and Breman 1990), and a distinction should be made between herbs with rapidly germinating seeds (RG) and those slowly germinating (SG). In order to germinate, SG species require moisture for a longer period than RG species. SG species therefore particularly germinate where the soil remains moist for a relatively long period.

In the open field situation, RG species appear rapidly with some rainfall at the onset of the rainy season and reach maturity if no long intermittent drought periods occur. Some SG species appear at a later stage but have a poor competitive power. If one or more drought periods occur, a large proportion of the RG species do not survive and few seeds remain for renewed germination. The proportion of SG species increases as drought periods at the beginning of the rainy season are more frequent, or early scattered showers occur (Cissé AM 1986). This process of variable response to rainfall distribution is an adaptation to erratic rainfall.

Under woody canopies average soil moisture contents are higher (Sect. 4.4.2.3), so that the proportion of germinating seeds at the beginning of the rainy season is higher than in the open field, and includes SG species (Fig. 4.5). It is common to observe a dense herbage cover under woody canopies in the early rainy season, when the open field is still bare. However, a long subsequent drought period may also lead to the elimination of all herbs under woody canopies, including those of SG species. Under woody canopies seedling mortality represents a larger proportion of all seeds than in the open field. With subsequent rainfall little vegetation appears under the woody canopy. This explains why in some years the herbage production under woody canopies in the Sahel zones may be poor, or even absent, whereas normally it is luxurious (Fig. 4.5).

Herbs with a C_4 photosynthetic pathway (Sect. 4.6) have a higher water use efficiency, are adapted to higher temperatures and generally show higher production than C_3 species. Herbs in the Sahel zones are usually C_4, growing under conditions of high temperature, low moisture availability and high radiation. Towards the south, the relative importance of C_4 species decreases. Under woody canopies the importance of C_3 species is higher due to shade conditions and lower temperatures. Some herbage species commonly found under woody canopies are *Achyranthes argentea*, *Blepharis maderaspatensis*, *Triumfeta pentandra*, *Pennisetum pedicellatum* and *Chloris pilosa* (Breman and de Ridder 1991). In the northern Sahel zone, C_3 species are more common under isolated woody plants than in the open field, although C_4 species still dominate (Fig. 4.5). Towards the southern Sudan zone, dense shade conditions are increasingly common, so that dominance of C_3 species under woody canopies increases. An increased proportion of C_3 species corresponds to a lower production. Since C_3 species have a higher minimal N concentration than C_4 species (10 and $5\,g\,kg^{-1}$ respectively; Penning de Vries and Djitèye 1991), the average N concentration of the herbage layer under a woody canopy is usually higher than in the open field.

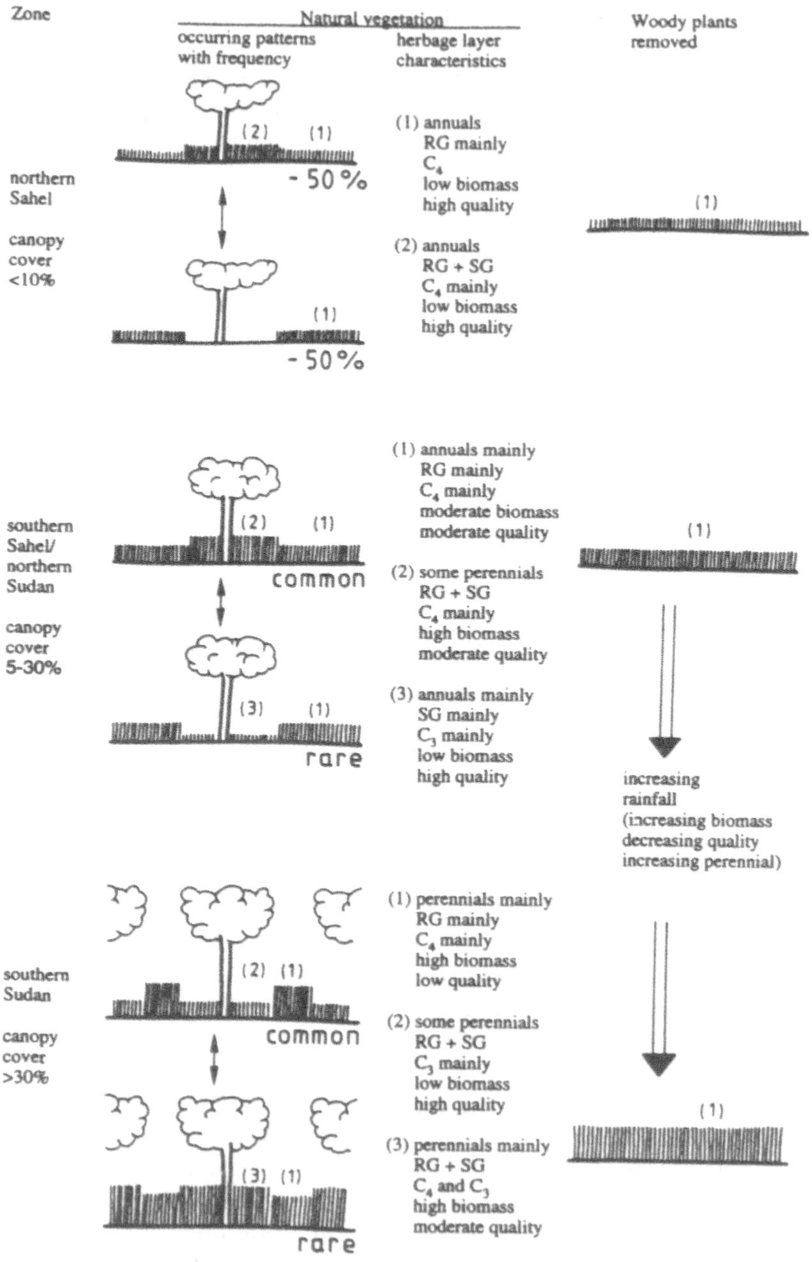

Fig. 4.5. A generalised schematic pattern of the woody and herbage layer in different climatic zones in the SSWA. *RG* Rapidly germinating; *SG* slowly germinating; *C_3* C_3 photosynthetic pathway; *C_4* C_4 photosynthetic pathway

4.2.5 Conclusions

The influence of woody plants on the herbage layer cannot be demonstrated by comparing different vegetation types. Variability of site conditions seems more important, both vegetation components showing a common response to the variation in the water and nutrient availability. Comparing sites within one vegetation type, there is some evidence for an inverse relationship between woody cover and herbage production. Tree felling or bush clearing generally leads to increased herbage production, although most studies considered short-term effects only.

Evidence from either natural or man-made variation in vegetations with woody plants suggests that up to a low woody cover herbaceous production is not suppressed, or is even stimulated (Figs. 4.2 and 4.4). Herbage biomass or production is highest where woody plant densities are low, as compared to the open field or a densely wooded vegetation. The critical canopy cover is the level below which other factors than woody cover are more important in determining herbage production and/or beneficial effects associated with woody plants are more important than competitive effects. If canopy cover exceeds the critical level, the relationship between herbs and woody plants becomes predominantly competitive. Competitive effects are most significant for woody plants and herbs with comparable resource exploitation patterns, e.g. shallow rooting shrubs and perennial grasses.

On the basis of scarce available data, the critical canopy cover in the SSWA is estimated at an average of 15%, for woody plants with two leaf layers. In the Sahel zones this level is probably lower than 15%, and it increases towards the south. Evidence is too scarce to further specify these assumptions.

The relationship between woody plants and herbs is determined by the possibilities for niche differentiation between the two vegetation components. Evidence was provided to assume that niche differentiation is less likely to occur as site production potential decreases (Scanlan and Burrows 1990). Site conditions influencing the critical canopy cover in a positive sense are those increasing production potential and increasing the possibilities for niche differentiation between woody plants and herbs (Sect. 2.3): deep soil, coarse soil texture, high soil fertility, high rainfall and well-distributed rainfall.

There seems to be a contradiction between, on the one hand, the commonly observed competitive relationship between woody plants and herbs and, on the other hand, the regularly encountered pattern of enhanced herbage production under woody plants. A higher herbage biomass is most common under not densely foliated woody canopies, under tall (mature) trees and under isolated woody plants.

Variation in herbage production and nutrient concentrations under woody plants is also related to variations in herbage species composition according to seed germination rate and photosynthetic pathway (Fig. 4.5). The most important determining factor seems water availability, in terms of total quantities and distribution in time and within the soil profile. Since differences in herbage production on bushed and cleared plots do not decrease in the years of highest rainfall, it seems that other factors such as shading and nutrient availability act as well (Dye and Spear 1982; Wilson et al. 1988).

The apparent contradiction is related to the fact that, particularly in situations of resource competition between woody plants and herbs, redistribution processes commonly lead to improved water and nutrient availability under woody plants. Isolated and tall woody plants allow most lateral incoming light to reach the soil surface under their canopies. Such a spatial concentration of production resources should be, to some extent, at the expense of moisture and nutrient availability in the surrounding open areas, leading to reduced herbage production in the open areas. The underlying processes and resulting patterns are analysed in the following sections.

4.3 Root Systems of Woody Plants

4.3.1 Root Systems of Woody Plants in Semi-Arid Regions

In order to analyse the influence of woody plants on water and nutrient availability, insight into root systems of woody plants is essential. On the basis of a summary of available data on root systems from woody plants in semi-arid regions, root dynamics in relation to environmental factors and resource availability are discussed.

Root biomass can be differentiated into fine and coarse roots. Although fine roots comprise only a small proportion of total root biomass, they provide the most important absorbing surface of a root system. Fine roots absorb water and nutrients, are associated with mycorrhizal transfer and incorporation of nutrients, and add organic matter and nutrients to the soil through turnover (Berish and Ewel 1988). Coarse roots provide support for woody plants and may penetrate to deep soil layers. Fine roots particularly develop at the extremities of coarse roots.

The root system is genetically determined as long as environmental conditions are favourable, but it is actually controlled by physical conditions in case of an unfavourable environment (Kummerow and Mangan 1981). Woody species vary in respect to the flexibility of their root system. Root systems of woody plants can be classified into three different types (Hellmers et al. 1955; Hairiah and van Noordwijk 1986; Fig. 4.6):

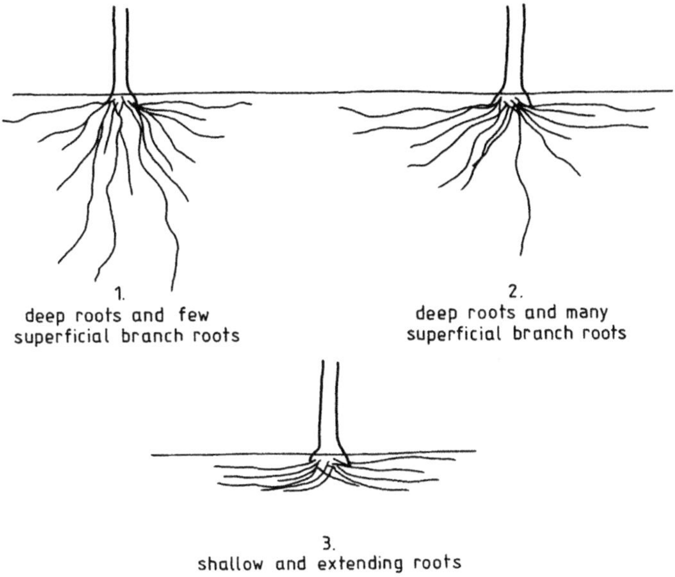

Fig. 4.6. Diagrammatic sketches of three types of root systems of woody plants

1. deep root penetration and few superficial branch roots;
2. deep root penetration and many superficial branch roots;
3. shallow roots only and a large horizontal spread.

A schematic presentation of root systems in arid, semi-arid and (sub-) humid tropical regions is given in Fig. 4.7. Woody plants in semi-arid regions generally have well-developed root systems, of type 1 mainly. In the humid tropics, forest species show a concentration of nearly all fine roots in the uppermost 0.3 m of the soil and have only few roots extending to deeper soil layers (Kerfoot 1963; Whitmore 1990). In a sub-humid savanna in Belize (1560 mm rainfall) more than 90% of the root biomass occurs in the first 40 cm of the soil (Kellman and Sanmugadas 1985).

Only recently it has been argued that in semi-arid regions roots are also predominantly shallow (Menaut 1983). In fact, systematic studies on root systems of indigenous woody plants in semi-arid regions demonstrate the predominance of root systems types 2 or 3 only, irrespective of soil depth (e.g. in Somalia: Glover 1951; in Egypt: Batanouny and Abdel Wahab 1973; in Australia: Story 1967; Pressland 1973; in California: Kummerow et al. 1977; in Israel: Noy-Meir 1979; in Zimbabwe: Dye and Spear 1982; in South Africa: Knoop and Walker 1985; in Argentinia: Montana et al. 1991). Root systems of exotic woody species show similar patterns. For example, in alley cropping systems with *Leucaena leucocephala*, at least 80% of the root biomass occurs in the

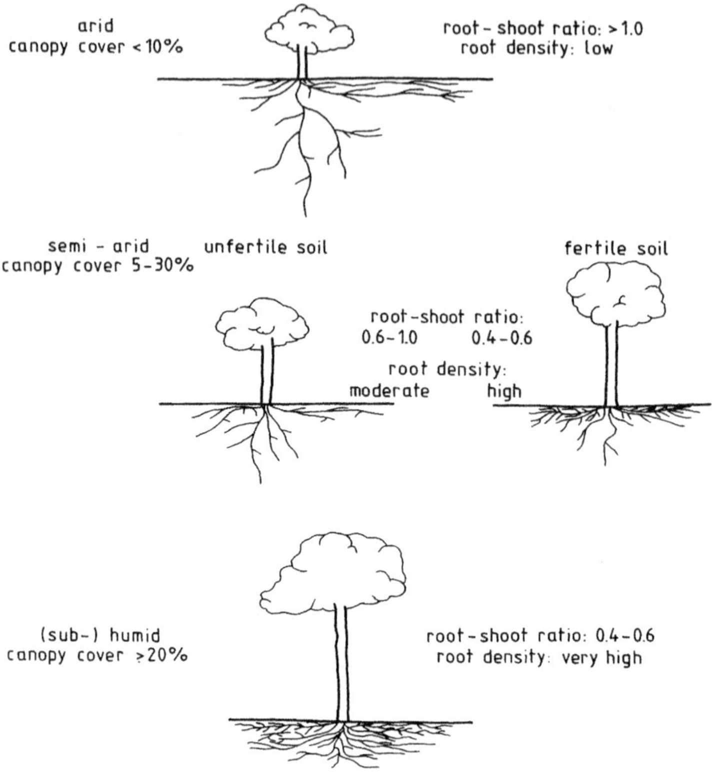

Fig. 4.7. Schematic pattern of root systems of woody plants in relation to climatic region, in a deep soil, with indication of root-shoot ratio and root density in the topsoil. (Based on data from various sources, see text)

first 50 cm of the soil whereas roots rarely extend to layers beyond 100 cm (Jonsson et al. 1988; Toky and Bisht 1988; Singh et al. 1989).

The lateral spread of roots from woody plants varies according to climatic zone (Fig. 4.7). In humid forests the lateral spread of shallow roots is limited; roots are concentrated in the crown canopy area (Kellman 1979). Everett et al. (1986) found fine root biomass to peak at the crown edge of *Juniperus* shrubs, coinciding with the area receiving large amounts of leaf drip (with nutrients). Although there is evidence that in semi-arid regions the density of fine roots is also highest under the crown canopy (e.g. Kummerow et al. 1977), the area covered by lateral roots generally extends far beyond the crown canopy area. Lateral roots may exceed crown canopy by a factor 10 (Australia: *Acacia pendula*, Story 1967) up to 40 (California, chaparral shrubs, Kummerow et al. 1977). In this way the woody plants effectively exploit the soil surface, in spite of low densities. Glover (1951) remarks that less than

10% woody cover in semi-arid Somalia may seem low, but he estimates on the basis of lateral root systems that the topsoil is fully exploited by roots from woody plants.

Most woody plants in semi-arid regions have one or few deep roots. Some woody plants having exceptionally deep roots, generally reaching deep water resources. Well-known examples are *Prosopis* spp. and *Atriplex* spp. (Le Houérou 1980a; Shankarnarayan et al. 1989) and *Leptadenia pyrotechnica* (Batanouny and Abdal Waheb 1973). Chaparral shrubs in California can penetrate joints of the bedrock, reaching depths of more than 10 m (Schlesinger and Gill 1980).

When woody plant densities increase, tree roots are more likely to exploit deeper water resources, if available. This was observed for apple tree plantations in temperate climates (Atkinson 1980), *Eucalyptus* plantations in sub-humid Australia (Eastham and Rose 1990) and *Eucalyptus* plantations in arid Israel (Lövenstein et al. 1991). Studies from Australia show that with increasing tree density total root length increases, but root length per tree decreases (Table 4.1).

Woody species with root systems of the first type (Fig. 4.6) generally show slow initial aboveground growth and high root to shoot ratios, whereas species of the second and third type may show rapid initial aboveground growth and lower root to shoot ratios (Hairiah and van Noordwijk 1986). Species of the third type are commonly pioneer species, for example those colonizing young fallowlands. Evidence from Ghana and Zaire shows that as the fallow period and/or plant succession proceeds, proportionally more roots are found in deep soil layers: the proportion of roots in the first 50 cm of the soil decreases from 98% two years after fallowing to 90% after five years (Hairiah and van Noordwijk

Table 4.1. Tree and pasture root system characteristics at different *Eucalyptus grandis* densities, including root to shoot ratios, in semi-arid Australia. (Eastham and Rose 1990)

Tree density ha^{-1}		82 (low)	304 (medium)	2150 (high)
Percentage of total tree root length at				
1.2 m from the stem, at 0.0–1.0 m soil depth		68.4	48.0	61.0
at 1.0–2.0 m soil depth		22.1	26.0	23.8
at 2.0–3.0 m soil depth		7.3	14.3	9.4
at 3.0–4.0 m soil depth		2.2	7.9	3.8
at 4.0–5.0 m soil depth		0.0	3.8	1.5
at 5.0–6.0 m soil depth		0.0	0.0	0.5
Total root length (10^3 m) per tree		42.2	35.3	32.9
Total root length per area (10^3 m m^{-2})	– Trees	1.3	1.0	7.3
	– Herbage	29.1	9.4	6.2
Root to shoot ratio of trees		0.8	0.3	0.2

Table 4.2. Characteristics of root systems of woody species occurring in the SSWA, including indication of the root system type according to Fig. 4.6

Country, rainfall + source	Woody species	Root system						Type
West Africa								
Senegal,	Depth (cm):	0–20	20–50	50–100	100–200	200–400	>400	
300 mm	*Adansonia digitata*	33%	33%	20%	13%	1%	–	2/3
(Bille 1977)	*Balanites aegyptiaca*	35%	23%	19%	15%	7%	1%	2
	Acacia senegal	33%	24%	19%	15%	6%	2%	2
	Commiphora africana	65%	21%	8%	5%	2%	–	3
	Guiera senegalensis	41%	27%	16%	11%	4%	1%	2
	Grewia bicolor	72%	10%	15%	3%	–	–	3
Mali,	*Acacia seyal*	Predominantly shallow, deep taproot (>5 m)						2
500 mm	*Bombax costatum*	lateral extension 5 × crown radius (5–10 m)						
(Penning de Vries,		Predominantly shallow, short taproot (<1 m)						3
unpubl.)	*Boscia senegalensis*	lateral extension 7 × crown radius (10–35 m)						
	Combretum ghasalensis	Predominantly at 0.5–1.5 m depth						3
	Guiera senegalensis	alike *Acacia seyal*						2
		Predominantly shallow, no taproot						3
	Pterocarpus lucens	lateral extension 5 × crown radius (4–6 m)						
		Very shallow roots (<0.5 m)						3
Sahel region (Wickens 1980)	*Acacia senegal*	Deep tap root reaching water table, extensive lateral roots						2
Sahel region (von Maydell 1986)	*Prosopis juliflora*	Deep tap root (up to 35 m)						1/2
Sahel region (CTFT 1988)	*Faidherbia albida*	Sandy soils: deep tap root reaching water table (40 m?)						1
		valleys: predominantly shallow roots						2
Sahel region (pers. observ.)	*Vitellaria paradoxa*	Predominantly shallow, large taproot not very deep (about 2 m)						2
	Parkia biglobosa	Predominantly shallow, short taproot (<2 m)						2

Other semi-arid African regions			
Sudan, 500 mm (Adams 1967)	*Acacia mellifera*	Shallow roots only (0–60 cm), extending laterally 8–15 m	3
	Acacia seyal	Predominantly shallow, extending laterally 8 m, taproot until 1.2 m depth	2
	Balanites aegyptiaca	Predominantly shallow, extending laterally 7 m, taproot until 1.3 m depth	2
Sudan, 700 mm (van Noordwijk 1984)	*Acacia seyal*	Taproot reaching 1–2 m depth; shallow	2
	Acacia senegal	Roots at 20–60 cm depth, laterally	2
	Balanites aegyptiaca	Extending to 8 m from trunk	2
Kenya, 450 mm (Belsky et al. 1989)	*Acacia tortilis*	Shallow roots with a deep taproot reaching groundwater	2
	Adansonia digitata	Very shallow root system	3
South Africa, 630 mm (Knoop and Walker 1985)	*Burkea africana*	Root biomass concentrated at depth 50–60 cm	2
	Ochna pulchra	Root biomass concentrated at depth 20–40 cm	2/3
	Terminalia sericea	Root biomass concentrated at depth 12–23 cm	2/3
Kenya, 500 mm (Fenner 1980)	*Adansonia digitata*	Maximal depth 1.8 m, maximal lateral extension 44 m	3

1986). Colonizing species in the humid tropics develop lateral roots mainly (Berish and Ewel 1988).

The summarised data on root systems from woody plants in the SSWA (Table 4.2) show patterns that correspond to the general picture presented above. Most root systems should be classified as types 2 or 3 (Fig. 4.6) with more than 80% of the root biomass in the top 1 m of the soil, and only 5% at a depth of more than 3 m (Bille 1977; Penning de Vries, unpubl.). On the basis of several excavated root systems of *Acacia seyal* trees in Niono (500 mm yr^{-1}; Mali) by the latter, the relation between root volume and soil depth has been graphically presented (Sect. 4.8.; Fig. 4.18). Regarding the lateral spread of roots, Fenner (1980) estimates that the shallow root system of one *Adansonia digitata* (baobab) tree may cover 0.7 ha, explaining their regular distribution at low densities. In the Mauretanian desert, shallow roots occur up to 50 m from the stem of *Acacia raddiana* and *Tamarix senegalensis* trees. These trees have some deep roots as well (Stocker 1970). The longest lateral root measured in Niono is from *Sclerocarya birrea*: 58 m. The average maximum extension is, however, 40 m, against 25 m for *Acacia senegal* and *Combretum ghasalense*, and 20, 10 and 5 m for respectively *Acacia seyal*, *Balanites aegyptiaca* and *Commiphora africana*. The maximum depths found for tap roots are 6, 4.5, 4, 3, 2.5 and 2 m for, respectively *Acacia seyal*, *Combretum ghasalense*, *Acacia senegal*, *Balanites aegyptiaca*, *Sclerocarya birrea* and *Commiphora africana* (Soumaré et al. 1994; Radersma 1994).

Deep roots of *Faidherbia albida* reach the water table at depths of more than 30 m. However, the root system of this tree probably varies according to site conditions. *Faidherbia albida* develops relatively deep roots on sandy soils, whereas a predominantly shallow root system develops in valleys with more fine-textured soils, where water is available in the topsoil throughout the year (CTFT 1988; Alexandre and Ouedraogo 1991). *Faidherbia albida* trees from southern Africa (ICRISAT 1990), as well as from India (Shankarnarayan et al. 1989), have more shallow roots and more rapid growth than those from the SSWA, but those from southern Africa show low survival during drought periods (ICRISAT 1990).

Deep rooting seems more frequent for (semi-) evergreen woody plants than for drought-deciduous woody plants, the first group requiring water (from deep soil layers) in the dry season (Belsky et al. 1989). For *Faidherbia albida*, with green leaves in the dry season, the necessity to reach dry season water reserves is most obvious.

The subsoil often presents physical or chemical barriers to root extension. Shallow soils are widespread in the Sudan zones (Table 2.2), often with a hardpan layer at less than 1 m depth. Chemical barriers to root growth include acidity, salinity, Al or Mn toxicity, and P and Ca deficiency. Such constraints are commonly encountered in semi-arid climates (Szott et al. 1991). Woody species differ in respect to their ability to penetrate fissures in the plinthite layer or bedrock. Penetrating woody roots lead to strings of trees in the landscape.

4.3.2 Root System Dynamics

Roots generally proliferate into soil layers offering the greatest moisture and nutrient supply if also well aerated (Pritchett 1979). Moisture and oxygen are determined by rainfall and soil texture in the first place.

Anaerobic conditions may occur at locations with fine-textured (heavy) soils where run-on water collects. Soil bulk density and root biomass of *Pinus radiata* in Australia are inversely related (Squire et al. 1978), due to aeration problems and mechanical impedance of root growth. Roots of woody plants vary considerably in their ability to penetrate heavy soils. Oxygen supply to deep roots can be a problem, but internal air channels in roots may overcome such problems (van Noordwijk 1984).

It is commonly stated that fine roots developing in depth follow the waterfront. Average depth of soil moisture is determined by rainfall and soil texture mainly. Nutrients are transported with the water, so that the influences of the two factors are difficult to distinguish.

Scarce evidence is available on the influence of soil texture on root systems. In temperate climates roots are more shallow on sandy soils and with plants have a higher root to shoot ratio than on fine (clay) soils (Atkinson 1980; Knoop and Walker 1985). In semi-arid regions, where the topsoil of sandy soils is prone to occasional drying out, the opposite pattern seems to prevail. In a South African savanna the depth of lateral root extension varies from 15–20 cm in heavier textured soils to 30 cm in sandy soils (Smith and Goodman 1986). Most observations of deep roots are from sandy soils. In fine textured soils in semi-arid regions soil moisture may be limited to the upper 40 cm (e.g. Glover 1950; Adams 1967; Seghieri 1990), but reaches more than 2 m on sandy soils in the southern Sahel zone (Penning de Vries and Djitèye 1991). Batanouny and Abdel Wahab (1973) showed how in a deep, predominantly sandy soil in an arid region fine roots develop in soil layers where soil moisture contents are higher than 1% (corresponding to silt layers). Tree root penetration into deep soil is related to variation in depth of soil moisture as a result of variation in ploughing and soil texture (Eyog-Matig 1982).

When the soil dries out, fine roots die. The attrition and regrowth of fine roots under suitable conditions occurs rapidly, but at considerable cost to the tree. In seasonal climates the effective root to shoot ratio shows seasonal variation (Landsberg and McMurtrie 1984). For root biomass of chaparral shrubs in California variation is sevenfold (Kummerow et al. 1977). Fine roots of young woody plants show a greater dry-season decrease in root biomass than mature woody plants (Berish and Ewel 1988).

Soil fertility, i.e. nutrient availability, also influences root dynamics. In humid tropical forests, roots follow nutrient distribution in both space and time. In early successional stages, when a litter layer has not yet formed, fine roots extend relatively deep into the soil. As the litter layer develops and nutrients are increasingly concentrated in the topsoil, fine roots occur in the topsoil mainly (Berish and Ewel 1988).

Fine root biomass increases with increased water and/or nutrient availability in the topsoil (Fig. 4.7). This is an indirect effect, via the increased aboveground production, and a direct effect on root growth in the soil (Squire et al. 1978; Atkinson 1980). For pines (Squire et al. 1978; Pritchett 1979) and fruit trees (Atkinson 1980) in the USA and Australia, a high supply of organic matter (e.g. by mulching) favours the development of an extending, but superficial root system. The same phenomenon occurs for millet in Senegal (Cissé L 1986): following the application of 10 t ha^{-1} of organic fertiliser, root biomass of millet in the soil layer of 0–50 cm increased threefold, but maximal rooting depth decreased from 1.7 to 1.4 m. In one specific year, as a result of the higher uptake of water by the better developed root-mat, drainage losses due to leaching decreased more than twofold.

The effect of water or nutrient stress on the root to shoot ratio is demonstrated in Fig. 4.8. High root to shoot ratios (more than 1.0) are

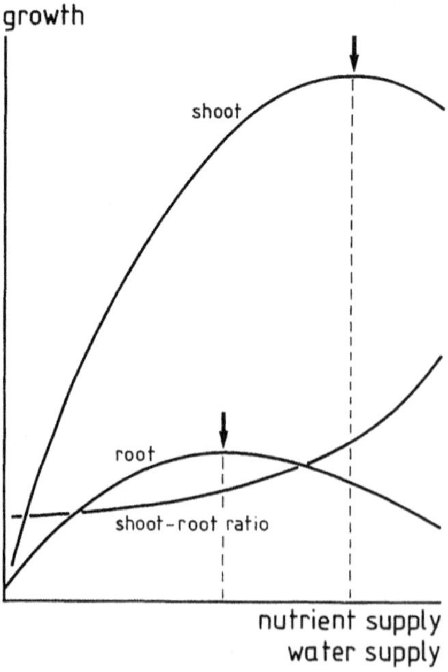

Fig. 4.8. Relationship between shoot growth, root growth and the shoot-root ratio, and nutrient and moisture supply. (van Noordwijk 1989a)

characteristic for stress situations although absolute root biomass is reduced by stress: stress-induced reduction in shoot growth exceeds the reduction in root growth.

Under stress conditions, the woody plant proportionally increases investments in the root system in order to obtain sufficient resources for shoot development (Atkinson 1980; Kramer 1983; Tilman 1985). Stress may be caused by drought, mineral deficiencies, pruning, cropping or defoliation of the woody plant, but also by competition with herbs. Some woody plants reduce root growth under stress conditions, others develop a shallow and laterally extending root system, or (a) deep root(s) reaching water resources through deep soil or rock fissures (Davies et al. 1989). Roots under the main trunk often penetrate deepest, a pattern that may be related to deep water infiltration by stemflow (e.g. Adams 1967; Young et al. 1984; Sect. 4.4.2). Woody plants can also develop water-storage organs as part of the root system, such as found for *Lannea humilis* in semi-arid Cameroon (Seghieri 1990) and for some shrubs in arid regions (Kemp 1983). Plants from infertile habitats maximise nutrient intake and storage through a high root to shoot ratio, as well as mycorrhizal associations (Sect. 4.5.4.9), rather than through a high root absorption capacity (Stuart Chapin 1980). The association with mycorrhiza increases both drought resistance and the absorption capacity of some nutrients (Kramer 1983). The exact nature of belowground chemical effects remains largely unknown (Coleman et al. 1989).

4.3.3 Comparison with Herbaceous Root Systems

Herbaceous root biomass in the topsoil is generally higher than of woody plants. Of total root lengths, Atkinson (1980) gives estimates of 35–100 cm cm^{-2} for tree roots and 100–4000 cm cm^{-2} for herbaceous roots in the topsoil in temperate regions. Root densities of woody plants, crops and grasses are generally estimated at 0.1–0.5, 1–5 and 5–50 cm cm^{-3} respectively (van Noordwijk 1989b; Young 1989). Eastham and Rose (1990) report values of 0.1–0.9 cm cm^{-3} for roots of *Eucalyptus* trees, compared to 2–40 cm cm^{-3} for herbaceous roots, the latter with highest values furthest away from the trees. In a natural savanna vegetation in southern Africa the density of herbaceous roots in the top 10 cm of the soil varies from a factor 10 to 100 times higher than that of woody plants, on sandy and loamy soils respectively (Knoop and Walker 1985).

In the SSWA, herbaceous root systems were investigated by Penning de Vries and Djitèye (1991). In the southern Sahel zone, root densities of herbs in the rainy season are 2.0–3.0 cm cm^{-3} in the top 0–10 cm of the soil, decreasing to 0.5 cm cm^{-3} at 50 cm depth, and negligible values beyond 120 cm. These root densities are sufficient to extract most available moisture until about 100 cm soil depth. Roots of perennial grasses may reach 200 cm depth. In the herbaceous vegetation the root to shoot ratio is 1.0 or more for aboveground biomass less than 1 t ha^{-1}, gradually decreasing to 0.5 or less when aboveground biomass exceeds 3 t ha^{-1}. The herbaceous root biomass in the rainy season averages 1.3 t ha^{-1} for aboveground production exceeding 1.5 t ha^{-1}, with 80% in the top 20 cm of the soil (Penning de Vries and Djitèye 1991). This may be compared with about 200 kg ha^{-1} of root biomass from woody plants in a woody plant community in the Sahel zones, with about 40% in the top 20 cm of the soil (Bille 1977).

The pattern of water absorption from soil differs between perennial and annual plants. Whereas in the rainy season annuals gradually develop a root system, perennials start the season with an extensive root system and immediately absorb water (e.g. Kramer 1983), and/or utilise water from storage organs (Seghieri 1990). This is a competitive advantage at the start of the rainy season, which is however soon taken over by the annuals due to their much denser root systems. Trees develop relatively more deep roots in the presence of a herbage layer (Atkinson 1980) and root to shoot ratios are higher when competition with annuals is more severe (Table 4.1). There is evidence that grass inhibits root growth of woody plants by depleting oxygen and increasing the carbon dioxide concentration in the soil. On the other hand, roots of some woody plants release toxic substances (Story 1967; Kramer 1983).

4.4 Influences of Woody Plants on Water Availability

4.4.1 Introduction

In this section are analysed the processes determining the influences of woody plants on water availability in the agro-ecosystem, and its consequence for primary production.

In Section 4.4.2 the following processes are discussed:

1. Redistribution of gross rainfall: how is rainfall reaching the soil surface and its distribution pattern influenced by a woody canopy?
2. Water infiltration and -storage: how are infiltration rate and soil water holding capacity influenced by woody plants?
3. Potential evapotranspiration (PET): how is the microclimate influenced by woody plants and what are the consequences for water partitioning into evaporation and transpiration?
4. Water use from deep soil layers: to what extent is niche differentiation possible by deep rooting?

Horizontal precipitation, a term used for atmospheric water supply to plants by the interception of dew or fog (e.g. Kramer 1983), is not discussed. Fog is rare in semi-arid regions and the occurrence of dew is limited to (the early morning at) the end of the rainy season, but very rapidly evaporates due to high temperatures.

Neither is discussed possible rainfall induction by woody plants (forests). The commonly claimed higher rainfall in afforested areas, as compared to similar areas without trees, remains unproven (e.g. Charney 1975). Too often micro- and macroscale are mixed up as far as this process is concerned.

The influences of woody plants upon temporal and spatial patterns of water availability for primary production are evaluated on the basis of available data on soil moisture contents in relation to woody plants (Sect. 4.4.3). The effects on both a small and large scale are discussed.

4.4.2 Processes Influencing Water Availability

4.4.2.1 Redistribution of Rainfall by Interception

Introduction

Rainfall interception definitions are based on the work of Helvey and Patric (1965). Gross rainfall, as measured in the open field, is redistributed by a woody canopy into several components (Fig. 4.9). Gross rainfall is partly intercepted by the woody canopy, the remaining part of the rainfall directly reaches the soil surface under the woody canopy (direct throughfall). Of the gross intercepted rainfall three components are distinguished: tree-surface adsorption and evaporation, drips from leaves and branches (indirect throughfall), and stemflow. Surface adsorption depends on the canopy and bark saturation value (storage capacity) for water, which is largely related to the leaf area index and bark roughness. Absorption of surface water by the foliage is probably negligible (Kramer 1983). Throughfall is gross rainfall which directly reaches the soil surface through spaces in the canopy, and indirectly through branch- or leaf drip. Stemflow is that part of gross intercepted rainfall that moves along the branches and the stem to the soil surface. Net interception is the proportion of gross rainfall lost by evaporation after having been intercepted by the canopy. Net throughfall or net rainfall is rainfall that effectively reaches the soil surface, equal to either:

- gross rainfall minus intercepted rainfall, or
- throughfall plus stemflow.

The factors regulating the importance of these components are both species-specific canopy parameters and climate specific rainfall parameters (Fig. 4.9).

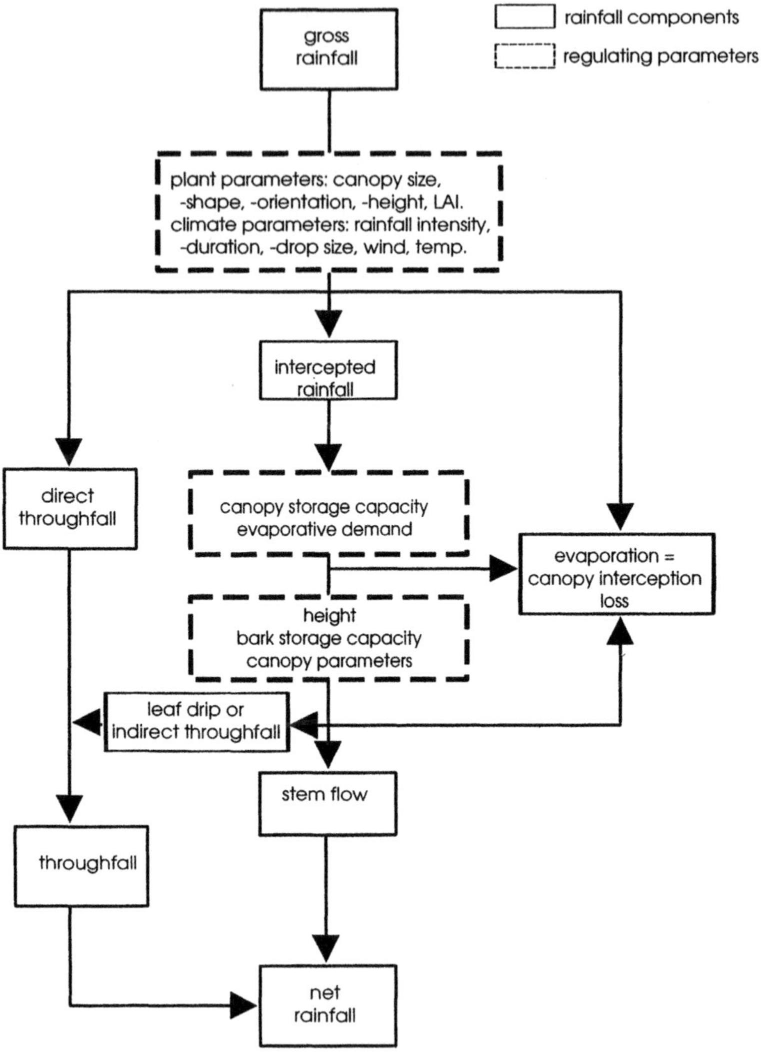

Fig. 4.9. The redistribution of gross rainfall by a woody canopy, with indication of the regulating parameters

For hardwoods in eastern USA, throughfall and stemflow require rainfall of at least 1.5 and 3.0 mm respectively to take place due to net interception by the canopy (surface of leaves and branches) and bark. Net interception is related to the storage capacity (saturation value) of the canopy and bark, which varies according to species, and is proportionally related to canopy volume. Interception losses per woody plant decrease with increasing density of woody plants (Teklehaimanot et al. 1991). The evaporation rate of intercepted rainfall (from leaf surfaces mainly) is high due to high surface roughness of a woody canopy and high advection in higher air layers (Jarvis 1985).

For woody plants without foliage (e.g. after leaf shedding) net interception is lower than fully foliated. Likewise, throughfall and stemflow increase following defoliation (Tiedemann et al. 1980). For small rainfall events (<15 mm) and low rainfall intensities, net canopy interception losses are highest (about 50% of rainfall); with high rainfall events (>15 mm) interception losses decrease to 10–20% of rainfall (Table 4.3). Net throughfall is lower in regions with low rainfall intensities (e.g. most temperate climates) than in regions with high rainfall intensities (e.g. most semi-arid regions). Due to high wind velocities and resulting shaking of canopies, interception losses may be much less. Thus, the proportions of stemflow and throughfall increase with increasing rainfall intensity and wind velocity (Helvey and Patric 1965), and the canopy storage capacity is not a constant (Jackson 1975).

Stemflow varies according to bark characteristics and branching patterns. Woody plants with smooth, dense bark usually have much stemflow, contrary to those with rough, corky bark (Voigt 1960). An upright, funnel-shaped branching pattern most efficiently leads water to the stem. Stemflow for mature beech trees reaches 10% of gross rainfall, assuming an even distribution of stemflow over the projected canopy zone, but reaches 275% of gross rainfall, assuming (more realistically) infiltration in a limited area (radius 30 cm) around the tree stem. With rough-barked woody plants stemflow actually drops in a diffuse pattern around the stem; increasing the width of collars catching stemflow from 2.5 to 50 cm increases stemflow catch on rough-barked trees by more than 50% (Voigt 1960; Helvey and Patric 1965). Coniferous trees, due to their dense canopy, show particularly low stemflow and high interception losses. Stemflow is generally less for large-diameter trees than for smaller-stemmed woody plants (Bruijnzeel 1989). This may be ascribed to differences in branching patterns and woody plant structure.

Another point of water concentration under a woody canopy is the crown limit. At the edge of the canopy of *Cassia siamea* trees, rainfall reaching the soil surface averages 106% of gross rainfall, an effect attributed to leaf drip at the edge of the canopy (Darnhofer et al. 1989). Under a woody canopy the spatial variation of rainfall reaching the soil surface is higher than in the open field, due to the structure of the woody canopy (Zinke 1962; Helvey and Patric 1965; Joffre and Rambal 1988). A positive correlation between the pattern of throughfall and fine root distribution has been observed for pine trees in temperate climates (Ford and Deans 1978).

Table 4.3. Redistribution of gross rainfall into different components, for individual trees or for the woody vegetation as a whole, in relation to climate, rainfall intensity and woody species

Vegetation, country + source	Average rainfall	Rainfall event + specifications	Interception losses (%)	Throughfall (%)	Stemflow (%)	Net throughfall (%)
Young spruce, England (Jackson 1975)	Temperate climate	2.5 mm	64			36
		17.8 mm	21			79
Hardwoods, eastern USA (Helvey and Patric 1965)	Temperate climate	2.5 mm	40	60	0	60
		20 mm	10	86	4	90
		Annual basis	13			87
Forest trees, Germany (Lunt 1934)	Temperate climate	Annual basis:			13	8
		– beech	22	65	6	79
		– oak	21	74	6	78
		– maple	23	72	1	41
		– spruce	59	40		
Tropical forest, Surinam (Jackson 1975)	High rainfall	2.5 mm	48			52
		20 mm	21			79
Rainforest, Tanzania (Jackson 1975)	High rainfall	2 mm	60			40
		20 mm	12			88
Tropical lowlands forests (Bruijnzeel 1989)	High rainfall	Annual basis	12–14	86	0.5–2.0	86–88
Semi-arid regions *Juniperus occidentalis* California (Young et al. 1984)	300 mm	Annual basis:				
		– edge of canopy	19			81
		– under canopy	51			49
		– near trunk	69			31
		Total canopy	42	58	0.1	58

Acacia holosericea Australia (Langkamp et al. 1982)	1200 mm	10 mm	12	84	4	88
		300 mm	6	67	27	94
		Annual basis	11	73	16	89
Acacia aneura Alice Springs (Slatyer 1965)	275 mm	1 mm	70	30	0	30
		>12 mm	5	55	40	95
Acacia aneura Charlesville (Pressland 1973)	500 mm	2 mm	~35	~60	<5	~65
		10 mm	10	68	22	90
		Annual basis	13	69	18	87
Eucalyptus melanophloia Australia (Prebble and Stirk 1980)	700 mm	5 mm	30	70	0	70
		15 mm	~13	87	0.6	~87
		Annual basis	~12	88	0.6	~87
Faidherbia albida Senegal (Dancette and Poulain 1969)	300 mm	<15 mm				95
		>15 mm				120
		Annual basis				110

Data from Semi-Arid Regions

In *Acacia aneura* woodlands in semi-arid Australia, stemflow as a proportion of gross rainfall increases to 40% for large rainfall events at Alice Springs (275 mm rainfall) whereas at Charleville (500 mm rainfall) stemflow is at most 22% (Table 4.3). Rainfall patterns being similar, the difference is related to different phyllode morphology. Phyllodes are longer and narrower at the drier location, thus reducing the canopy saturation value and increasing stemflow. Both this phyllode morphology and the funnel-like shape of the tree are adaptations to dry conditions. At Charleville, stemflow water infiltrates the soil within 45 cm of large tree stems, and within 15 cm for small trees. From a 25-mm shower an equivalent of 140 mm rainfall can infiltrate this area due to stemflow. For water to percolate in a dry loamy soil beyond 200 cm depths, a 75-mm rainfall event, or accumulated rainfall of 75 mm in a short period, would be required (Pressland 1973).

For *Eucalyptus melanophloia* trees in Australia (700 mm rainfall), stemflow is 0.6% of annual gross rainfall only (Prebble and Stirk 1980). For these trees, 7 mm of rain was required to offset stemflow, probably related to the high adsorption capacity of the thick layer of bark. The largest measured stemflows for this tree represent only a 15–30 mm addition to a soil zone 0.5 m around the tree stem. The canopy shape of this tree species is characterised by a low leaf density and many open spaces.

Throughfall under *Acacia aneura* trees increases with rainfall intensity (Table 4.3) and is lower near the stem (40%) than at the canopy edge (60%) (Pressland 1973). At Alice Springs, net throughfall on an annual basis is 95% of gross rainfall, but on several occasions net throughfall exceeds gross rainfall. At Charleville, net throughfall is lower (87%) and only on two occasions exceeds gross rainfall.

Net throughfall under a woody canopy exceeding gross rainfall can be explained by the lateral interception of rain by the tree canopy during intensive showers, particularly when associated with strong winds. Throughfall under the canopy therefore varies according to wind direction: for rainfall events of 40–50 mm, throughfall varies between less than 40 and more than 60 mm on the leeward and windward sides of the canopy respectively (Belsky et al. 1989). Darnhofer et al. (1989) found a rainshadow area at the leeward side of a row of trees, at a distance equal to 50% of tree height, where only 73% of gross rainfall reaches the soil surface. Neumann and Pietrowicz (1989) found rainfall in the open areas of an alley cropping system to be 90% of gross rainfall in open, untreated fields.

For *Eucalyptus melanophloia* trees in Australia throughfall is only 88% of gross rainfall on an annual basis. Here throughfall is lowest for intermittent rainfall events, as compared to continuous rainfall events.

For small rainfall events net throughfall is lower near the stem (40%) than at the edge of the canopy (60%); for high rainfall events it is just the opposite: 110% near the stem and 100% at the edge of the canopy. These differences are consistent with the greater phytomass near the centre of the canopy, absorbing an equivalent amount of water (Prebble and Stirk 1980). For most *Eucalyptus* trees in semi-arid regions, net rainfall under the canopy is lower than gross rainfall: interception losses average 11–20%, whereas stemflow averages 2–5% (Poore and Fries 1985).

Interception data from the SSWA are limited to studies of net throughfall under the canopy of *Faidherbia albida* trees in Senegal. Results show a 5% decrease and a 20% increase as compared to gross rainfall, with fine rains ($<15\,\mathrm{mm\,h^{-1}}$) and heavy rains ($>15\,\mathrm{mm\,h^{-1}}$) respectively. On an annual basis, there is an increase of 10% under the woody canopy (Dancette and Poulain 1969). This increase should be assciated with a similar reduction of net rainfall in the open areas between the trees. The leafless status of this woody species during the rainy season contributes to particularly low interception losses.

It can be concluded that throughfall as a proportion of gross rainfall increases from zero for very small rainfall events to almost 100% for large rainfall events (Belsky et al. 1989; Darnhofer et al. 1989). The minimum rainfall required to offset throughfall is related to the canopy saturation value (Sect. 4.2.1.1). The canopy saturation value, averaged on an annual basis, is about 2 mm for mature *Eucalyptus melanophloia* trees, 1.5 mm for mature *Acacia aneura* trees at Alice Springs and 2 mm for mature *Acacia tortilis* trees in Kenya (Belsky et al. 1989). In the Sahel zones, interception losses are probably less than in tropical Australia, as a result of the predominance of small woody plants and their low total canopy cover. For the herbage layer, an average of 1 mm net interception is commonly applied (Breman and de Ridder 1991). The canopy saturation value increases with increasing woody plant size due to an increase of intercepting foliage. In Australia, net throughfall (X in mm) is related to precipitation of a rainfall event (Y in mm) and to mean basal area of *Acacia aneura* trees (Z in $\mathrm{cm^2}$) by the equation:

$$X = 0.885Y - 0.0015Z \text{ (Pressland 1973).}$$

As woody plant densities increase, and the canopy gradually closes, interception losses for the woody plant community as a whole increase. Ovalle and Avendaño (1988) found the proportion of rainfall reaching the soil surface to average 75 and 90% for a woody canopy of about 80 and 50% respectively.

On the basis of the preceding data, it can be concluded that in the SSWA stemflow probably varies between 0 and 20%. The highest value is applicable to woody species with a typical funnel-shaped branching pattern (e.g. *Acacia seyal*), and to years with a high proportion of

Table 4.4. Rainfall required for water to infiltrate beyond 2 m soil depth in an area around the trunk of the woody plant equal to 3% of the canopy cover, assuming 10% interception (evaporation) losses and infiltration of all stemflow (no run-off), for three different soil textures characterised by their water content at field capacity

	Soil texture		
Water content at field capacity ($cm^3\,cm^{-3}$)	Sand 0.05–0.10	Sandy loam 0.15–0.20	Loamy clay 0.30–0.35
Infiltrated water (mm) required to reach 2 m soil depth[a]	250–400	550–750	1000–1150
Rainfall (mm) required to reach 2 m soil depth in tree-trunk area, assuming:			
– no stemflow	280–440	610–830	1110–1280
– 10% stemflow	70–110	150–205	270–310
– 20% stemflow	40–60	85–115	155–180

[a] Based on Fig. 2.5.

intensive rainshowers. High stemflow leads to intensive soil wetting of an area around the tree stem and water infiltration in the subsoil, increasingly beyond the reach of herbaceous roots (>2 m soil depth). The area around the tree stem receiving additional moisture mainly depends on the diameter of the tree trunk and the canopy size. For small trees, dominating in the Sahel zones, this area is more than 3% of the total canopy area; for large trees, dominating in the Sudan zones, this area is at most 2%. As a result, the concentration factor of water in this area receiving stemflow varies from less than 30 to more than 50. The minimum rainfall required for water to infiltrate to 2 m soil depth varies for different soil textures (Table 4.4). Interception losses of 10% are considered as independent of variation in total rainfall. Stemflow provides woody plants a competitive advantage in relation to herbs (Sect. 4.8.3.3) and may even recharge subsoil water reserves if percolation exceeds the reach of tree roots.

4.4.2.2 Water Infiltration and Moisture Retention

Processes by which woody plants influence water infiltration and moisture retention capacity of the soil are presented in Fig. 4.10. Water infiltration and moisture retention capacity of the soil are largely determined by soil structure and texture. Regarding soil structure, soil porosity is a crucial factor: macropores reaching the soil surface particularly enhance water infiltration, whereas micropores are most important in respect to

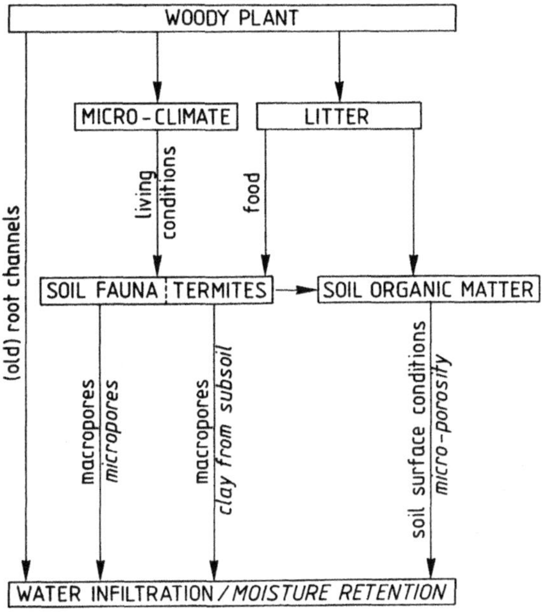

Fig. 4.10. Processes by which woody plants influence water infiltration rate and moisture retention capacity of soil (the latter in *italics*)

moisture retention capacity. Porosity is especially related to vegetation cover, improving soil porosity by plant roots, and providing feeding material for soil fauna, which, in their turn, improve soil porosity. Maintenance of soil organic matter seems crucial in two ways: providing suitable conditions for soil fauna and maintaining a favourable soil structure (Sect. 4.5.5). This is particularly important where soil crusting may occur (mainly on loamy soils). There is an inverse relationship between infiltration rate and moisture retention capacity for different soil textures.

Old root channels of woody plants improve water infiltration, particularly to greater depths. In the humid tropics in Indonesia and in Nigeria water does not infiltrate homogeneously, but preferably follows old root channels of woody plants (van Noordwijk et al. 1991). It is unknown to what extent living roots have a similar function. This effect probably has limited importance in semi-arid regions; in the SSWA, water infiltration patterns in different soils can be predicted by models that do not take into account infiltration by root channels (Breman and de Ridder 1991).

Soil biological activities improve water infiltration and moisture retention capacity by increased clay content and improved soil porosity mainly (Lal 1987; Lavelle et al. 1990). In humid tropical regions in Nigeria, infiltration rates of alfisols are 30–70% higher with than without earthworms (Lal 1989). In semi-arid savanna grasslands termites generally improve water infiltration, although termite mounds without vegetation (Sect. 2.3.3) have low water infiltration rates (Arshad 1982; Lal 1987). In savanna woodlands in Ivory Coast, the total surface of pore openings due to the activities of termites is 2–4 $m^2 ha^{-1}$ (Lavelle et al. 1990). Although under woody canopies termite activities seem to be enhanced (Sect. 4.5.3.4), Elkins et al. (1986) assume that other biological or physical processes are more important regarding the influence on water infiltration. In semi-arid New Mexico (USA), extermination of termites under shrubs does not reduce water infiltration rate, contrary to the situation in the open field.

Glover et al. (1962) observed that (1) moisture penetration in the soil from a 20 mm shower is considerably deeper under a *Balanites aegyptiaca* tree than in the surrounding grassland, and that (2) soil moisture profiles are closely related to the (scattered) presence of the herbage layer.

Data on water infiltration under woody canopies in semi-arid regions are summarised in Table 4.5. Water infiltration rates are presented in relative terms, since absolute values are difficult to compare due to different methods used. The data show that under woody canopies water infiltration rates are relatively high, which is generally acknowledged

Table 4.5. Relative water infiltration rates in relation to soil texture and presence of a woody canopy, from studies in semi-arid regions

Country + source	Soil texture	Canopy specifications	Relative infiltration rate (%)
Zimbabwe (Kennard and Walker 1973)	Sandy	Closed canopy	100
		Open canopy	84
		Open grassland	55
Zimbabwe (Kelly and Walker 1976)	Variable	Complete litter cover	100
		Partial litter cover	33
		No litter cover	12
Kenya (Belsky et al. 1989)	Loamy	Under canopy *A. tortilis*	100
		Open field	25
		Under canopy *Adansonia*	100
		Open field	20
Kenya (Scholte 1989)	Loamy	Under shrub	100
		Open field	5
Australia (Slatyer 1961)	Sandy	0.5 m from tree stem	100
		2.0 m from tree stem	60
		Open field	40
New Mexico (Elkins et al. 1986)	Variable	*Carrea tridentata* canopy	100
		Medium cover grassland	92
		Low cover grassland	87

(e.g. Wiersum 1984). The reduction of the infiltration rate in the open field as compared to the canopy zone varies from 50 to 95%, depending on the soil conditions and the vegetation cover. Both herbage and litter biomass are commonly higher under a woody canopy than in the open field (Sect. 4.2). Presence of a herbage or litter layer improves water infiltration by 10–20% (e.g. Kelly and Walker 1976; Knoop and Walker 1985; Jackson et al. 1989; Scholte 1989). By a litter or a mulch layer (the latter as deliberately deposited organic material) covering the soil, soil organic matter is improved, water infiltration is increased and soil erosion reduced (Lal 1975). The effect largely depends on the amount of accumulated organic material. For most soils, runoff and erosion show an exponential decrease with increasing mulch biomass (Lal 1976). A mulch or litter layer also provides shade, and protection from mechanical impacts (Stigter 1985). The reduction of run-off from slopes by woody plants (e.g. when planted in hedges) is probably more related to the increased litter cover and the improved soil structure than to actual physical barriers.

The available moisture for plant growth is estimated by the difference between moisture contents at field capacity (pF 2.0–2.5) and at permanent wilting point (at least pF 4.2). Comparing the topsoil under woody canopies with open areas, moisture content at permanent wilting point of the topsoil may be a factor 2 higher, moisture content at field capacity up to a factor 1.4 higher (Kennard and Walker 1973; Joffre and Rambal 1988). Due to this difference, available moisture under a woody canopy in the mediterranean region is 40–80% higher than in the open field, corresponding to 50–100 mm additional water in a soil layer of 150 cm (Joffre and Rambal 1988). Scarce available evidence from tropical semi-arid regions indicates that the absolute changes of the two components (moisture content at field capacity and at wilting point) are similar, so that the available moisture for plant growth is little influenced (e.g. Kennard and Walker 1973). Available moisture under *Faidherbia albida* canopies is similar to that in the open field, although organic matter content under the tree is almost twice as high (Dancette and Poulain 1969). This finding corresponds to an experiment in the same region, where doubling of organic matter contents in the topsoil on croplands did not influence the moisture retention curve (de Ridder and van Keulen 1990). The 40% decrease of available moisture following continuous cropping cannot be attributed solely to the reduction in organic matter contents (up to a factor 3) because soil texture had also changed (increased sandiness).

Improved moisture availability under woody canopies as compared to the open field should be related to improved soil structure by a combination of increased organic matter and reduced soil bulk density, and possibly also to the improved soil fauna activities and the increased proportion of clay under woody canopies (e.g. Koechlin et al. 1986;

Joffre and Rambal 1988). The presence of woody plants is probably most important on fine-textured soils that easily form a crust in order to maintain a sufficiently high water infiltration capacity of the soil at the onset of the rainy season mainly. On the other hand, where woody plants suppress the herbage layer and/or where tree litter is removed, water infiltration under woody canopies may be less. Pressland and Fisher (1982) found water losses by streamflow to be lowest for a water catchment dominated by a grassland vegetation, as compared to a vegetation with *Eucalyptus* trees in Australia.

It has been observed that, shortly after a rainfall event, water pools form under isolated woody plants. This may be due to the woody plants selectively establishing in micro-depressions where water availability, prior to woody plant establishment, is (slightly) higher.

In Section 4.2 it was shown that when woody cover exceeds a certain value, availability of limited resources is likely to lead to a reduced herbage cover in the open field, depending on the climatic zone and the soil type (Fig. 4.5). The tiger bush pattern in the Sahel zones is one example of this phenomenon. Here, an improved water infiltration under the woody canopy may lead to a reduced water infiltration in the open areas. This effect is particularly important where water is the most important plant production limiting factor.

4.4.2.3 Evapotranspiration

Introduction

Water moves from the soil to the atmosphere by the process of evapotranspiration, involving water transport directly from the soil surface (evaporation) and indirectly through the plant (transpiration). Both are diffusion processes, driven by concentration gradients and controlled by resistances, while radiation provides the required energy (Lövenstein et al. 1992).

Regarding soil evaporation, resistance for vapour transfer is built up as the area of transfer from liquid to vapour moves deeper into the soil. Eventually a layer of dry soil may effectively prevent further evaporation. For coarse soil textures this soil layer is less than for fine soils, since in the latter case water diffuses upwards through micropores.

Transpiration from plant tissues takes place via the stomata mainly. The vapour pressure deficit is defined as the difference between water vapour concentration in the stomatal cavity and the outside air. Temperature is a major determinant of the vapour pressure deficit. The combined effect of vapour pressure deficit and windspeed is referred to as the drying power of the air. When transpiration is not restricted by stomatal aperture, transpiration rate is determined by radiation and the drying power of the air, together constituting the evaporative demand (Goudriaan 1989). One common way to reduce water losses in case of an increasing vapour pressure deficit is by closure of stomata in order to reduce water conductance. Under water shortage conditions, stomata may close (reduced conductance), curbing water losses by transpiration, but leading to reduced photosynthetic activity. In order to adequately respond to continuously changing climatic circumstances, different mechanisms of stomatal regulation operate on the basis of light, CO_2 in the stomatal cavity and vapour pressure deficit. A low water availability in the soil generally leads to low moisture contents in plant tissues and a reduction of water conductance. In the case of high temperatures, some woody plants, with sufficient water available, increase transpiration in order to reduce the temperature, but cooling by advection is probably more important (Sutcliffe 1979).

While water demand determines the potential transpiration rate, water supply is a modifying factor leading to ultimate water use, i.e. the actual transpiration rate. The relation between production and transpiration rate is expressed as the water-use efficiency (Lövenstein et al. 1992). Water-use efficiency (WUE) of woody species is also influenced by weather conditions: WUE can vary a factor 2 in relation to air humidity only. Plant production is related to WUE divided by potential evapotranspiration (PET) (de Wit 1958). In this relationship the drying power of the air predominates as the determining factor of PET (Tanner and Sinclair 1983).

In the following the influence of woody plants on the climatic factors determining potential evapotranspiration (PET) is investigated first: temperature, air humidity and wind speed. Subsequently potential evapotranspiration is analysed and data on actual transpiration by woody plants are summarised.

Microclimate and Woody Plants

Analysing the influence of a woody canopy on microclimate factors, one should consider where the measurements were taken (soil depth or air stratum), at what time (season and time of the day), and for which plants (species, structure, condition of herbs and woody plants). Also, it is important to describe the reference vegetation type with which a comparison is made (e.g. the vegetation in the open areas). Very commonly, such information is not provided.

The following selection of available evidence from semi-arid regions serves as illustrations of the prevailing trends. A distinction is made between the influences of a woody canopy (in comparison to an open area) and those of a wooded area (in comparison to a grassland).

Table 4.6. Temperatures (°C) in a semi-arid tree savanna in Botswana, at four different locations, annual averages and standard deviations. (Veenendaal 1991)

	Bare area		Grassland[a]		Outer canopy[b]		Inner canopy[c]	
	x	σ	x	σ	x	s	x	σ
5 cm above soil	36.5	12.5	37.7	13.3	35.9	8.6	27.2	4.8
Soil surface	43.0	15.1	32.3	8.5	29.8	6.5	25.8	4.6
2 cm soil depth	40.3	11.8	28.8	5.9	28.2	5.7	22.7	2.0

[a] Dense grass vegetation.
[b] Average value under the edge of the canopy of eight *Acacia tortilis* and three *Dichrostachys cinerea* trees.
[c] Average value next to the trunk of the same trees.

The reduction of mean annual *temperature* under a woody canopy depends on air stratum and soil depth where measurements took place (Table 4.6). In comparison to the adjacent annual grassland, temperature reduction under a woody canopy is more than 10 °C in the topsoil, but at 30 cm soil depth temperatures are not any more influenced (Tiedemann and Klemmedson 1973; Kelly and Walker 1976; Shankarnarayan et al. 1989). The temperature differences are greatest for seasons with intense radiation. Under the woody canopy, soil and air temperatures gradually increase from the vicinity of the trunk towards the outer canopy area, but also vary up to 10 °C between the sunny and the shadow side of a tree or alley (Tiedemann and Klemmedson 1977; Malik and Sharma 1990).

The reduction in temperature by woody cover is influenced by the vegetation cover of the open area. Data from Table 4.6 demonstrate the influence on temperature of vegetation cover, and of the location under the canopy. Belsky et al. (1989) found mean soil temperatures to be lowest in the shade of a tree (24 °C), intermediate in a grassland with 90–100% cover (29 °C) and highest in a grassland with 50–70% cover (33 °C). A reduction of the cover of annual grasslands from 65% to 6% causes an increase in mean temperatures of 14 °C at 2.5 cm soil depth (Tiedemann and Klemmedson 1977). A mulch or litter layer buffers and dampens the effects of environmental factors on the soil. The stabilising effect is due to radiation absorption, change of albedo and physical barriers. While albedo is a function of mulch colour, the barrier effect depends on thickness and cover of the mulch layer (Coleman et al. 1989). Under a layer of woody litter applied at 5 t ha^{-1}, soil temperature at 5 cm soil depth is 3–7 °C lower than at bare sites (Budelman 1989).

Alley cropping versus sole groundnut reduces maximum air temperatures by 3–10 °C whereas alley cropping versus sole millet reduces maximum air temperatures by at most 1 °C (Corlett et al. 1989). Air temperatures recorded under the canopy of *Faidherbia albida* and in the millet field are at maximum 38 and 41 °C respectively, but minimum temperatures do not differ (Dancette and Poulain 1969). For the same tree, maximum temperatures differ by 10 °C in comparison to the adjacent uncultivated field, in the dry season at the time of full tree foliage (ICRISAT 1991).

One conclusion is that in tropical regions both mean temperatures and the variation in temperatures are reduced by woody canopies because maximum temperatures are influenced more than are minimum temperatures (Lawson and Jeník 1967; Belsky et al. 1989).

Air *humidity* under a woody canopy and in a woody vegetation is generally higher than in the open field. This is particularly so during the rainy season and early dry season since in the full dry season soil and air humidity is low everywhere. Higher air humidity under woody canopies is a consequence of reduced air temperatures and higher soil moisture in the topsoil under woody canopies. In Senegal, at the end of the rainy season (October), average daily air humidity under *Faidherbia albida* trees is 59% as compared to 51% in the open field (Dancette and Poulain 1969). In semi-arid India, during the rainy season, air humidity in a wooded area is 7% higher than in a tree-less area (Shankarnarayan et al. 1989).

On a regional scale similar effects may be expected, in relation to temperature effects, but no data are available.

Wind speed is always reduced by woody plants in comparison to a vegetation without woody plants. Thus, the evaporative demand is reduced: moist air under a woody canopy is not transported and a higher air humidity is maintained. Windbreaks may serve to optimise this suppressive effect on a regional scale. Several consecutive windbreaks have a regional effect on windspeed, as well as a local effect in relation to each individual windbreak (Guyot 1989). A decrease of wind velocity on a regional scale may also result from evenly spread woody plants, such as occurring in parklands (Sect. 2.4.4). This effect has, however, never been specifically demonstrated (Stigter 1985).

Windbreaks on croplands are supposed to lead to increased crop yields by increasing moisture availability due to lower PET, but recent studies throw doubts on this simplistic explanation (Stigter 1985). In arid zones the beneficial effect on crops by the reduction of wind velocity can also be explained by the decrease of the abrasive effect on plant seedlings by the wind blown soil (e.g. Lyles and Woodruff 1960). Protection against mechanical stress should also be considered (Stigter 1988). Both primary and secondary wind injury effects (influence of air movement on physiology) may occur and depress plant productivity.

Potential Evapotranspiration

It is difficult, on the basis of the influence on microclimate factors as cited above, to determine the overall effect of woody plants on PET. Observations are scarce.

At 0–6 m distance from a row of 12-m-high *Eucalyptus tereticornis* trees, mean daily temperature at the shade (cool) side is 18 °C and pan evaporation 8.9 mm d^{-1}, whereas on the sunny (hot) side of the trees mean temperature is 28 °C and pan evaporation 15.5 mm (Malik and Sharma 1990). Pan evaporation in the open field is twice as high as on the leeward side of a bush, although mean temperatures differ only by some degrees (Lawson and Jeník 1967). In some cases the influence of a woody canopy on potential evaporation seems insignificant. For instance, in Senegal, in the rainy season, there are no differences in pan evaporation at varying distances from *Faidherbia albida* trees (Dancette and Poulain 1969). In Chile, during the cold and rainy season, Ovalle and Avendaño (1988) could demonstrate no significant differences in PET between plots with and without *Acacia caven* trees. In both cases, measurements were taken in the rainy season; temperature reduction under the woody canopy is small or insignificant, whereas wind speed under isolated woody plants is probably not lower than in the adjacent open field.

From these studies, the apparently contradictory findings are due to the fact that woody plants influence, partly independent, both radiation and drying power of the air. Simulation helps to reinforce insight. A model simulating the influence of woody plants on PET in the southern Sahel zone is based on a year with average rainfall: average radiation during the rainy season is 21 350 kJ m^2 d^{-1}, air temperature 27.4 °C, vapour pressure 2.73 kPa (Pa = J m^{-3}) and wind speed 1.2 m s^{-1}. From the previous evidence it is concluded that the temperature difference between the canopy zone and the open field is at most 10 °C, air humidity under the canopy can be at most 15% higher, wind speed may be reduced by 50% and radiation intensity under isolated trees may be reduced by 50% or more (Sect. 4.7). Calculating PET according to the Penman formula, the maximum difference of PET between woody canopies and the open field is a factor 3 (1.9 and 5.6 mm d^{-1} respectively), but the mean reduction is much lower. The variation of the drying power of the air is much more. In the open field, the average contribution of the drying power of the air to PET is 26%. Resulting from the influence of the woody plant it can increase to 38% where the woody plant only intercepts radiation, but it can also decrease to almost nil where the interception of radiation is accompanied by strongly increased air humidity and decreased temperature. The importance of the factors involved is as follows:

- 10% variation of radiation intensity leads to 0.35 mm d^{-1} PET variation;
- 1 °C variation of temperature leads to 0.15 mm d^{-1} PET variation, largely related to the drying power of the air;
- 10% increase of air humidity leads to 0.35 mm d^{-1} decrease of PET, strongly related to the reduced drying power of the air;
- 50% reduction of wind speed leads to 20% reduction of PET, entirely resulting from the reduced drying power of the air.

Hence, woody plants intercepting radiation always cause reduced PET under their canopies; but where the drying power of the air is little influenced, WUE of the vegetation under the canopy decreases, in spite of reduced radiation and evaporation. These conditions occur where no moisture is available to increase air humidity or to reduce air temperature, and can be found where water reserves are available, beyond the reach of herbs or in soils with effective barriers for evaporation (coarse sandy soils).

If the reduction of the drying power of the air exceeds the reduction of the radiation component, WUE of the vegetation under the canopy increases, but at a low growth rate. These conditions occur where both evaporation and transpiration are high, leading to reduced temperatures and increased air humidity, and where PET is high. These conditions can be found where soils are moist and where PET is high: for instance, valleys in the Sahel zones, particularly during the dry season. This conclusion corresponds to that of other authors: most benefits of woody plants on water availability are expected when PET is high, for dense woody canopies, and where soil and/or plant moisture is sufficient to reduce temperature and increase air humidity, leading to prolonged crop growth (Gupta et al. 1984; Guyot 1989). Under too dry conditions, the opposite pattern may appear: crops protected by woody plants (wind breaks) may produce less due to reduced water-use efficiency (FAO 1986).

Transpiration by Woody Plants

Data on WUE from woody plants in semi-arid regions are scarce. Estimates of WUE for aboveground production from arid Israel vary from 770 l kg^{-1} for *Acacia salicina* to 340 l kg^{-1} for *Eucalyptus occidentalis*, both tree species in their juvenile growth phase (Lövenstein et al. 1991). Chaturvedi et al. (1988) in semi-arid India found estimates varying from 1400 to 500 l kg^{-1} (highest value for *Eucalyptus* species), also for young plants. Values are lower when both above and belowground production are considered (Sect. 3.2.2.2). For annual plants, the amount of water transpired per total (above- and belowground) net production varies from $250\,l\,kg^{-1}$ for C_3 species to $160\,l\,kg^{-1}$ for C_4 species (Penning de Vries and Djitèye 1991). Thus, WUE of woody plants is lower. Part of the explanation is that woody plants, due to their perennial status, require more assimilation products for maintenance than annual plants. On the other hand, the perennial status of woody plants enables utilisation of water resources (if available) during the entire year. This provides a higher production than by annual plants with a growth period limited to the rainy season.

The low WUE of a woody plant is also related to the evaporative demand per unit of leaf area of a woody plant, due to its position and structure, exceeding that of a herb. However, to offset this disadvantage, a woody canopy can exercise greater control of water through morphological, phenological and physiological adaptations to limit transpiration losses (e.g. Connor 1983). Most woody species in semi-arid areas have adopted a combination of morphological structures such as a thick leaf cuticle or a small leaf area, phenological strategies such as leaf drop in the dry season, and physiological adaptations such as stomatal resistance and increased reflectance of leaves. Open canopies, in combination with small leaves, enhance convective cooling, reducing leaf temperatures and transpiration (Shmida and Burgess 1988), but also lead to a higher evaporative demand.

Survival and production are conflicting goals under water-stressed conditions. Adaptations are firstly aimed at improved survival during the dry season, but usually lead to reduced growth rates at the same time. In perennial plants, adaptation to drought means a combination of avoidance of, and tolerance to dehydration. If these plants exist in non-competitive habitats, a high WUE is beneficial because it provides higher productivity per amount of water. However, in semi-arid regions competition is the rule and any water saved by one individual is used by competitors. Here, woody plants are more likely to close stomata at higher water potentials than in non-competitive situations (Pereira and Pallardy 1989).

In Niger, transpiration of woody species was measured by the porometer method (Ullmann 1985). The average daily transpiration in March and April (dry season) of fresh leaves of various *Acacia* species is $53–105\,mg\,m^{-2}$ leaf surface s^{-1}. Transpiration losses are highest for *Acacia* trees rooting in groundwater as compared to those that do not, and are highest for young leaves as compared to old leaves (Table 4.7). The palm tree *Phoenix dactylifera* shows lower transpiration losses than the *Acacia* species.

Based on the above data and assuming an average canopy cover of the *Acacia* trees of $20\,m^2$ and 2 transpiring leaf layers, transpiration losses during the dry season would be 100 to $200\,l\,d^{-1}$ per tree. The high estimate applies to *Acacia* trees rooting in groundwater (not restricting water losses), the low estimate to trees without additional groundwater

Table 4.7. Transpiration losses by some woody species in the SSWA during the dry season

Country + source	Rainfall	Woody species	Transpiration rate		Details
			Max.	Mean	
			(mg m^{-2} s^{-1})		
Niger,	500 mm	*Faidherbia albida*	90	~ 53	
Niamey		*Acacia senegal*	118		Young leaves
(Ullmann 1985)		*Acacia seyal*	153		id.
			(mg m^{-2} s^{-1})		
Niger,	126 mm	*Acacia ehrenbergiana*	179	~105	Rooting in groundwater
Indoudou		*Acacia ehrenbergiana*	53		Old leaves
(Ullmann 1985)		*Acacia ehrenbergiana*	90		Young leaves
		Acacia leata	161		
		Acacia nilotica	134	~ 75	Rooting in groundwater
		Acacia tortilis	161	~105	
		Phoenix dactylifeca	41	~ 20	
Burkina Faso	600–900 mm		(g g^{-1} d^{-1})		
(Tolman 1989)		*Vitellaria paradoxa*	4.8	4.5	
		Parkia biglobosa	5.1	4.0	
		Faidherbia albida	6.3	6.1	
		Khaya senegalensis	3.4	3.2	
		all trees		4.2	

resources. Ullmann (1985) states that transpiration losses may be even further reduced in the case of severe water stress. Transpiration losses in the dry season are generally higher than in the rainy season, due to a lower air humidity and resulting high vapour pressure deficit. Drought deciduous woody plants shed leaves in the dry season to reduce transpiration losses, (semi-)evergreen woody plants have physiological adaptations to reduce transpiration losses. Transpiration losses by *Faidherbia albida* in the dry season (with leaves) are also relatively low (Table 4.7).

In Burkina Faso, Tolman (1989) estimated transpiration losses of large trees in valleys during the dry season. Use was made of the rapid weighing method of cut twigs and small branches. Transpiration losses vary between 3.2–6.3 $g\,d^{-1}$ per g of leaf biomass (Table 4.7), equivalent to 150–300 $l\,d^{-1}$ per tree, with an average of 200 $l\,d^{-1}$. Acknowledging the limitations of the method used, this would be a transpiration loss of about 4 $mm\,d^{-1}\,ha^{-1}$ for a closed canopy. Such transpiration losses are low in comparison to those of herbs growing under conditions of sufficient water and nutrient availability: in the dry season 6 to 12 $mm\,d^{-1}$ for a closed herbage cover with a leaf area index of 4 (respectively with and without a regulatory mechanism of stomata; Penning de Vries and Djitèye 1991). In most cases, tree cover is much less than 100% so that transpiration losses are less. Transpiration from woody plants averages 0.1 $mm\,d^{-1}\,ha^{-1}$ in valleys in the northern Sahel zone (with 2.5% woody cover), and 1.3 $mm\,d^{-1}\,ha^{-1}$ in valleys in the Sudan zones (with 30% woody cover). When rooting in the subsoil with available water, woody plants can continue to transpire during the dry season. Total transpiration losses by woody plants remaining in foliage during the dry season (e.g. in valleys) thus reaches 30 mm during the 10-month dry season in the northern Sahel zone (equal to about 15% of annual rainfall), and almost 300 mm during the 7-month dry season in the southern Sudan zone (equal to about 30% of annual rainfall).

There are both reasons to assume that transpiration by woody plants in the rainy season is less than in the dry season (lower evaporative demand) as the opposite (higher water availability in rainy season, no stomata closure). Therefore, in valleys in the southern Sahel zone, assuming transpiration by a mature woody plant of about 200 $l\,d^{-1}$ during the entire year, and based on the average aboveground production of woody plants (Table 3.10), WUE of a woody plant with a canopy cover of 150 m^2 averages 555 $l\,kg^{-1}$. This value is comparable to the observations cited above.

On the basis of the previous analysis, 240 $l\,kg^{-1}$ is used as the average WUE for woody plants with a rapid growth in their juvenile stage, and 480 $l\,kg^{-1}$ as the average WUE for mature woody plants and those showing slow growth (taking into account above- and belowground production).

The influences of the exploitation of woody plants on its transpiration are variable. Defoliation of the woody plant may lead to reduced transpiration losses due to reduced leaf area, which may enable prolonged growth and water use by the woody plants (van Duivenbooden 1985), or by the associated herbage. The opposite effect can also occur e.g. due to vigorous regrowth of the woody plant following defoliation (Stuart-Hill and Tainton 1989). Pruned *Leucaena* shrubs show transpiration rates similar to unpruned plants, probably as a result of regrowth of the pruned plants, with actively transpiring young leaves, in contrast to mature plants with old leaves and reproductive organs (Ong et al. 1992).

Grazing of the herbage layer may be beneficial for woody plants since it reduces moisture competition and may eventually lead to bush-encroachment (Sect. 2.4.2).

4.4.2.4 Deep Roots

By deep roots, woody plants obtain water from deep soil layers, beyond the reach of herbaceous roots. In the SSWA, water uptake by deep roots becomes important with rainfall exceeding 500 mm yr^{-1} (Sect. 4.8.3). In the northern Sudan zone, the additional water availability by this process is 50–170 mm yr^{-1} (variation according to soil texture); in the southern Sudan zone it is 240–340 mm yr^{-1} (Table 4.33). The additional water may be used for either woody or herbaceous plant growth (Sect. 4.8.3).

With closed stomata at night, passive water transport may take place through woody roots, from the moist subsoil to the dry topsoil, by difference in vapour pressure. Indications for this process, called hydraulic lift, were found by Baker and Bavel (1986) and by Alexandre (1990). In a sagebrush (*Artemisia tridentata*) community in semi-arid USA Richards and Caldwell (1987) observed a regular pattern of drying of the soil at 35–80 cm depth during day-time followed by wetting at night-time. The wetting can not be explained by water movements through the soil and is assumed to result from water efflux from roots that take up water in deep soil layers. The quantities of water implicated are at most 0.1 mm per night, so not more than 20–30 mm during one annual dry season. Such small quantities are not sufficient for plant growth but can be important for plant survival and may avoid dying of fine roots in the dry season. It is hypothesised that hydraulic lift and parasitism of water lost to the topsoil by shallow-rooted plants can result in greater resource competition than root distributions would suggest. Hydraulic lift may be particularly important for *Faidherbia albida*, explaining its growth during the dry season, when the topsoil is dry.

4.4.3 Water Availability and Water Use

4.4.3.1 Water Availability for Primary Production

Woody plants firstly influence the availability of water for plant production within an ecosystem, and secondly the spatial distribution of gross rainfall into soil moisture availability. In Table 4.8 are summarised the relative importance of the discussed processes associated with woody plants influencing water availability for primary production, in comparison to the situation without woody plants. The data refer to a situation of unexploited woody plants in an equilibrium situation with canopy covers corresponding to canopy cover in the SSWA before the drought periods, and also refer to water availability for primary production in the ecosystem as a whole.

Water losses from the ecosystem occur by run-off, percolation beyond the root zone and evaporation. Woody plants reduce such losses by improving water infiltration (directly by stemflow, or indirectly by improving soil structure), by deep rooting and by reducing evaporation in favour of transpiration, whereas increased water losses may occur by water interception and evaporation from woody canopies subject to a high evaporative demand, and by decreased WUE. An optimal density of woody plants occurs where interception losses are still relatively low whereas evaporation losses are already considerably reduced. Results of an experiment in Australia (1100 mm rainfall) show topsoil water contents (residual moisture) averaged over a year to be highest at medium *Eucalyptus* tree densities (304 ha^{-1}) due to reduced evaporation in comparison to low tree densities (82 ha^{-1}), but less interception losses in comparison to high densities (2150 ha^{-1}) (Eastham and Rose 1988). By removal of all woody plants, the reduced interception losses would be more than offset by increased evaporation from the open field (Tiedemann and Klemmedson 1977).

Table 4.8 shows that the spatial distribution of available water is also influenced by woody plants. Processes of spatial concentration related to the presence of woody plants lead to an increased water availability under the woody canopy (horizontal redistribution) and in the subsoil (vertical redistribution), particularly around the woody trunk. Such redistribution occurs at the expense of water availability in the open areas and in the topsoil. This redistribution is one factor leading to patchiness of vegetation in semi-arid regions. The roughly quantified effects of woody plants on water availability (Table 4.8) indicate that most important are the processes increasing water infiltration, particularly towards the subsoil. In the most arid parts of semi-arid regions (e.g. the Sahel zones), rainfall is insufficient for these processes to increase water availability by more than 50 mm ha^{-1} yr^{-1}; in less arid parts (e.g. the Sudan zones) benefits may exceed 200 mm ha^{-1} yr^{-1}. Woody plants with

Table 4.8. The maximum effects of various processes associated with woody plants in the Sahel and Sudan zones on water availability for primary production of a mixed vegetation, considering the effects in the canopy area and canopy cover in the vegetation as in the pre-drought situation, in comparison to a vegetation without woody plants

Process	Effect on water availability	Zone	
		Rainfall: 150–600 mm canopy cover: 2–20% (Sahel zones)	Rainfall: 600–1200 mm canopy cover: 15–35% (Sudan zones)
Rainfall interception[a]	Losses by evaporation from foliage, gains by net throughfall exceeding gross rainfall under woody canopy	– 0	– 0
Stemflow[b]	Redistribution to deep soil around stem[c]	+	++
Improved soil structure[b]	Increased water infiltration leading to reduced run-off[c]	+	+++
Increased soil organic matter content[b]	Increased soil moisture retention capacity[c]	0	+
Transpiration by woody plants[a]	Decrease of water use efficiency[c]	–	–
Micro-climatic changes[b]	Reduced PET, relative increase of transpiration and increase of water use efficiency	+	+
Hydraulic lift[b]	Less percolation losses	0	0
Uptake by deep roots[a]	Less percolation losses[c]	+	+++

0 = negligible effects.
+, ++, +++ = positive effects on water availability of 10–50, 50–100 and >100 mm yr^{-1} respectively.
– = negative effects on water availability of <50 mm yr^{-1}.
[a] Processes that improve water availability for the woody plants only.
[b] Processes that might improve water availability for the herb layer.
[c] Effects decreasing rapidly through exploitation of the woody species.

deep roots are required to utilise water in the subsoil which may originate from distant areas, by subsoil water flows (e.g. the *dallols* in Niger).

The improved water availability by woody plants is roughly related to woody cover in the vegetation, so is highest on loamy soils with redistribution of rain water and in depressions with run-on. For a certain rainfall, the improvement of water availability is inversely related to PET (on an annual basis) and the length of the dry season. Both are extremely high in the SSWA; in many other semi-arid regions woody plant densities are higher and beneficial effects on water availability can be more important (Sect. 5.3.4). Of course, the importance of the effects listed in Table 4.8 decreases as woody plants are exploited.

4.4.3.2 Water Use by Woody Plants and Herbs

Data from semi-arid Africa on soil moisture contents in relation to woody canopies are summarised in Table 4.9. Figure 4.11 shows the development of soil moisture contents under an Acacia tortilis canopy, in comparison to an open grassland, for a semi-arid savanna in Kenya.

These data suggest that once the soil under the woody canopy is wetted, soil moisture contents during the rainy season remain higher under the woody canopy than in the open field, particularly in the topsoil. Even after the rainy season, the topsoil under the canopy generally remains moist for a longer period (Prebble and Stirk 1980; Table 4.9; Fig. 4.11). In the Sudan zones in the SSWA, herbage growth under a woody canopy continues for 1 month longer than in the open field, corresponding to higher soil moisture contents under the woody canopy at the end of the rainy season (Fig. 4.3). In Chile (semi-arid mediterranean climate), after the rainy season soil moisture remains available for about 1 month longer at the site with woody plants than at the open site (Ovalle and Avendaño 1988). In Zimbabwe Kennard and Walker (1973) found moisture content in the top 15 cm of the soil to remain highest in the closed tree canopy site, and to decrease most rapidly in the open tree canopy site.

These data show that interception losses from woody canopies plus PET under woody canopies is less than PET in the open field. On the basis of the analysis presented in Section 4.4.2.3, it can be calculated that, in comparison to the open field, under isolated trees the reduction of radiation intensity by 50% reduces PET by 40–50%. Since interception losses from woody canopies are 10%, a 30–40% reduction of PET under woody canopies can be expected. Section 4.5 shows that nutrient concentrations under woody canopies may be twice as high as in the open field. As a result, transpiration increases from 10–15% of moisture availability (Penning de Vries and Djitèye 1991) to 20–30%. The overall result is a reduction of PET of at most 20%, which corresponds to a 25%

Table 4.9. Soil moisture contents in relation to the presence of woody plants in semi-arid Africa

Country + source	Canopy specifications	Season	Soil depth + humidity (vol%)				
Senegal			0–10 cm				
(Charreau and Vidal 1965)	Under *F. albida* canopy	End of	1.4				
	Edge of canopy	rainy season	0.8				
	Open field		0.7				
Senegal			10 cm	30 cm	50 cm	70 cm	100 cm
(Bille 1977)	In a woody vegetation	End of	0.4	0.7	1.0	1.4	1.5
	In a herbage vegetation	rainy season	0.5	1.1	1.3	1.8	1.9
	Herbage vegetation removed		1.6	1.7	1.8	2.1	2.2
South Africa			0–12 cm	12–48 cm			
(Hirst 1975)	Under woody canopy	Dry season	2.0	2.0			
	Edge of canopy		2.2	2.3			
	Open field		1.6	2.1			

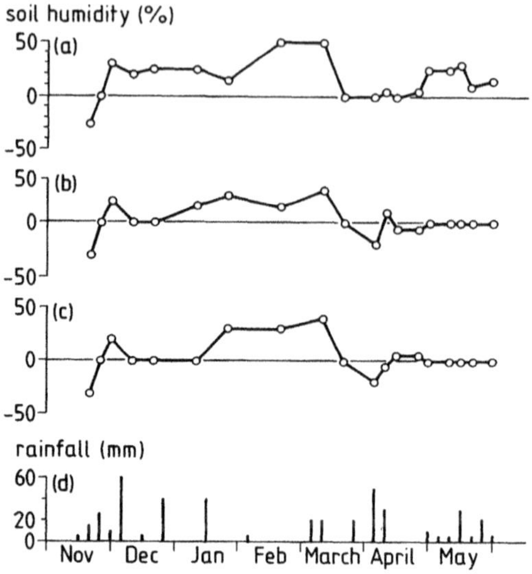

Fig. 4.11a–d. Soil humidity under the canopy of an *Acacia tortilis* tree in semi-arid Kenya, expressed as a proportion of soil humidity in the open field, at **a** 5–10 cm soil depth, **b** 15–20 cm depth, **c** 25–30 cm depth, **d** in relation to rainfall. (After Belsky et al. 1989)

increase of the period during which soil moisture is available for plant growth. This result varies according to the importance of the determining factors: interception losses, light interception, microclimate changes and soil fertility improvement. Since these factors are at least partly independent, the effects on the water balance of the felling of woody plants are highly variable. It may lead to increased amounts of drainage water leaving the ecosystem (Roose 1981), waterlogging (Menaut et al. 1985) or increased soil moisture in the subsoil (Tiedemann and Klemmedson 1977).

The available evidence shows that subsoil water is more rapidly used in the presence of woody plants (e.g. Prebble and Stirk 1980), as compared to herbs only. Particularly at high tree densities, much water is taken from deep soil layers, and the more so in dry periods when the topsoil is dry (Eastham et al. 1990). This is related to high root densities at greater depths (Sect. 4.2). There is evidence to assume that woody plants prefer utilisation of subsoil water resources, if available, before depletion of topsoil water (Australia: Pressland and Fisher 1982; Israel: Lövenstein et al. 1991). On the other hand, low topsoil moisture contents in *Pinus sylvestris* plantations in a mediterranean semi-arid region are caused by a shallow rooting pattern (Koechlin et al. 1986). Ultimately, woody plants may utilise all subsoil water reserves, possibly (partly) created through their own influences. Malik and Sharma (1990) found that additional soil moisture under *Eucalyptus tereticornis* trees is completely utilised by the trees themselves, whereas crop growth under the canopy is minimal.

Subsoil moisture is often further depleted under the woody plant than in the open field (Table 4.9; Menaut et al. 1985). Sole cowpea shows high residual moisture contents at more than 45 cm soil depth whereas intercropping with *Leucaena leucocephala* considerably reduces available soil moisture at more than 45 cm but does not influence moisture contents in the topsoil during the dry season (Singh et al. 1989). The pattern is probably different if subsoil moisture is scarce. In Burkina Faso it was estimated on the basis of measurements of water use by woody plants (Sect. 4.2.3.3), that in the Sahel zones during the dry season woody plants transpire most available water stored in the subsoil of valleys. Similarly, in the groves of the *Acacia aneura* woodlands in Australia, during the dry season all available soil moisture is gradually transpired by the woodland vegetation, and most rapidly nearest to woody trunks. In spite of initially highest moisture losses from the intergroves (open areas), water use is most continuous in the groves (Slatyer 1961).

In Senegal, Grouzis et al. (1990) found PET to be highest for a vegetation with *Acacia senegal* trees, intermediate for a vegetation with *Balanites aegyptiaca* trees, and lowest for herbs only. The additional transpiration by the woody plants, in comparison to herbs only, is obtained from soil layers beyond 60 cm depth. Likewise, total transpiration by an alley cropping system is more than double that of crops only, although transpiration by the crops only is reduced by the introduction of woody plants (Ong et al. 1992). Plant production of the alley cropping system is highest. Similar results were obtained for perennial *Andropogon* grasses intercropped in croplands (Renard and Vandenbeldt 1990).

The presence of woody plants reduces recharge of subsoil water reserves and deep rooting trees can reduce ground water levels when recharge is low. Dense *Populus* windbreaks in the Turpan oasis in China reduced the water table from 1 to 10 m depth (Dregne 1988). This can be expected in arid regions and with rapidly growing trees. With increasing rainfall, recharge of the groundwater increases (IWACO 1990) and the presence of woody plants will have less influence on water available in the subsoil (Roose 1981).

A general pattern of water use and soil moisture of a mixed vegetation in semi-arid regions is based on the previous information. At the end of the dry season the topsoil is dry, and the subsoil has probably dried out as well, particularly where woody plants are abundant. At the beginning of the rainy season topsoil moisture content under a woody canopy is often lower than in the open field because initial small rainfall events do not, or only partly, reach the soil surface under the canopy due to interception losses. In a similar way, after a dry period during the rainy season, recharge of soil moisture under woody canopies is relatively slow

(Fig. 4.11). However, if the first rains are intense, the soil is also wetted under the woody canopy. During the rainy season, improved water infiltration and reduced PET under woody canopies lead to water reaching deeper soil layers than in the open field. However, during the dry season, water uptake by deep roots from woody plants causes more intense drying out of the subsoil.

During most of the growing season, woody plants are poor competitors. with herbs for soil moisture in the topsoil because here the density of fine roots from woody plants is generally much lower than that of herbage plants (Sect. 4.3). Only at the onset of the rainy season woody plants have an advantage due to their perennial root system.

4.4.4 Conclusions

Several processes associated with woody plants play a role in reducing water losses by run-off, evaporation, and percolation to deep soil layers, and/or by enhancing water inputs from surrounding areas by run-on and interception of lateral rainfall, and by hydraulic lift from deep soil layers. Table 4.8 summarises the effects of these processes. Many of these processes lead to an increased water availability for plant production and a redistribution of the available water resources to the area under the woody canopy from the surrounding areas. It is important to realise that the improvement of soil moisture under woody canopies may be partly due to more favourable site conditions prior to woody plant establishment. Whether such a redistribution of water under the woody canopy is significant depends on the woody plant and site conditions. Most important are the capacity of the woody plant to generate stemflow, the improvement of soil structure and deep rooting. Regarding site conditions, maximum effects are expected on loamy soils.

Both the variation in (micro-)climatic factors and the variation in topsoil moisture contents are lower under woody canopies than in the open field or after clearing of the woody plants (Tunstall and Walker 1975; Roose 1981). The reduction in soil moisture variation by a woody canopy is estimated at 50% in the semi-arid mediterranean area (Koechlin et al. 1986). Peak flows from wooded catchments in semi-arid regions are 1.5 to 2.0 times lower than from similar catchments without woody plants (Roose 1981; IWACO 1990). Woody plants in semi-arid regions prolong the availability of topsoil moisture after the rainy season. A similar effect occurs during prolonged dry periods in the rainy season, a phenomenon particularly common in the Sahel zones. In the mixed vegetation the transpiration-evaporation ratio changes in favour of transpiration. The overall productivity of the mixed vegetation is therefore higher. Woody plants improve water infiltration to the subsoil (e.g. by stemflow), but at the end of the dry season subsoil reserves are generally exhausted. The increased water availability can lead to increased plant production if water is the main plant production limiting factor. This is the case in arid regions, but may also occur under woody canopies in semi-arid regions, as a result of the relatively important improvement of nutrient availability.

4.5 Influences of Woody Plants on Nutrient Availability

4.5.1 Introduction

Soil fertility is defined as the nutrients in the soil available for plants, in the absence of artificial inputs. The concentration of nutrients in the soil, as determined by chemical soil analysis, does not necessarily indicate the availability for plants. Nutrients may be present in a form or condition inaccessible for take-up by plant roots. Nutrients most commonly limiting plant production in semi-arid regions are nitrogen and phosphorus (Penning de Vries and Djitèye 1991). Particularly in semi-arid regions with seasonal climates, concentrations of available nutrients fluctuate throughout the year (e.g. Lambert et al. 1980; Penning de Vries and Djitèye 1991). With many studies no specifications are given regarding the season of soil sampling.

It is commonly assumed that the presence of woody plants improves soil fertility (e.g. Nair 1984; Young 1989). Two types of arguments are most commonly used. In the first place, reference is made to the higher soil fertility of forests in relation to croplands. However, it is difficult to draw firm conclusions from the comparison of entirely different ecosystems, varying in several aspects other than the presence of woody plants. For instance, higher soil fertility in a densely wooded vegetation in comparison to a sparsely wooded vegetation may also result from varying exploitation intensity, or from initial differences in soil properties. Some natural perennial grasslands show a higher soil fertility than forests (Janssen 1986).

In the second place, reference is made to the process of soil fertility restoration during a fallow period or by reafforestation. Improved soil fertility is related to biomass buildup and increased nutrient cycling. Rapidly growing woody plants rapidly build up biomass with consequences for the nutrients available in other compartments of the plant-litter-soil pool. Many nutrients are released when the woody vegetation is cleared, prior to another cultivation period. It is important to analyse to which extent soil fertility restoration is associated with woody plants, e.g. during fallowing. The role of perennial grasses in soil fertility restoration remains largely to be investigated.

In this section are evaluated the processes associated with woody plants that influence nutrient cycling in an agro-ecosystem. First, the available evidence for the presence and development of islands of soil fertility under and in the vicinity of woody canopies is summarised (Sect. 4.5.2). In the second place insight is provided in the buildup of soil fertility in the process of forest development and fallowland regeneration (Sect. 4.5.3). On the basis of available data the processes implicated are analysed and evaluated in Section 4.5.4. Fundamental for a proper functioning of the soil is a sufficiently high organic matter content, in view of its role regarding soil physical aspects (e.g. soil structure) and soil chemical aspects (e.g. cation exchange capacity) (de Ridder and van Keulen 1990), both generally improving the availability for plants of nutrients in the soil. The role of organic matter is analysed in Section 4.6.

4.5.2 Patterns of Soil Fertility in Relation to Woody Plants

4.5.2.1 Soil Fertility Patchiness

On a large scale and over several years, water and nutrient availability have been determined in the SSWA (Penning de Vries and Djitèye 1991). Considerable variation and uncertainty exist at both spatial and temporal scales (Shmida and Burgess 1988). The high spatial variability in semi-arid regions is generally referred to as patchiness of resources (Tongway and Ludwig 1990; Brouwer et al. 1993). Spatial heterogeneity results from the following factors:

1. geological and geomorphological heterogeneity (Sect. 2.1.3);
2. redistribution of rainfall water by run-off, and of soil particles by wind erosion, leading to concentration in some areas (at the level of both geomorphological units and microdepressions);
3. rainfall patchiness, which is predictable to some extent (e.g. related to variation in landscape elevation);
4. biologically induced patchiness, e.g. related to soil faunal activities, microhabitats around plants, differential leaching through old root channels etc.

This section demonstrates that in semi-arid regions the presence of woody plants is commonly correlated with patches of nutrient availability. It is difficult to determine whether the woody plant has improved soil fertility, or whether an initial higher soil fertility was at the basis of a successful establishment of the woody plant at this specific site. Spatial variation may occur at a small scale (plant site) or on a large scale (e.g. valleys and "groves"). Spatial variation in soil fertility parameters has only rarely been addressed by research. Brouwer et al. (1991) found that, within a flat, sandy area, early establishment and growth of young *Faidherbia albida* trees is particularly successful at microsites with favourable soil conditions.

It can be hypothesised that a combination of both options is probably most realistic: woody plants selectively establish at favourable sites and by doing so maintain and further enhance redistribution of nutrients and water, towards this microsite. Spatial concentration of resources is particularly beneficial for plant production in resource-poor environments (Sect. 5.2).

4.5.2.2 Nutrient Enrichment in Relation to Woody Plants

A summary of quantitative data from semi-arid regions, mainly from Africa, on soil nutrient concentrations in relation to the distance from a woody plant is presented in Table 4.10. Some other studies provided qualitative information only (Bosch and Van Wijk 1970; Garcia-Moya and McKell 1970; Christie 1975; Parker and Muller 1982). Most study locations are characterised by relatively low densities of woody plants.

The general pattern which appears from these data is that of a concentration of nutrients in the topsoil under a woody plant, gradually decreasing from the trunk towards the open field, and from the topsoil towards deeper soil layers (e.g. Charley 1971). Most studies consider the topsoil as the first 10 cm only. Islands of fertility, as shown in Fig. 4.12, are a general phenomenon in semi-arid regions. In the islands of fertility differences in nutrient concentrations between the topsoil and the subsoil are larger than in the open areas.

In the SSWA studies on soil fertility in relation to woody plants focussed on *Faidherbia albida*. Only recently, other trees were also investigated. The results suggest that higher soil fertility under isolated woody canopies is a general phenomenon (Table 4.10). Contrary to earlier assumptions, processes causing concentration of nutrients under isolated woody plants are not unique to *Faidherbia albida*.

Indirect evidence for higher nutrient concentrations under woody canopies is provided by the fact that under isolated woody plants nutrient yields from the herbage layer are often higher than in the open field. Indirect evidence was obtained by using reference crops grown in pots with soil material from different locations and depths on the woody plant gradient. Maize and sorghum grow two to three times faster in topsoil from under trees than from the open field (Verinumbe 1987). Reference grasses grown in pots with soil from under *Prosopis juliflora* trees yield 15 times more nitrogen and 1.5 times more phosphorus than with soil from the open field (Tiedemann and Klemmedson 1973).

In temperate regions, soil fertility parameters can vary up to twofold at different sides of a tree trunk, such as for a beech tree (Gersper and

Table 4.10. Soil nutrient concentrations in relation to the distance from a woody plant, in semi-arid regions, with specification of site conditions and soil depth

Country + source	Site conditions (rainfall, soil texture + depth)	Location on woody plant-open field transect	Soil nutrient concentrations				
			C $g\,kg^{-1}$	N $g\,kg^{-b}$	C/N	P_{av}[a] $mg\,kg^{-b}$	K_{av}[a] $mEq\,kg^{-1b}$
a) West Africa							
Senegal (Charreau and Vidal 1965)	670 mm Sandy soil 0–10 cm	Under *Faidherbia albida*	5.32	0.60	9	3.5	6.3
		Edge of canopy	4.80	0.52	9	1.9	4.9
		Open field	3.29	0.31	11	1.5	5.2
						0–10	10–20 cm
Senegal (Dancette and Poulain 1969)	670 mm Sandy soil 0–20 cm	Under *Faidherbia albida*	3.7	0.4	9	3.3	2.4 1.0
		Open field	2.7	0.3	9	2.9	2.2 0.7
Senegal (Jung 1969)	670 cm Sandy soil 0–5 cm	Under *Faidherbia albida*		0.82		9.2	7.4
		Open field		0.32		5.8	5.1
Senegal (Bernhard-Reversat 1982)	300 mm 0–10 cm	Under *Commiphora africana*	4.2	0.38	11		
		Under *Balanites aegyptiaca*	5.8	0.52	11		
		Under *Acacia senegal*	6.2	0.56	11		
		Under *Boscia senegalensis*	7.0	0.55	13		
		Under *Adansonia digitata*	8.1	0.69	12		
		Open field – sanddune	2.2	0.18	12		
		– valley	3.7	0.30	12		
Nigeria (Radwanski 1969)	787 mm Sandy soil 0–15 cm	Under *Azadirachta indica*	3.6	0.31	12	P_{total}	
		Open field	1.2	0.13	9	116	
						195	

Mali	900 mm 0–20 cm	Under parkland trees	6.6	0.61	11	8.0	6.9
(Kater et al. 1992)	Loamy soil	Open field	5.1	0.51	10	8.1	4.1
	20–40 cm	Under parkland trees	4.9	0.48	10	3.9	6.1
		Open field	4.1	0.44	9	4.3	3.8
Togo	1180 mm	Under *Azadirachta indica*	19.8	1.7	12	6.1	7.1
(Drechsel et al.	Sandy-loam soil	Under *Albizia lebbeck*	14.2	1.3	11	4.9	3.6
1991)	0–10 cm	Under *Cassia siamea*	13.7	1.1	13	7.5	4.6
		Under *Acacia auriculiformis*	12.7	1.1	12	3.0	3.5
		Grassland	9.1	0.7	13	4.6	2.4
b) Other semi-arid regions in Africa							
Sudan	450 mm	Under *Acacia senegal*	2.0	0.30	7	4.4	1.3
(Gerakis and	Sandy soil	Open field	1.4	0.13	11	4.3	0.8
Tsangarakis 1970)	0–20 cm						
Zimbabwe	630 mm	Under closed canopy	22	0.22		4.6	5.7
(Kennard and	Sandy soil	Open field	19	0.28		6.4	5.8
Walker 1973)	0–15 cm	Open grassland	6	0.20		6.1	1.9
				Mineralised $g\,kg^{-1}$			
Kenya	450 mm	Under trees	~14	~40		13.6	11.5
(Belsky et al. 1989)	Loamy soil	Open field	~ 8	~10		5.0	7.7
	0–10 cm						
Kenya	500 mm 0–2.5 cm	Under *A. raddiana*	4.7	0.54	9		
(Weltzin and	Loamy soil	Open field	2.7	0.31	9		
Cougheor 1990)	7.5–10 cm	Under *A. raddiana*	2.0	0.26	8		
		Open field	1.5	0.28	5		
	47.5–50 cm	Under *A. raddiana*	0.9	0.26	4		
		Open field	1.0	0.17	6		

Table 4.10. (*Contd.*)

Country + source	Site conditions (rainfall, soil texture + depth)	Location on woody plant-open field transect	Soil nutrient concentrations				
			C $g\,kg^{-1}$	N $g\,kg^{-b}$	C/N	P_{av} [a] $mg\,kg^{-b}$	K_{av} [a] $mEq\,kg^{-1b}$
Botswana (Tolsma 1989)	300 mm 5–30 cm	Under *Acacia erubescens*					6.4
		Under *Acacia karroo*					7.9
		Under *Acacia tortilis*					2.8
		Under *Dichrostachys cinerea*					5.1
		Open field					2.8
c) Other semi-arid regions							
Chile (Ovalle and Avendaño 1988)	650 mm Clay soil 0–10 cm	Under *Acacia caven*, 6 yrs old	7.7	1.4	6	1.8	3.6
		Open field	7.7	1.2	7	1.5	2.2
		Under *A. caven*, 14 yrs old	12.7	2.3	6	2.5	4.9
		Open field	8.3	1.5	6	1.7	3.8
		Under *A. caven*, mature	14.9	2.7	6	6.0	4.7
		Open field	11.6	2.2	6	5.4	3.7
Australia (Ebersohn and Lucas 1965)	350 mm 0–3 cm	Under various trees				5.2	7.4
		Inter-tree areas				3.2	5.0
Argentinia (Braun Wilke 1982)	350 mm 0–10 cm	Under various shrubs	7.7	1.1	7		
		Inter-shrub areas	3.3	0.8	4		
	40–50 cm	Under various shrubs	4.0	0.9	4		
		Inter-shrub areas	4.3	0.7	6		
	80–100 cm	Under various shrubs	3.0	0.8	4		
		Inter-shrub areas	3.3	0.5	7		

				kg ha^{-1}		kg ha^{-1}	kg ha^{-1}
India	250 mm 0–15 cm	Under *Prosopis cineraria*		250		22	633
(Aggarwal 1980,		Under *Prosopis juliflora*		203		10	409
cited by Young		Open field				8	370
1989)	15–30 cm	Under *Prosopis cineraria*		193		10	325
		Under *Prosopis juliflora*		212		5	258
		Open field		196		4	235
USA	300 mm 0–5 cm	Under *Atriplex* shrubs	14.5	1.7	9	3.6	
(Charley and		Open field	11.0	1.3	8	2.7	
West 1975)	10–15 cm	Under *Atriplex* shrubs	7.5	1.0			
	150 mm 0–5 cm	Under *Coleogyne* shrubs	12.1	0.9	12	1.6	
		Open field	4.1	0.6	7	0.6	
	10–15 cm	Under *Coleogyne* shrubs	5.0	0.6			
USA	200 mm	Under *Prosopis juliflora*	7.3	0.8	9	P_{tot}	7.6
(Tiedemann and	0–4.5 cm	Open field	2.5	0.3	8	370	7.5
Klemmedson 1973)						310	

[a] av = available.
[b] Unless indicated otherwise.

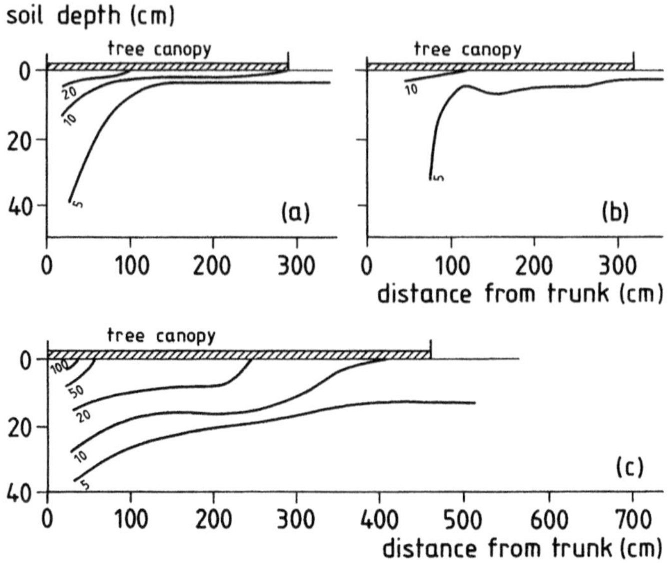

Fig. 4.12a–c. Horizontal and vertical gradients of soil nitrogen ($g\,kg^{-1}$) under **a** *Acacia senegal*, **b** *Balanites aegyptiaca* and **c** *Adansonia digitata*. (After Bernhard-Reversat 1982)

Holowaychuk 1971). Such differences are related to differences in stemflow quantity, resulting from variable inclination of the trunk. From the SSWA no such evidence was obtained (Kater et al. 1992; Kessler 1992). Under the canopy itself, nutrient concentrations are not always highest nearest the trunk.

Nitrogen, phosphorus, carbon (organic matter) and potassium generally fit to the pattern of higher concentrations under the woody canopy. Sometimes, P concentrations under woody canopies are lower than in the open field (Table 4.10). Differences in available P may also result indirectly from variation in pH under the woody canopy (Drechsel et al. 1991). Mueller-Harvey et al. (1985) provide evidence to assume that variability in P concentration under woody plants is overshadowed by large-scale variations (correlated with clay content), contrary to variability in C and N, which is small-scale and predominantly correlated with vegetation patterns. Both the CEC and concentrations of cations Mg^{2+}, Ca^{2+} and K^+ are generally higher under the woody canopy. However, whereas C, N and P usually show a surface concentration decreasing with depth, the cations may be concentrated in the subsoil (more than 30 cm) (e.g. Everett et al. 1986). Data presented in Table 4.10 show that P concentrations decrease more rapidly with depth than N concentrations. This is probably related to the higher mobility of nitrogen.

Only one study, from the Mediterranean region, shows an increase in clay content under woody canopies (Joffre and Rambal 1988). Soil bulk density is commonly lower under woody canopies than in the open field. This may be caused by either increased soil fauna activities loosening the soil (Joffre and Rambal 1988), roots of woody plants loosening the soil and creating air-filled pores (Sanchez et al. 1985), and/or improved soil organic matter under woody canopies. There is evidence to assume that the latter process dominates: extermination of soil fauna in semi-arid New Mexico does not significantly influence soil bulk density under bushes ($1.7\,g\,cm^{-3}$), nor soil porosity (36%), whereas in the open field extermination of soil fauna increases soil bulk density from 1.7 to $2.0\,g\,cm^{-3}$, and reduces soil porosity from 36% to 25% (Elkins et al. 1986). Soil bulk density decreases from 1.4 to $0.9\,g\,cm^{-3}$ with increasing forest maturity on fallowed croplands (Koechlin et al. 1986) and decreases from $1.6\,g\,cm^{-3}$ in the open field to $1.2\,g\,cm^{-3}$ near the stem of *Pinus contorta* trees in the USA (Zinke 1962).

4.5.2.3 Factors Correlated with Nutrient Enrichment

The enrichment factor is defined as the ratio, for a specific nutrient, of its concentration under a woody canopy and in the open field. The enrichment factor has been suggested to depend on woody plant parameters, such as plant density, woody species (Barth and Klemmedson 1978) and age of woody plants (Zinke 1962). What evidence is available from semi-arid regions to support these assumptions?

In semi-arid regions in the USA, the enrichment factor for nitrogen varies between 0.2 and 5.5 for different shrub species (Charley and West 1977). In Senegal the enrichment of C and N in relation to woody plants also varies according to woody species (Fig. 4.12). Tolsma (1989) found enrichment factors of potassium, magnesium and calcium to vary according to woody species (Table 4.10). However, Belsky et al. (1989) found similar nutrient enrichment for *Acacia tortilis* and *Adansonia digitata*.

Nutrient enrichment under woody canopies increases with increasing age of woody plants, generally represented by size (trunk diameter). In Senegal, both C and N enrichment of the topsoil under *Acacia senegal* and *Balanites aegyptiaca* canopies increase with age. Similarly, soil enrichment with C and N in the topsoil under *Acacia caven* trees in semi-arid Chile increases with age (Table 4.10). Kellmann (1979) also found nutrient enrichment under woody canopies in a natural vegetation to depend, among other factors, on the size of the tree.

The summarised data from semi-arid regions do not permit firm conclusions regarding a relationship between enrichment under woody canopies and woody plant densities. With increasing density of woody plants, the enrichment patterns of woody canopies are expected to gradually merge, until an evenly distributed topsoil fertility occurs. Vertical nutrient gradients are expected to become less extreme with increasing humidity (cf. Charley and West 1975) because biological activities in humid regions are less limited to the topsoil.

Thus, it appears that the enrichment factor varies per woody species, increases with plant age, and decreases with plant densities. However, from the evidence cited, it cannot be concluded that differential enrichment is associated with woody plant parameters only. Variation between species, or between plant densities, may be related to site conditions because woody species show preference for specific site conditions.

On the basis of available evidence in semi-arid Africa, the relationship between the enrichment factor and annual rainfall is investigated for nitrogen (N) and phosphorus (P) in the topsoil (0–20 cm). Averaged for all available data, the enrichment factor for N is 1.9 (variation 1.1–2.8), and for P 1.3 (variation 0.6–2.3). For both nutrients there is a tendency for the enrichment factor to decrease with increasing rainfall (Fig. 4.13), which is certainly related to increasing plant densities. This finding corresponds with findings from northern America, showing an increasing enrichment factor with increasing aridity (Charley and West 1977). The enrichment factor decreases with soil depth. In the subsoil (20–50 cm) enrichment of P is negligible (1.05) whereas for N it is 1.25 (Table 4.10).

No relationship was found between the enrichment factor and the initial soil fertility, as reflected by nutrient concentrations in the open

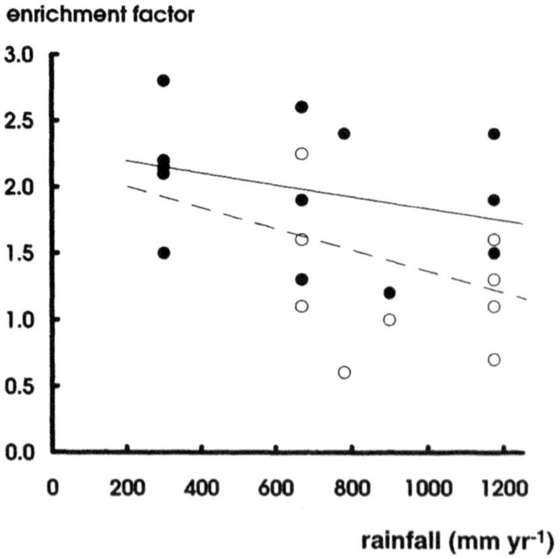

Fig. 4.13. The relationship between the enrichment factor under woody canopies (see text) and annual average rainfall, for nitrogen (•) and phosphorus (– –, ○), based on data from semi-arid Africa

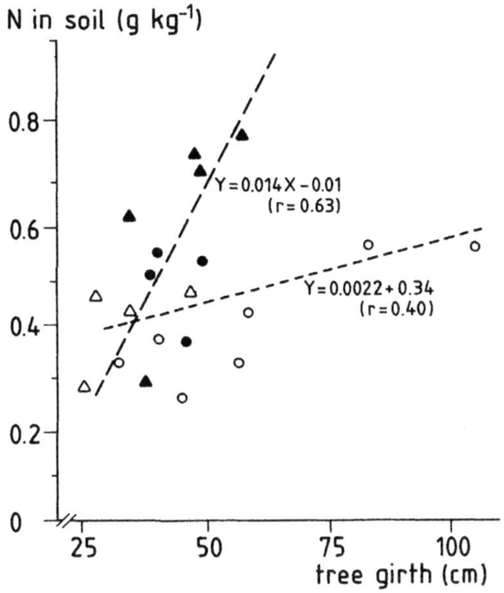

Fig. 4.14. Nitrogen concentration in the topsoil (0–10 cm) under *Acacia senegal* (△▲) and under *Balanites aegyptiaca* (○●), in relation to tree girth, at two different sites: sandy dunes (○△) and the lower part of a slope (●▲). (After Bernhard-Reversat 1982)

field. However, Fig. 4.14 shows that enrichment takes place more rapidly at the more fertile lower part of slopes than on the sand dunes, which is also associated with differences between the two species.

4.5.3 Soil Fertility Changes During Fallowing

In the tropics, fallowing is the predominant method of reconstituting soil fertility after a period of cropping. During fallowing the vegetation regenerates, gradually developing to a vegetation structure that resembles the initial vegetation before first clearing took place. Changes.
in nutrient concentrations in the topsoil during fallowing result from the relative importance of the following processes:

1. uptake of soil nutrients by the developing perennial (woody) vegetation;
2. addition of soil nutrients from the subsoil and from atmospheric origins;
3. loss of nutrients by erosion, leaching, exploitation and through fire (volatilisation).

In the process of vegetation development during a fallow period, three phases may be distinguished (Nye and Greenland 1960). Growth

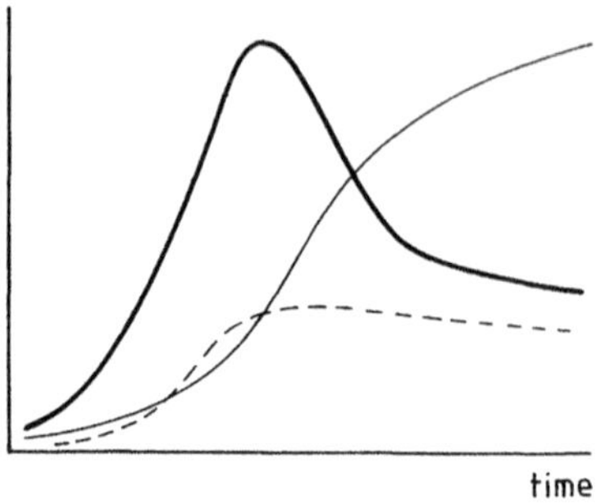

Fig. 4.15. The relative rate of accumulation of biomass (–), uptake of nutrients (—) and release of nutrients by litterfall (- - -) of a maturing woody plant in semi-arid regions. (After Miller 1989)

and nutrient uptake characteristics of a woody plant (Fig. 4.15) are related to these three phases. During the first phase of primary succession, uptake of nutrients by the rapidly growing fallow vegetation predominates. During this phase, a net loss of nutrients occurs from the (already impoverished) topsoil to the vegetation biomass. Uptake of nutrients by (woody) perennials includes sources from parent material in the subsoil (if available). In semi-arid regions, in the first years after fallowing, the herbage layer usually is poorly developed, whereas woody plants grow rapidly. During this phase there is a net increase of the nutrient pool for the vegetation as a whole. Nakano and Syahbuddin (1989) found that in forest fallows in the humid tropics in Southeast Asia, where fire does not occur, concentrations of most nutrients in the topsoil are lowest at 3 to 4 years after the beginning of fallowing. Vegetation at that moment corresponds to the so-called "bush" stage. For phosphorus, however, the decline may continue for many more years.

As fallowing proceeds, accumulation of nutrients from the subsoil by the developing perennial vegetation gradually terminates. During the second phase, nutrient inputs to the topsoil by litter fall exceed storage of nutrients in standing biomass (Fig. 4.15), and losses by leaching and erosion decrease as both the above- and belowground biomass develops. As a result topsoil fertility gradually increases, particularly under woody canopies.

During a third phase, an increase of total nutrient cycling within the ecosystem increasingly depends on atmospheric inputs. However, when dealing with plantations of trees selected for their rapid growth, the depletion of soil nutrients may continue (Sanchez et al. 1985). After clearing of the vegetation for a renewed cropping cycle, nutrients are made available for plant growth by burning of the cleared (woody) vegetation mainly.

During fallowing, soil structure improves: soil bulk density decreases, soil porosity, soil moisture retention and soil organic matter increase. These improvements are not related to nutrient concentrations (Sanchez et al. 1985). Improved soil physical properties contribute to improved conditions for cropping after fallowing.

Most data and theoretical insights on vegetation development and soil fertility restoration during fallowing are from the humid tropics (summaries provided by, e.g. Nye and Greenland 1960; Sanchez et al. 1985; Ewel 1986; Nakano and Syahbuddin 1989). Available data from the Guinean zone (Areola 1980; Aweto 1981; van Reuler and Janssen 1989), located south of the southern Sudan zone, correspond to the general theory for humid tropics. The annual uptake of nutrients, equal to the sum of the annual net increase in the biomass and the soil nutrient pool, gradually increases during fallowing until an equilibrium is reached at forest maturity. However, variability in soil nutrient concentrations is large: in spite of a general trend of increasing fertility, some samples from mature forests show lower nutrient concentrations than samples from young fallowlands.

From the SSWA, quantitative data are provided by Ohler (1982) from a densely populated area in Mali, by Leenaars (1990) from a sparsely populated area in Burkina Faso, and by van Reuler and Janssen (1989) from a region just south of the southern Sudan zone in Ivory Coast (Table 4.11). These studies include measurements of biomass development. Contrary to the above theory, from these data no trend of decreasing or increasing soil fertility can be detected during distinct phases of fallow development (Table 4.11). Ohler (1982) concludes that systematic changes related to fallowing are overshadowed by site-specific variability in soil fertility and exploitation during fallowing. Leenaars (1990) could detect no systematic variations in soil fertility of fallowlands within distinct sets of samples classified according to soil texture. A gradual increase of aboveground biomass was recorded (Sect. 3.3.1), but biomass increase varies in relation to exploitation intensity by a factor 3 (Ohler 1982). A major condition for soil fertility improvement during fallowing is that no, or little exploitation takes place of the fallow vegetation. The extent to which woody plants contribute to improvement of the soil fertility depends on the extent to which organic matter and nutrient inputs from litter are maintained (Sanchez et al. 1985). Because of bush fires and exploitation by grazing and firewood cutting, the average fraction of N annually lost from the aboveground biomass in the natural vegetation increases from 0.3 in the northern Sahel to 0.5 in the southern Sudan zone. Due to variation in exploitation intensity, the availability of N for primary production in the southern Sahel zone varies from 5 to 35 $kg\,ha^{-1}\,yr^{-1}$ (Penning de Vries and Djitèye 1991). Nowadays, fallowlands are increasingly used for exploitation of firewood and grazing, so that little soil organic matter and nutrient buildup can be expected.

There are other reasons to assume that soil fertility improvement by fallowing may be of limited importance in semi-arid regions. In the first

Table 4.11. Nutrient concentrations in the topsoil in relation to the period of fallowing, at three locations in West Africa

Country + source	Site conditions				Period of fallowing (years)	Nutrient concentrations		
	Rainfall (mm)	Soil[a]	Fire[b]	Exploitation[b]		C ($g\,kg^{-1}$)	N ($g\,kg^{-1}$)	P_{av}[c] ($mg\,kg^{-1}$)
Mali (Ohler 1982)	700	SL	+	0	0	7.7	0.5	2.3
					1	4.1	0.4	2.3
					2–4	7.3–11.0	0.5	2.3–3.6
					5–10	8.0–9.5	0.6	2.3–4.2
					>10	7.5	0.6	2.9
	700	SL	0	+	0	4.0	0.5	2.7
					1	5.7	0.4	–
					4	5.1	0.3	–
					>10	4.5	0.4	2.3
Burkina Faso (Leenaars 1990)	900	LC	–	+	2–4	16–26	0.8–1.3	0.2–0.6[d]
					5–10	14–24	0.7–1.1	0.2–0.7[d]
					>10	19–27	0.9–1.3	0.3–0.4[d]
	900	SL	+	0	1	9	0.5	0.1[d]
					2–4	9–12	0.5	0.1[d]
					5–10	10–18	0.5–0.9	0.1–0.3[d]
					>10	10–17	0.4–0.7	0.1–0.2[d]
Ivory Coast (van Reuler and Janssen 1989)	1800	SL	?	0	0	9.0	0.7	9.0
					1	10.0	0.9	9.5
					2–4	10.3–11.9	0.9	9.5
					5–10	11.4	1.0	9.5
					>10	13.5	1.1	13.5

[a] S = sand, L = loam, C = clay.
[b] – = absent, 0 = moderate intensity, + = high intensity.
[c] av = available.
[d] P_{total} in $g\,kg^{-1}$.

place, vegetation recovery and biomass increment, fixing the majority of the annual uptake of nutrients during the first phase of fallowing, is low in comparison to humid tropics. This is the result of poor rainfall and the low availability of nutrients in the often highly weathered soils in semi-arid regions (Allen 1985). Moreover, croplands left for fallowing have very low organic matter concentrations and cation exchange capacity, and are often degraded to such an extent (including removal of tree trunks) that vegetation regeneration is slow.

In the second place, in semi-arid regions leaching of nutrients is limited (Sects. 4.5.4.3 and 4.5.4.4) and so is the potential of improving topsoil fertility by deep rooting. Additionally, woody plants in semi-arid regions seem particularly efficient in maintaining a closed (largely internal) nutrient cycling (Sect. 4.5.4.6). Therefore, the increase of the nutrient pool largely depends on atmospheric inputs, which are low (Table 4.13). Hence, the second and third phases of fallowland development merge.

The data from *Acacia caven* plantations in semi-arid Chile (Ovalle and Avendaño 1988: Table 4.10) show that as the plantation matures, both enrichment of the soil under the trees and in the open field takes place, but the latter at a lower rate. This is as expected. Under mature woody plants soil fertility is improved, but for the fallowland as a whole, where woody cover is low and mature woody plants occur isolated, soil fertility regeneration is not significant. Most nutrients are fixed in aboveground biomass.

On the basis of the enrichment factor of soil fertility under woody plants in relation to annual rainfall (Fig. 4.13), and the average canopy cover per climatic zone, the expected enrichment of soil fertility by woody plants can be calculated (Table 4.12). It is assumed that enrichment under the woody canopy does not occur at the expense of the fertility of the open field, which is, however, unlikely in the SSWA (Sect. 4.5.4.4). The results show that for the region as a whole, fertility improvement by woody plants is very limited. Most important are probably the buildup of organic matter in the soil (Sect. 4.6) and protective aboveground biomass.

4.5.4 Woody Plant Processes Influencing Nutrient Availability

4.5.4.1 Introduction

Various processes associated with woody plants influence nutrient availability under and in the vicinity of the woody canopy, and may be responsible for the improved nutrient availability under mature woody plants. The processes influencing nutrient availability can be classified

Table 4.12. The impact of soil enrichment by woody plants on nutrient concentrations under the woody canopy, and in the region as a whole, based on open field nitrogen concentration of 0.4 $g\,kg^{-1}$ and phophorus concentration of 2.5 $mg\,kg^{-1}$

Climatic zone	Enrichment factor[a]		Woody cover[b] (%)	Nutrient concentrations			
				Under woody canopy		Average wooded area	
	N	P		N ($g\,kg^{-1}$)	P ($mg\,kg^{-1}$)	N ($g\,kg^{-1}$)	P ($mg\,kg^{-1}$)
Northern Sahel	2.2	1.9	6	0.88	4.75	0.43	2.7
Southern Sahel	2.1	1.7	14	0.84	4.25	0.46	2.7
Northern Sudan	2.0	1.5	26	0.80	3.75	0.50	2.8
Southern Sudan	1.8	1.3	30	0.72	3.25	0.50	2.8

[a] According to Fig. 4.13.
[b] Before drought period according to Fig. 2.4.

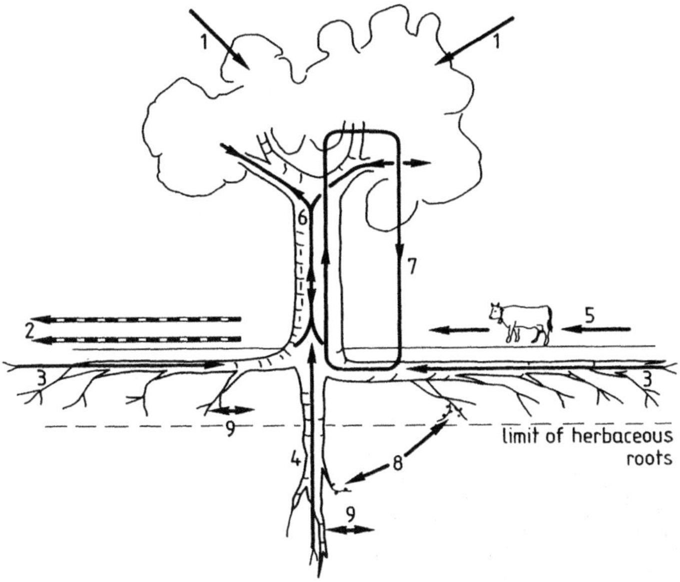

Fig. 4.16. Woody plant related processes influencing nutrient availability: *1* atmospheric deposition and interception; *2* losses by wind and water; *3* uptake by lateral roots; *4* uptake by deep roots; *5* deposition by animals; *6* internal recycling; *7* external recycling; *8* N_2 fixation, *9* rhizosphere interactions

into two categories (Switzer and Nelson 1972; Garner and Steinberger 1989):

1. physical (geochemical) processes of nutrient import or export: deposition by rainfall and wind (Sect. 4.5.4.2) and losses by erosion and leaching (Sect. 4.5.4.3);
2. biological (biochemical) processes, some of which are basically internal (Sect. 4.5.4.6), but most of which are related to the interaction between woody plants and soil: nutrient uptake by roots (Sect. 4.5.4.4), deposition by animals (Sect. 4.5.4.5), external recycling (Sect. 4.5.4.7), nitrogen fixation (Sect. 4.5.4.8) and rhizosphere interactions (Sect. 4.5.4.9). The various processes are illustrated in Fig. 4.16.

In the following sections the processes are classified into categories of:

1. redistribution of available nutrients within the ecosystem (leading to a spatial concentration),
2. reduction of nutrient losses, or
3. an enrichment of the nutrient pool of the ecosystem as a whole.

This differentiation is fundamental to evaluate the potential for increasing primary production in semi-arid regions by the introduction or protection of woody plants (Kessler and Breman 1991).

4.5.4.2 Atmospheric Deposition

Woody plants have a greater potential of filtering the air than herbs because of increased leaf area surface, height and the depressing effect on wind velocity. Wet deposition includes depositions from rain, mist or dew. Nutrients in rainfall are intercepted and redistributed heterogeneously by throughfall and stemflow, similar to rainfall partitioning (Sect. 4.4). Dry deposition includes interception and filtering by woody plants of aerosol particles, ashes (human or volcanic), dust and sea spray. The interception of dust or soil particles by woody canopies has never been specifically demonstrated (Young 1989). Atmospheric conditions and canopy structure basically determine the amounts of nutrients trapped. The deposits are largely washed down to the topsoil (Sect. 4.5.4.7).

Nutrients in rainfall originate from water surfaces, dust, lightning and (bush) fires mainly, by evaporation and volatilisation. In industrialised areas there may be high inputs from industrial or intensified agricultural activities. There is a trend to somewhat higher cation deposition in regions with seasonal climates, as compared to other tropical regions (Kellman et al. 1982). Relatively high amounts of air-bound particles in seasonally dry climates are related to the occurrence of bush fires (air-bound ashes and volatilisation of nitrogen and sulphur), and wind intensity in combination with sparse vegetation cover. Here, the major influx of nutrients from the atmosphere probably takes place in the early rainy season, with the first rains. Thus, incidental measurements of atmospheric inputs are not reliable.

Stemflow directly funnels intercepted nutrients towards the soil. It has been suggested that stemflow is particularly important for small-diameter (young) woody plants as an adaptation to low nutrient availability (Bruijnzeel 1989). There is no evidence to support this hypothesis. Atmospheric accession is more important as the soils and parent materials are weathered and nutrient poor (Richards and Charley 1983).

For West Africa, some estimates of nutrient concentrations in rainfall are presented in Table 4.13. Note that the variability of such parameters is large, between sites and throughout the year, particularly for P (Vitousek and Sanford 1986). Concentrations of Ca, Cl, K and Na in rainfall decrease with increasing distance from the Atlantic Ocean (Roose 1981). A similar pattern was found in Middle America (Kellman et al. 1982). Estimates from West Africa on P concentrations in rainfall are high compared to data from other semi-arid regions (e.g. $0.01\,kg\,ha^{-1}\,yr^{-1}$ in Australia: Attiwill 1981). This may be related to the more frequent occurrence of fires in West Africa, leading to high P concentrations in wind-blown ashes. Throughfall (under a woody canopy) is also enriched by leaching of nutrients from leaf and bark surfaces (Sect. 4.5.4.7), as is demonstrated by the difference between nutrient concentrations in rainfall in open areas and under woody canopies (Table

Table 4.13. Nutrient concentrations in rainwater, at bare soil sites and under forest, at three locations in West Africa (after Roose 1981) and at open areas in West Africa; estimates of nutrient deposition by Sahara dust, from various sources

Location and specifications	Rainfall ($mm\,yr^{-1}$)	Nutrient concentrations in rainfall ($g\,ha^{-1}\,mm^{-1}$)					
		N	C	P	K	Ca	Mg
Adiopodomé	1700						
– bare soil		14	13	1.0	3	18	4
– under forest		25	74	2.9	39	32	22
Korhogo	1200						
– bare soil		9	12	1.0	3	19	1
– under forest		11	38	1.2	13	24	4
Saria	700						
– bare soil		6	13	2.5	4	21	3
– under forest		9	39	3.8	12	26	12
Rainfall at open areas							
– Penning de Vries and Djitèye 1991		6.5		0.5–2.5	1–5		
– Stoorvogel and Smalink 1990		2–6					
		Nutrient concentrations in dust ($kg\,ha^{-1}\,yr^{-1}$)					
Dust at open areas							
– Niger: Jones (1960)				2.6	37	1	3
– Niger: McTainsh and Walker (1982)				1.2	27	13	8
– Ivory Coast: Janssen et al. (1990)				0.9			
– Niger: Stahr et al. (1993)				1.1	17.5	10	6

4.13). Between woody canopies and bare soil, the differences in nutrient inputs reaching the soil surface amount to 19 kg N ha^{-1} and 3.2 kg P ha^{-1} in Adiopodomé (1700 mm rainfall), decreasing to 2.1 kg N ha^{-1} and 0.9 kg P ha^{-1} in Saria (700 mm rainfall). Note that only a small proportion of these P inputs is available for uptake by plants. Other nutrients also show decreasing inputs with decreasing rainfall. For nitrogen, the rate of decrease is larger than expected on the basis of nutrient inputs from rainfall only. Other sources of N inputs, e.g. dust deposition and leaching from plant tissues, should be important as well.

The importance of atmospheric nutrient inputs for plant production in semi-arid regions was argued by Gorham (1961). Estimates of deposition of wind-blown light soil particles (dust) in Kano (Nigeria), originating from the Sahara at 1000–1200 km distance (McTainsh and Walker 1982) and in northern Nigeria (Jones 1960), show high values for all nutrients, particularly for P and K (Table 4.13). In the SSWA, southeasterly winds prevail during the dry season, causing transport of dust (soil particles with nutrients) from bare areas in the north to vegetated areas in the south. Monsoon winds during the rainy season, predominantly in the other direction, are not likely to have a high particle load, since the soil is covered and humid. Interception of dust may also explain higher clay contents under some tree canopies (Szott et al. 1991), though the opposite effect may occur when wind-blown sand is trapped. Whereas dust, constituted of fine particles, may be transported over large distances, transport and deposition of sand, due to the weight of the particles, occurs on a local scale only.

By obstruction and reduced wind speed, soil particles and organic material, moving over the soil surface by the action of wind and/or water, may accumulate where vegetation is dense, e.g. under shrubs with low branches. Charley and West (1975) suggested that the lower P concentrations under desert shrubs, as compared to the open field, result from the deposition of large amounts of nutrient-poor (sandy) wind-blown material. Strips of perennial grasses can be equally effective in accumulating wind-blown material. A strip of *Andropogon gayanus*, 100 m long and 10 m wide, was found to intercept 225 t of sand in a period of 3 years (ICRISAT 1990), equivalent to 750 t ha^{-1} yr^{-1}. Obviously, in the vicinity of the study area bare areas with coarse-textured topsoils predominated.

Organic material densities are a factor 5 to 20 higher under desert shrubs than in the open field, and are particularly concentrated around the trunk (Barth and Klemmedson 1978; Young et al. 1984). On the other hand, in arid and sandy regions, depressions can be formed around isolated trees without undergrowth due to whirling of winds around the stems. In that case, litter is blown away from under large trees with no or few low branches. These highly variable effects have never been quantified for woody plants.

4.5.4.3 Reduction of Losses

Losses of nutrients occur by wind erosion, water erosion, leaching or volatilisation (the latter particularly following fire). These physical transport mechanisms have as agents water, wind and gravity. By such actions nutrients are dispersed, possibly beyond the reach of plants and their roots, reducing the resource availability for primary production.

In arid zones *wind* is the most important transport mechanism. Losses by wind erosion are rarely quantified, being extremely variable. By reducing wind speed and/or increasing interception of wind blown material, losses of nutrients by wind erosion can be reduced. A direct soil cover is particularly effective in reducing transport (uptake) of soil particles by wind and water, whereas a woody canopy, due to its effect on the microclimate (Sect. 4.4.2.3) and its volume, is most effective in intercepting wind blown material. Soil particles and organic material accumulate in dense vegetation, particularly shrubs, also resulting from the filtering of run-off water.

The incidence of *fire* is important in all semi-arid regions (Sect. 2.4.2). Although most of the dense natural vegetation in the SSWA is annually burnt, internal translocation of nutrients to plant components that are either not or less susceptible to fire damage, results in less nutrient losses than expected on the basis of foliage nutrient concentrations. Provided that only part of the foliage from the woody plants burns, N losses by fire are lower from woody plants than from herbs. Nutrients lost by volatilisation may be redistributed by rainfall and dust deposition. This is particularly important for nutrients lost by smoke. Losses of nitrogen, besides sulphur and carbon susceptible to volatilisation, are estimated for the SSWA (Table 4.14).

Although it has never been quantified, bush-fires seem less destructive to the herbage layer under woody canopies than in the open field. The wooded parts of the vegetation and the canopy zones of large trees often remain intact, contrary to open areas. This is due to reduced evaporation from the soil and the vegetation in wooded areas, and due to higher moisture contents of woody foliage than of herbs in the dry season. Since vegetation is concentrated in run-on areas, reduced fire damage as a result of the presence of woody plants has a large impact on nutrient losses from the entire ecosystem.

In the majority of tropical regions *water* is the most important transport agent of nutrients. Kellman et al. (1985) conclude, for a sub-humid savanna on a granite plateau in Belize, that nutrient losses by run-off are more important than those by leaching. Data from the natural vegetation in West Africa (Table 4.15) show that erosion becomes more important than leaching (drainage) as rainfall decreases from the south to the north (Roose 1981). In Niono (500 mm rainfall), run-off equals 12–57% of annual rainfall, the variation being related to soil charac-

Table 4.14. The effect of annual burning of the herbage and woody layer on nitrogen losses, per climatic zone in the SSWA

	Herbage layer[a]		Woody layer[b]		Proportion of area burnt annually (%)[c]	Loss of N by fire ($kg\ ha^{-1}$)[d]
	biomass ($t\ ha^{-1}$)	N ($g\ kg^{-1}$)	biomass ($kg\ ha^{-1}$)	N in foliage ($g\ kg^{-1}$)		
Northern Sahel	1.0	12	50	16	0	0
Southern Sahel	2.0	7	100	15	15	2.2
Northern Sudan	3.0	6	350	15	50	4.7
Southern Sudan	4.0	5	750	14	80	8.6

[a] Based on Breman and de Ridder (1991).
[b] Based on data from Fig. 3.9 (1980 situation), Table 3.13 (early dry season) and Table 3.15 (minimum N concentrations).
[c] Within land-use type "old fallowlands, rangelands and forests": see Section 2.4.2.
[d] Assuming 25% of foliage burnt in the Sahel zones, and 15% in the Sudan zones (in relation to tree height); and assuming 50% of all N in the burnt plant material lost by volatilisation (Sect. 2.4.2.2).

Table 4.15. Run-off and leaching beyond 180 cm soil depth, and their nutrient concentrations, at three sites with sandy-loam soils, in West Africa (Roose 1981)

		Adiopodomé	Korhogo	Gonsé
Average annual rainfall (mm)		~1700	~1200	~700
Average run-off (% of rainfall)				
– natural vegetation		0.5	2	7.6 (1.5–24)
– cropland		22	20	16.5
Average drainage at 180 cm depth (mm)				
– natural vegetation		78.5	23	15.4 (0–49)
– cropland		39	25	1.5
Nutrient concentrations of run-off/drainage				
water in natural vegetation	N ($mg\,l^{-1}$)	4.5/3.3	2.0/1.5	2.8/2.4
	P ($mg\,l^{-1}$)	0.5/0.2	0.2/0.2	0.3/0.2

teristics. Here, leaching beyond 2 m depth is 25% of annual rainfall on coarse textured soils, but decreases as soil texture becomes finer, so that on most soil types little or no leaching occurs (Breman and de Ridder 1991). For a given rainfall (quantity and distribution), run-off and leaching mainly depend on soil conditions, slope and vegetation characteristics. The relative increase of water erosion due to clearing of the natural vegetation is highest in humid tropical regions, but clearing leads to a reduction of leaching losses (Table 4.15).

Nutrients are associated with run-off, leading to enrichment at run-on sites, according to their concentrations and mobility. Data on soluble nutrient concentrations in run-off and drainage water have an indicative value only, as nutrient concentrations vary considerably according to season, location (upslope or downslope, depth of soil profile) and soil and vegetation characteristics. Nutrient concentrations in run-off are generally higher than those in drainage (Roose 1981). Nutrient losses by run-off are higher than expected on the basis of nutrient concentrations only, since additional losses occur by non-soluble nutrients in soil particles. With increasing rainfall, free available nutrients originating from leaching of nutrients from dense vegetation increase more than proportionally to rainfall (Roose 1981). With increasing rainfall the nutrient cycle becomes significantly more "open" (Sect. 4.5.6).

Soil organic matter has a dual positive effect on reducing nutrient losses, via the increase of the cation exchange capacity of the soil (reducing the potential to lose nutrients) and the improved water infiltration and moisture retention capacity of the soil (Sect. 4.6).

Here, the dilemma of soil fertility preceding or resulting from woody plant establishment (Sect. 4.5.2.1) can be addressed. Where woody plants selectively establish at favourable sites, redistribution and concentration of nutrients and water towards this microsite is maintained, and possibly enhanced, because nutrients are fixed and recycled by the concentrated vegetation. Woody plants, due to their perennial status,

Table 4.16. Mean annual N and P yields from herbaceous above-ground peak biomass, in relation to site conditions: uplands (run-off) and depressions (run-on), canopy cover (%) and exploitation intensity. (Penning de Vries and Djitèye 1991)

	Uplands		Depressions	
Canopy cover (%)	0.5	0.4	47	66
Exploitation intensity:	low	high	low	high
N yield (kg ha^{-1})	10	5	12	23
P yield (kg ha^{-1})	1.0	0.3	1.5	4.3

reduce losses of nutrients by effective internal and external recycling (Sects. 4.5.4.6 and 4.5.4.7). Once established, woody plants contribute to sustain suitable conditions for water infiltration and nutrient accumulation, thus reducing losses and enhancing uptake of run-on water, with associated soil nutrients and organic matter. Due to this process of redistribution, the vegetated parts of the tiger-bush vegetation pattern in the Sahel zones are, and remain, more fertile than the open areas (Penning de Vries and Djitèye 1991). In a similar way, the groves of *Acacia aneura* in Australia are more fertile than the intergroves (Tongway and Ludwig 1990). The enrichment is particularly enhanced in the vicinity of degraded uplands, from which run-off is intensive. An indication of the enrichment by run-on is based on the comparison of the annual yield of nutrients in the herbage layer at run-on and at run-off sites on a loamy plain in the southern Sahel zone (Table 4.16). Where heavily grazed, microdepressions receive higher amounts of run-off water and associated nutrients than with moderate grazing (Penning de Vries and Djitèye 1991). The availability of P seems more influenced by run-off than N availability.

4.5.4.4 Nutrient Uptake by Roots

Lateral Uptake

The uptake of nutrients from the topsoil by lateral roots of woody plants is a spatial concentration process of nutrients, from which the woody plants benefit in the first place. Charley and West (1975) consider redistribution to be of predominant importance for woody plants and suggest that accumulation under woody canopies is associated with concentration decline in the surroundings. This is particularly important for non-mobile nutrients (e.g. Ca, P, Mg); for mobile nutrients such as N, K and C other processes may be of at least equal importance (Sect. 4.5.4.7). Stuart-Hill and Tainton (1989) showed the reduction of herbage

Table 4.17. Nutrient concentrations in the topsoil and subsoil at some locations in the SSWA

Location, rainfall + soil texture		C ($g\ kg^{-1}$)	N ($g\ kg^{-1}$)	P_{tot}[e] ($mg\ kg^{-1}$)	P_{av}[f] ($mg\ kg^{-1}$)	K_{av}[f] ($meq\ kg^{-1}$)
Niono (500 mm)	[a]					
Sandy	0–50 cm	2.5–5.6	0.10	65–200	5	1.3
	>50 cm	<2.0	<0.05	100–200	<2	<1.3
Loam	0–50 cm	3.0	0.12	70–100	1	1.0–1.3
	>50 cm	<1.0	<0.12	80–100	1	1.5–2.5
Clay	0–50 cm	1.3–3.9	0.10	>100	0–3	2.5–3.8
	>50 cm	1.5	0.06	83	1	3.1
Fété Olé (300 mm)	[b]					
Sandy	0–50 cm	1–4	0.14–0.40	70–190	n.a.	n.a.
	>50 cm	<1	<0.14	40–90	n.a.	n.a.
Saria (700 mm)	[c]					
Sandy-clay-loam	0–100 cm	1.6–4.2	0.2–0.4	200–420	n.a.	1.0–3.3
	>100 cm	<0.7	<0.1	<300	n.a.	1.5–2.5
id.	[d]					
	0–100 cm	1.6–9.6	0.4–0.8	350–750	n.a.	1.0–1.4
	>100 cm	<1.0	<0.2	<250	n.a.	2.4–7.2

[a] Penning de Vries and Djitèye (1991); natural rangelands.
[b] Bille (1977); natural rangelands.
[c] Roose (1981); young fallowland.
[d] Roose (1981); old fallowland.
[e] tot = total.
[f] av = available.

biomass to increase from about 3 m from the trunk for small woody plants (≤1 m height) to more than 9 m for woody plants exceeding 2 m height. This observation suggests a lateral spread of roots with increasing tree maturity. There is, however, no quantitative evidence to support the redistribution hypothesis, due to the difficulties in experimental research set-up and analyses associated with the subject. Experiments with root barriers do not suffice since growth and production of the woody plant are influenced, and so will be its effect on microclimate.

A number of arguments can be postulated supporting the importance of the process of lateral uptake for woody plants in semi-arid regions.

1. Moisture during the rainy season is concentrated in the topsoil and as a result nutrients, in a state available for uptake by plants, are also concentrated in the topsoil (Table 4.17; Fig. 4.18). Even at run-on sites nutrients are available in the topsoil mainly, since here, due to the predominance of heavy soils, water infiltration is slow and mobility of nutrients is limited. Obviously, a root system would be most effective when concentrated in the topsoil. Exceptions should be made for two situations:
   - locations with a relatively shallow groundwater table;
   - woody species with an effective redistribution of intercepted rainfall towards stemflow; stemflow enhancing deep water infiltration (Sect. 4.4.2.1) may be associated with water uptake by a taproot, but it remains doubtful whether, besides water, nutrients are obtained from deep soil layers.
2. Woody plants in semi-arid regions have predominantly lateral spreading of shallow roots (Sect. 4.3). The lateral root extension of woody plants in the SSWA varies at least between two and seven times the canopy diameter. Thus, an evenly distributed woody plant population with total canopy cover of e.g. 10% would be able to exploit all topsoil. In that case the concentration factor of nutrients would be a factor 9. Since the topsoil root biomass of herbs generally is much higher than of woody plants (Sect. 4.3.3), the processs of nutrient concentration by woody plants is probably rather slow (Sect. 4.5.3). For desert shrubs in the USA, Garcia-Moya and McKell (1970) found N concentration under the canopy to be highest for the shrub species with the most extensive, lateral root system.
3. The build-up of an increased nutrient pool under a woody canopy is most rapid for sites with high overall topsoil nutrient concentrations, as compared to less fertile sites (Fig. 4.14). It may be expected that the process of nutrient concentration by lateral uptake under a woody canopy proceeds throughout the life-time of the woody plant, provided that the chemistry of the element is such that the rate of diffusion back to the surroundings is less than the rate of continuing accretion (Charley and West 1975).

Uptake from Deep Soil

It is generally assumed that deep rooting woody plants play a crucial role in the nutrient balance through "pumping" of nutrients from the subsoil to the topsoil (e.g. Nye and Greenland 1960; Young 1989). Relatively immobile nutrients such as P and basic cations would be made available by weathering in the subsoil, whereas mobile nutrients would be captured from subsoil layers following leaching. Only recently, doubts have been cast on this opinion, in the first place for humid tropical regions (e.g. Grubb 1989; van Noordwijk 1989b). Preliminary results using ^{32}P injected at different soil depths and distances from tree trunks suggest that uptake of P takes place from the topsoil mainly (IAEA 1975). Leaching of nutrients is limited by the high organic matter contents and the high density of fine roots in the topsoil under woody canopies. A dense root layer in the top 15 cm may be as efficient in tapping a downward flow of nutrients as a perennial root system down to several metres. The leaching rate of nutrients is related to their apparent adsorption constant, which increases in the order NO_3^-, NH_4^+, K and P (van Noordwijk 1989b). Under humid tropical conditions, leaching beyond the topsoil root zone may be expected for NO_3^- and NH_4^+ in particular.

The arguments to doubt the importance of the nutrient pumping mechanism are particularly valid to semi-arid regions for a number of reasons.

1. In semi-arid regions, water infiltration to the subsoil is limited, whereas physical and chemical root barriers are widespread (Szott et al. 1991). Roots of some trees only are claimed to break up compact or indurated soil layers (Young 1989).
2. Nutrient concentrations in the subsoil are low (Table 4.17; Fig. 4.18) and few nutrients are made available to the subsoil by weathering. Kellman (1979; 1985) found that in a sub-humid savanna in Belize soils are so deeply weathered, that the source of P and bases in the subsoil, awaiting transfer by deep roots, is indeed very limited. Similar observations are from other areas (Grubb 1989) and are particularly applicable to deeply weathered soils. Phosphorus, as the nutrient most often referred to as being obtained from the subsoil after weathering, is easily fixed in organic and ferroaluminium compounds in the soil (Bruijnzeel 1989). Most semi-arid regions with low nutrient availability can be characterised as having a predominantly "closed" nutrient cycle, where nutrient input from weathering materials is negligible (Burnham 1989). Besides P, non-mobile nutrients such as Ca and Mg have low concentrations in most subsoils.
3. If in semi-arid regions low contributions by weathering dò occur, these nutrients will accumulate in the topsoil, via litterfall, as leaching is limited. The apparent adsorption constant is highest for P, reducing its mobility to the subsoil in case of downward water flow.

If woody plants significantly increased P availability in the topsoil by uptake from the subsoil, one would expect, in the majority of the locations where N and P limit primary production, a correlation between P yield in the herbage biomass and woody cover. This relationship is tested on a transect from the northern Sahel to the southern Sudan zone in the SSWA, where vegetation sampling took place at 36 distinct sites at the end of the rainy season during 2–4 years. A multiple regression analysis shows that water availability at each site accounts for 73% of the variance in P yields of the herbage layer. This can be expected, as more water causes increased depth and period of soil wetting and thus P availability in deeper soil layers. Total woody cover, or woody cover of leguminous woody plants only, does not show any relationship with P yield and does not improve the explanation of remaining variation (Fig. 4.17). In fact, the correlation between herbage P yield and woody cover is negative (level of significance = 0.10). This might be explained by woody plants competing with the herbage layer for limited P resources.

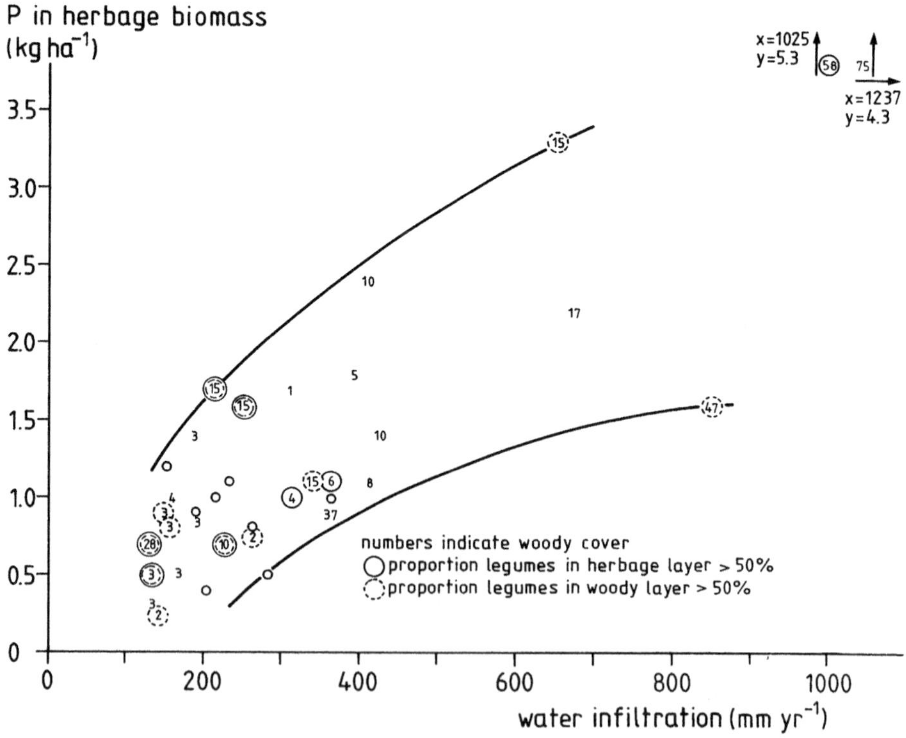

Fig. 4.17. The relationship between annual phosphorus yield in the herbage biomass and water infiltration, for sites with variable woody cover and variable contribution of legumes in the woody and herbage layer. *Lines* indicate outer limits of the relationship

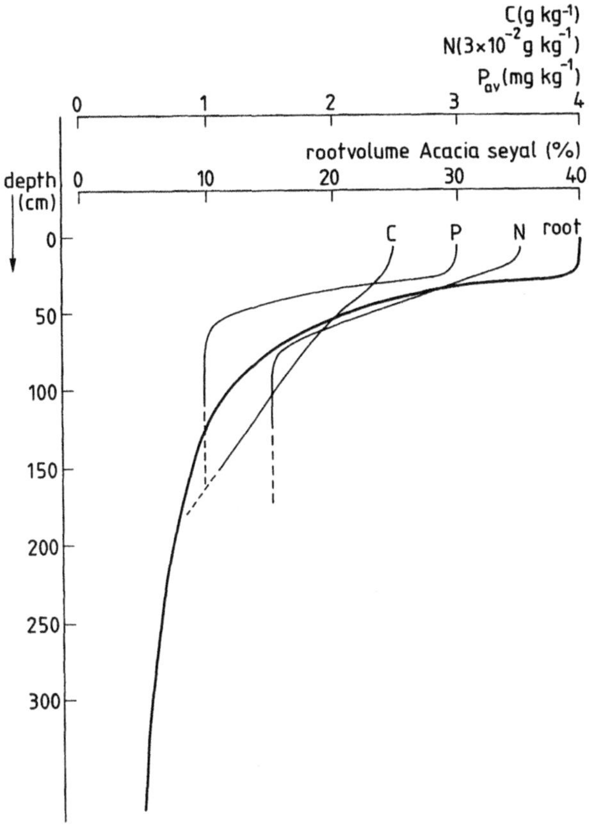

Fig. 4.18. Nutrient concentrations (carbon, nitrogen, available phosphorus) and root volume (as a proportion of total root volume) of an *Acacia seyal* tree in the southern Sahel zone, in relation to soil depth

Similar conclusions are from Kenya, where the introduction of various leguminous woody plants does not lead to increased P uptake by intercropped millet (Nyami 1991).

The potential uptake of nutrients from the subsoil is estimated for an *Acacia seyal* tree on a sandy-loam soil in the southern Sahel zone. The root system of this tree was entirely excavated and measured in detail in relation to soil depth (Fig. 4.18). The analysis is based on concentration patterns of available phosphorus, and of nitrogen and carbon (Fig. 4.18), and on the availability of water in the soil profile. It is assumed that P uptake takes place in a layer of 0.1 cm around the roots. The resulting uptake of P from the soil layer exceeding 1 m depth is thus calculated to be about 0.5 mg per tree. For an estimated density of 150 woody plants in the area, total uptake of P from the subsoil would be 75 mg ha^{-1}, which is very little (0.005%) in relation to total P requirements of woody plants, and which is little compared to the annual P uptake by the herbage layer (2 kg ha^{-1}). Nutrient uptake from the subsoil might be improved by stemflow, increasing P availability around the roots, and by rhizosphere interactions (Sect. 4.5.4.9), increasing the area around the roots from which P can be absorbed (Sect. 4.8.4.3).

4.5.4.5 Deposition by Animals and Influence of Soil Fauna

Nutrients are deposited with dung or droppings of animals resting in the shade of, or in the trees. Spatial concentration of nutrients by this process is particularly important if tree density is low. Belsky et al. (1989) found a concentration of bird droppings under woody canopies; bird droppings are especially rich in P. Large trees attract livestock by offering both shade and highly palatable pods, particularly in the dry season. It has been suggested that high soil fertility under *Faidherbia albida* trees largely results from dung deposition (Robinson 1986).

Herbivores produce 50–70% of their manure during rest periods. Cattle dung contains about 12 g N and 2.4 g P kg^{-1} dry matter (van Duivenbooden et al. 1991; Breman and de Ridder 1991). Assuming for the southern Sahel zone a woody cover of 10% (Fig. 2.4; 1980 situation), a livestock density of 0.12 TLU ha^{-1} and 50% of manure being deposited during resting in the shade of trees, the annual nutrient deposition under the woody canopy equals 1.6 g N m^{-2} and 0.4 g P m^{-2}. These quantities are about twice as high as nutrient inputs by rainfall under woody plants (Table 4.13). Manure deposition may be concentrated most under leguminous trees. For the northern Sahel zone, similar estimates are obtained; for the Sudan zones, quantities of nutrients deposited under woody canopies are much lower due to lower livestock densities and higher woody cover.

The soil fauna in semi-arid regions is dominated by termites, whereas earthworms are relatively unimportant (Lamotte 1979; Lavelle et al. 1990). Termites concentrate organic matter and nutrients towards termite mounds, which therefore have high nutrient concentrations. Termites are rare or absent on sandy soils where mounds cannot be built. Transport of soil particles between the subsoil and the topsoil is caused by termites and other soil macrofauna (Trapnell et al. 1975; Arshad 1982; Lal 1987). The effect of vertical transport on nutrient concentrations is unclear. Regarding mineralised nutrients, a flux from the topsoil to the subsoil will predominate; on the other hand, weathered nutrients are brought to the soil surface. Water infiltration is improved by termites, but may lead to leaching of nutrients, through pores, to the subsoil. Extermination of termites leads to accumulation of N in the topsoil (Elkins et al. 1986).

Termites are generally more active in a mixed vegetation and under woody canopies than in monocultures and in open grasslands (Lal 1987). During the 1970s drought period Lepage (1974) found termite activities in the northern Sahel zone to be limited to locations with woody plants, dead or alive. Termites may benefit from woody plants by high litter inputs and a large proportion of poorly decomposable material, and the constant supply of organic matter (Sect. 4.4.6). Several woody species show a preference for termite mounds, establishing either on or around termite mounds (Sect. 2.3). *Faidherbia albida* seedlings showed significantly better survival and growth at locations with previous termite activity (Brouwer et al. 1991). Thus, there is a relationship between soil fauna and woody plants, with mutual advantages, both improving and stabilising soil fertility in the topsoil mainly.

4.5.4.6 Internal Recycling

Internal recycling (translocation) is an efficient way of reducing nutrient losses and accumulating a nutrient pool. Studies on internal recycling of nutrients by woody plants show variable results. Summaries provided by Attiwill (1981), Gray (1983) and Vitousek and Sanford (1986) show that estimates of N, P and K translocation from foliage in forest ecosystems vary between 10–75%, irrespective of rainfall. The estimates are based on the comparison of nutrient concentrations on a dry matter basis between young leaves and leaf litter. Such estimates are unreliable for two reasons. Firstly because variable amounts of both organic matter and nutrients are translocated (Vitousek and Sanford 1986). Secondly because nutrient losses also occur by leaching. Leaching is the predominating process for soluble nutrients such as K. To overcome these problems, Vitousek and Sanford (1986) proposed using the nutrient/Ca ratio, assuming that Ca is immobile once it reaches leaves.

On the basis of the available evidence, it appears that P translocation, as a proportion of P uptake, is more important in less fertile sites (70%) than at fertile sites (50%) in moist tropical regions (Vitousek and Sanford 1986; Whitmore 1990). Leguminous N_2-fixing woody plants in Botswana show a lower proportion of N translocation and leaching from foliage (about 50%) than do non-leguminous woody plants (more than 50%) whereas translocation of P is important (50–85%) for all species (Tolsma 1989). Translocation is probably most important for those nutrients that are most limiting. Translocation seems to increase with age, so that nutrient concentrations in litter are less for mature trees (Gill et al. 1987).

Translocation of nutrients also occurs from wood. An almost complete withdrawal occurs for certain nutrients, particularly P and K, from older wood in some woody species, e.g. *Eucalyptus* in Australia (Beadle and White 1968, cited by Grubb 1989).

During early foliage growth, the nutrients N, P and K are taken up in excess of requirements, to be translocated immediately after growth ceases. Early accumulation and storage of nutrients is aimed at avoiding nutrient limitations to growth as the soil dries out and/or nutrients become exhausted. Luxurious consumption, like seasonal storage, is essential to the success of perennial species in low-nutrient environments (Stuart Chapin 1980). This was demonstrated for evergreen chaparral shrubs in California (Gray 1983), where translocation (as a proportion of nutrient uptake) amounts to 56% for P, 44% for N and 12% for K. Thus P is most efficiently accumulated and shows the lowest turn-over rate. Flowers and fruits, incorporating large amounts of nutrients with little or no translocation, constitute a potential source of large nutrient "losses". Translocated nutrients from leaves are stored in plant tissues (e.g. inner wood) or are allocated to flowers and fruits or to new leaves. For tropical trees a relation was found between nutrient concentrations in roots and wood, and nutrient concentrations in foliage at the end of the growing season (Noij et al. 1990). Such data support the statement made by Janssen et al. (1990) that translocation forms the major process by which the woody components of woody plants receive nutrients.

In actively growing plant tissues N and P concentrations are much higher than concentrations of less mobile nutrients (Evans 1979). Less mobile nutrients, such as Ca and Mg, are accumulated in foliage after growth. They are not translocated and therefore may have high turn-over rates (Stuart Chapin 1980; Gray 1983; Tolsma 1989). The accumulation of these nutrients varies considerably according to species (Attiwill 1981).

As regards nutrient recycling evergreen species have advantages in comparison to deciduous species, particularly under conditions of nutrient-stress (Stuart Chapin 1980). High nutrient efficiency of evergreen species is due to the highly efficient and comparatively self-regulated nutrient cycle. Deciduous leaves show higher nutrient turn-over rates and are more susceptible to leaching and other nutrient losses. Key factors implicated are the possibility to translocate nutrients from old leaves directly to young leaves (related to the phenological cycle), and the degree of translocation of nutrients (Gray 1983).

Estimates of translocation and leaching of N can be obtained from an analysis of the variation in N concentration in foliage (Tables 3.14 and 3.15), representing young and old leaves. Translocation and leaching of N appears slightly less important for nodulating (N_2-fixing) woody legumes (about 50%) than for non-N_2-fixing woody plants (55–65%). No data are available on P concentrations of young and old leaves.

The importance of translocation explains why in semi-arid regions nutrient concentrations in woody plant tissues are only slightly correlated with rainfall or soil condition (Sect. 3.6.2), contrary to the situation for annual herbs (Breman and de Ridder 1991).

The proportion of (semi-)evergreen species in the natural vegetation (Table 3.12) is higher in the Sahel zones (60–80%) than in the Sudan zones (30–40%). A high proportion of (semi-)evergreen species contributes to an efficient translocation of nutrients, supporting the hypothesis that translocation is most important under nutrient poor conditions. On the basis of these data, and assuming 10–20% of N leached from plant tissues (Sect. 4.5.4.7), 40% of N is estimated to be translocated. Translocation of P is estimated at 70%, based on the data presented above.

4.5.4.7 External Recycling

Two components of external nutrient cycling can be identified: leaching of nutrients from plant tissues, and decomposition of litter originating from the woody plant.

Leaching from Plant Tissues

Rainwater is generally enriched with nutrients after interacting with the woody plant, by washing out nutrients from intercellular space in plant tissues (Parker 1983). This process is less effective with sclerophyllous leaves. There are also reports of leaves in infertile habitats removing N and P from rainwater (Stuart Chapin 1980). The origin of the nutrients in rainwater infiltrating into the soil may therefore be threefold: atmospheric inputs (Sect. 4.5.4.2), leaching from woody plant tissues (foliage and bark mainly), or leaching from litter.

Nutrients in net rainfall (Fig. 4.9) are important, even though generally in low concentrations, because their availability, in solution, is greater than that of nutrients in litter. Few data are available from tropical regions, where leaching of nutrients from plant tissues seems an often underestimated factor influencing soil fertility (Bruijnzeel 1989).

The contribution of nutrients in net rainfall to total nutrient return to the soil is variable (Table 4.18). Variation is according to the specific nutrient, climatic zone and season. In semi-arid Australia, the contribution of nutrients in rainfall to total nutrient return is higher than in temperate climates. The following comments per nutrient are based on a summary of data from temperate regions (Gersper and Holowaychuk 1971; Parker 1983) and tropical regions (Mathieu 1976; Roose 1981; Vitousek and Sanford 1986; Whitmore 1990; Okeke and Omaliko 1991).

1. Nitrogen. Negative net throughfall may be found for NH_4, apparently due to uptake by leaf and leaf-dwelling organisms, but is generally

Table 4.18. The contribution of nutrients in throughfall and stemflow to total nutrient return to the soil, for specific woody species at different locations

Location, climate and source	N	P	K	Ca	Mg
Abies grandis temperate climate USA (Tiedemann et al. 1980)	13%	18%	42%	24%	22%
Douglas fir id. (Cole et al. 1967, cited by Tiedemann et al. 1980)	11%	67%	82%	29%	n.d.
Eucalyptus melanophloia semi-arid Australia (Prebble and Stirk 1980)	n.d	50%	86%	37%	57%
Acioa barter sub-humid Nigeria (Okeke and Omaliko 1991)	33%	76%	21%	19%	12%
Montane rainforest New Guinea (Whitmore 1990)	25%	33%	72%	17%	37%

positive for organic and total N (e.g. Langkamp et al. 1982). Leaching from plant tissues is relatively low but due to high foliar N levels, some N in net rainfall originates from leaching. As a whole, N in net rainfall probably contributes little to total N return.

2. Phosphorus. The contribution of P in net rainfall to P return to the soil is variable but may be important. Small amounts of P in the atmosphere (rainfall), relatively large amounts in dust, and the low mobility of P in plant tissues, suggest that leachates from plant tissues do not contribute much to P in net rainfall.
3. Potassium. K in net rainfall generally contributes greatly to total K return, and is derived almost entirely by leaching from plant tissues as a result of its extreme mobility.
4. Calcium. Airborne Ca may be important but is contained largely in large soil particles. Probably little Ca originates from leaching from plant tissues, being rather immobile; Ca in net rainfall contributes moderately to total Ca return.
5. Magnesium. Small amounts of Mg in the atmosphere and low foliar levels of Mg suggest that leaching from plant tissues supplies at least 50% of Mg in net rainfall.

It can be concluded that susceptibility to leaching from plant tissues (mobility) generally decreases in the order K > N > Mg > P > Ca.

Nutrient concentrations in stemflow are generally higher than in throughfall, contributing to nutrient concentrations in the soil around the trunk (e.g. Tiedemann et al. 1980; Young et al. 1984). Soil nutrient gradients in relation to the distance from the trunk are therefore greatest for woody plants with much stemflow. Differential nutrient supply by net rainfall, defined as the biohydrologic factor, is an important soil-forming factor (Gersper and Holowaychuk 1971).

Data from West Africa show that concentrations of nutrients in rainwater under woody canopies are higher than in open areas (Table 4.13). The differences vary according to the solubility of the nutrient. However, the origin of the enrichment under woody canopies can be either or both leaching of plant tissues and atmospheric deposition. Bernhard-Reversat and Poupon (1980) estimated that, during a relatively moist year (330 mm) in the northern Sahel zone in Senegal, N inputs from rainfall under woody canopies averaged $2.2\,g\,m^{-2}$. Since N in rainfall in the open field averaged $0.4\,g\,m^{-2}$, the *Acacia senegal* canopy contributed $1.8\,g\,m^{-2}$. Assuming a tree with two leaf layers, N contribution from the canopy would be $7.5\,g\,kg^{-1}$ of foliage. In a dry year (170 mm) the tree contributes only $0.6\,g\,m^{-2}$. The contribution to N return by stemflow was reported to be very low.

The importance of leaching of nutrients from plant tissues can be expected to decrease from south to north, due to decreasing plant productivity, decreasing rainfall and increasing sclerophylly and ever greenness. Towards the north, nutrients in rainfall under woody canopies increasingly originate from dust deposition (Sect. 4.5.4.2) and the low woody cover.

Decomposition of Litter

Woody plants represent a high biomass per unit of area, both above- and belowground. A proportion of the biomass is annually turned over, constituting organic (litter) inputs to the soil. Whereas during early growth stages foliage production is the major component of aboveground organic inputs, at maturity woody components (twigs, branches and large wood) become more important (Sect. 3.3).

For the woody plant litter fall is in the first place a loss of nutrients, in spite of efficient translocation (Sect. 4.5.4.6). Average turn-over rates (as a proportion of total nutrients in biomass) for woody plants in the south-African savanna are 17% for N, 19% for P, 31% for K and 33% for Mg (Frost 1985b).

During mineralisation of the organic matter, nutrients are released and can be taken up by roots from the woody plant. Which proportion of the mineralised nutrients is taken up by roots depends on various loss factors (fire, erosion, herbivory) in the first place. Breman and de Ridder (1991) assumed that, amongst others, due to internal recycling N losses of a woody vegetation are 50% lower than of a herbaceous vegetation (Sect. 4.8.4). Other factors to consider are competition with other plants and micro-organisms, and synchronisation of nutrient release and nutrient demand by the plant. Litter fall of a woody plant is relatively well spread throughout the year (probably belowground as well as aboveground), and only to a limited extent corresponds to the period of growth and nutrient uptake of herbs (Sect. 3.5). Litter inputs from woody plants have components with variable susceptibility to decomposition (wood, twigs, foliage etc.) and the release of nutrients into the soil is therefore more spread throughout the year than nutrient release from herbage litter. Under the woody canopy the soil remains moist for a longer period than under a herbage layer, enabling decomposition processes to continue during (part of) the dry season. Thus, the woody plant, due to its perennial root system, may take up a considerable proportion of the released nutrients. In semi-arid regions, below-ground litter inputs are probably important due to seasonal drying out of the soil and corresponding root decay.

In Section 4.6 decomposition processes related to litter inputs from woody plants are further analysed.

4.5.4.8 Nitrogen Fixation

Which Woody Legumes Can Nodulate?

Not all woody legumes show symbiotic N_2 fixation, as characterised by the presence of nodules on roots. Knowledge on nodulation of woody legumes in semi-arid regions is still fragmentary, particularly in Africa (Ducousso 1991). Available data on the potential for effective nodulation of woody legumes in semi-arid Africa are summarised in Table 4.19. Woody legumes of the sub-family Caesalpinioideae generally do not nodulate. Of the genera *Bauhinia*, *Afzelia* and *Tamarindus* there have been only single positive reports of doubtful reliability (Sprent and Sutherland 1990). Of the tribe Mimosoideae most genera show nodulating species. However, *Parkia biglobosa* seems consistently non-nodulating (de Faria et al. 1989). Most woody species of the Papilionoideae nodulate but among the African species only the genera *Pterocarpus* and *Dalbergia* nodulate (Sprent and Sutherland 1990). For the SSWA it is remarkable that, although average N concentrations in foliage of N_2-fixing woody plants are slightly higher than those of non-N_2-fixing woody legumes (Table 3.15), species of the non-leguminous family Capparidaceae have the highest foliage N concentrations.

N_2 fixation is not restricted to nodules; microbial N_2 fixation occurs in stomata of tree leaves and is reported from stems (Sipkens 1984), but are insignificant. Additionally, some N_2 fixation may take place by microbial activity in the woody canopy (Gosz 1980).

Table 4.19. Nodulation in genera of woody legumes occurring in semi-arid Africa; indicated are per genera the total number of species and the number of species showing nodulation per genus. (de Faria et al. 1989)

	No. of species	Nodulation		
		Yes	No	Doubtful
Caesalpinioideae				
Afzelia Sm.	13	–	4	1[a]
Bauhinia L. (including *Piliostigma*)	~250	–	38	1[a]
Brachystegia Benth.	~30	–	8	–
Burkea Benth.	1	–	1	–
Cassia L. (including *Senna*)	~30	–	11	–
Daniellia Benn.	9	–	1	–
Delonix Raf.	10	–	2	–
Detarium Juss.	3	–	1	–
Isoberlinia Craib and Stapf.	5	–	–	–
Parkinsonia L.	~15	–	4	–
Tamarindus L.	1	–	–	1[a]
Mimosoideae				
Acacia Mill.	~1200	204	12	–[a]
Albizia Durazz.	~150	41	–	–
Dichrostachys Wight and Arn.	~12	3	–	–
Entada Adans.	~30	5	–	–
Faidherbia Chev.	1	1	–	–
Parkia R. Br.	~40	2	7	2[a]
Prosopis L.	44	7	3	–
Papilionoideae				
Cordyla Lour.	5	–	1	–
Dalbergia Linn. f.	~100	24	–	1
Pterocarpus Jacq.	~20	17	3	–[a]

[a] See comments in text.

Woody species with the potential to nodulate may not do so in the field. Evidence on natural occurrence of nodules with woody plants in semi-arid regions is limited. In the SSWA 6 *Acacia* species, *Piliostigma reticulatum* and *Pterocarpus lucens* were investigated for the presence of nodules, with positive results for the latter species only (Penning de Vries and Djitèye 1991). Most *Acacia* species occurring in the SSWA can be successfully nodulated with the less ubiquitous *Rhizobium* strains only. Their N_2-fixing potential is therefore probably limited (Dommergues 1987). Of the numerous indigenous *Acacia* species in Senegal, Ducousso (1991) found nodules on roots of *Acacia raddiana* and *Acacia senegal* only. In the Sudan nodulation on roots of *Acacia senegal* is limited, and with *Acacia seyal* irregular (Miettinen et al. 1992). Nodulation on roots of *Faidherbia albida* is common for young seedlings, but is rare for mature trees, suggesting that the N_2-fixing potential of this species is low

(Dommergues 1987). Under arid conditions, nodules on roots of woody plants may be found metres down only, possibly near the water table (Sprent and Sutherland 1990). On roots of *Prosopis* spp. in California, nodules were not observed closer to the surface than at 2.7 m soil depth (Felker and Clark 1982), suggesting that moisture requirements of the bacteria are met in the subsoil only.

Factors Determining Nitrogen Fixation by Woody Legumes

It is problematic to estimate N_2 fixation by woody plants, particularly to know if woody plants continue to fix N_2 throughout their growth. Most estimates of the amounts of nitrogen fixed by woody plants are obtained by the acetylene (C_2H_2) reduction technique. This method provides unreliable results since the assumption of a constant ratio of C_2H_2 to N_2 is unjustified. More recently, N_2 fixation has been estimated using ^{15}N isotopes. However, all values so far obtained are approximates, which do not concern semi-arid regions. Experiments concentrate on woody legumes used for alley cropping. Highly variable results, even for one species, range from 10–200 kg ha^{-1} of fixed nitrogen (Giller and Wilson 1991).

N_2 fixation by legumes is often limited by insufficient P-availability. Application of phosphorus may greatly stimulate N_2 fixation (e.g. Hingston et al. 1982). Due to an increased availability of P following fire, burning may stimulate N_2 fixation (e.g. Hingston et al. 1982). A high N availability in the soil, in turn, limits N_2 fixation by a feed-back mechanism in the plant (Vitousek and Sanford 1986; Högberg 1989). The absence of nodules on roots in the topsoil under mature woody legumes may be related to the gradually improved N concentrations in the soil under mature trees. Woody legumes are particularly successful as pioneer species at poor sites due to their N_2-fixing ability (Langkamp et al. 1982). *Acacia senegal* seedlings, which normally develop root nodules, do not nodulate in soil obtained from under mature trees (Bernhard-Reversat and Poupon 1980). The application of N fertiliser may increase the biomass of non-leguminous plants while depressing that of cooccurring leguminous plants (Grove 1988).

N_2 fixation is also depressed by moisture stress (Högberg 1986; Langkamp et al. 1982; Gray 1983; Sprent 1986), by low temperatures until about 20 °C (Langkamp et al. 1979), and by deficiencies of minerals such as molybdenum. In the rainy season, root nodules appear shortly after the appearance of young leaves (Beadle 1964). There is evidence to assume that N_2 fixation already declines during the rainy season (Langkamp et al. 1982), possibly due to a reduced supply of assimilate caused by leaf fall and increased competition for assimilate because of wood growth.

Estimates of N_2 fixation by naturally occurring leguminous herbs in the Sahel zones range from 0.3 to 5 $kg\,ha^{-1}\,yr^{-1}$ (Penning de Vries and Djitèye 1991). For cultivated legumes in semi-arid areas with sufficient P and moisture available, estimates of N_2 fixation range from 10 to 40 $kg\,ha^{-1}\,yr^{-1}$ (Greenland 1985).

Tables 2.4 to 2.7 indicate the canopy cover of nodulating woody legumes in the SSWA, per climatic zone and per geomorphological unit. The proportion of nodulating woody legumes within the woody layer decreases with increasing rainfall, from about 75% in the Saharan Sahel zone to about 10% in the southern Sudan zone (Table 4.20). Most woody legumes in the Sudan zones do not nodulate. The proportion of nodulating woody legumes is lowest on sandy soils.

Average N concentrations in foliage of nodulating woody legumes and non-nodulating woody legumes differs by approximately 3 $g\,kg^{-1}$ in comparison to non-leguminous woody plants (Table 3.15). Similar data are available from eastern Africa (Table 3.16). This difference need not result from N_2 fixation by woody legumes. It may well be that woody legumes more effectively incorporate N into woody plant tissues. Assuming the average difference of 3 $g\,N\,kg^{-1}$ to be due to N_2 fixation, based on the proportion of nodulating legumes and on the foliage production per climatic zone (Fig. 3.8), the total of N_2 fixed per hectare is estimated per climatic zone (Table 4.20).

Högberg (1989) hypothesises that the occurrence of legumes is particularly favoured when the P/N ratio in foliage of non-N_2-fixing plants is above 0.11, indicating a sufficient P availability. N_2-fixing plants in tropical Africa seem to maintain a low 0.04–0.06 P/N ratio despite large variations in soil P/N among the sites, whereas non-N_2-fixing plants have a higher P/N ratio, indicating a limited N availability. In Zambia, N concentrations in foliage of N_2-fixing and non-N_2-fixing woody plants differ about 9 $g\,kg^{-1}$ (Högberg 1989). In Botswana, the average N concentration of the foliage of *Acacia* species averages 30 $g\,kg^{-1}$, whereas

Table 4.20. Annual foliage production per climatic zone (Fig. 3.9), the proportion of woody legumes in the woody layer, and the estimated amount of N_2 fixed by woody legumes

Climatic zone	Northern Sahel	Southern Sahel	Northern Sudan	Southern Sudan
Foliage production ($kg\,ha^{-1}\,yr^{-1}$)				
before drought	110	340	1100	1800
after drought	40	100	900	1800
% Nodulating woody legumes	50	40	25	10
Estimated N_2 fixed ($kg\,ha^{-1}\,yr^{-1}$)				
before drought	0.2	0.4	0.8	0.5
after drought	0.05	0.1	0.7	0.5

Table 4.21. The relation between the proportion of woody legumes, herbage production and herbage quality at the end of the growing season, at different sites in the southern Sahel zone in Mali. (Breman and de Ridder 1991)

Soil type[a]	SL/LS	LS/LS	LS/LC	LS/LC	LS/LC	L/LC	C/C
Canopy cover (%)	15	15	7	34	24	33	56
of which nodulating legumes (%)	57	31	30	38	58	64	96
Annual herbage							
production ($t\ ha^{-1}\ yr^{-1}$)	3.0	2.8	3.4	5.7	2.4	3.0	2.1
of which legumes (%)	1	?	>5	?	5	1	0
N in herbage layer ($g\ kg^{-1}$)	6.3	6.5	6.6	5.0	6.5	5.3	5.2
($kg\ ha^{-1}$)	18.9	18.1	22.3	28.5	15.7	17.6	12.3
P/N ratio of herbage	0.10	?	0.11	?	0.14	0.18	?

[a] Topsoil/subsoil; S= sand, L = loam, C = clay.

the P/N ratio is a low 0.03 (Tolsma 1989). A decrease of the occurrence of nodulating woody species, commonly observed when moving from dry to moist savannas, corresponds to a trend of increasing P deficiency (Högberg 1989).

In the SSWA, the P/N ratio does not differ according to N_2-fixing ability of woody plants, nor according to climatic zone (Table 3.15). P/N ratios are, however, consistently lower than 0.11, suggesting insufficient P available. P/N ratios for annual herbs vary between 0.06 and 0.19 (average 0.10), but show no consistent trends according to climatic zone, nor according to soil texture (Breman and de Ridder 1991). However, detailed studies from the southern Sahel zone in Mali show a positive correlation between P/N ratio of the herbage biomass and the proportion of woody legumes in the woody layer (Table 4.21). This observation seems to support the hypothesis of Högberg (1989). Note that highest P/N ratios of the herbage biomass occur at loamy-clay soils, where P availability is likely to be high due to inputs by run-on (Sects. 4.5.4.3 and 4.5.4.4).

How is Fixed Nitrogen Utilised?

It is important to know to what extent fixed nitrogen contributes to an increased N availability in the soil and thus an improved production or quality of vegetation components other than the woody plant itself.

Nitrogen availability in the topsoil is commonly similar under N_2-fixing and non-N_2-fixing woody plants in semi-arid regions (Garcia-Moya and McKell 1970; Ahuja et al. 1978; Nair 1984; Belsky et al. 1989; Drechsel et al. 1991) as well as in humid tropical regions (Daccarett and Blydenstein 1968; Montagnini and Sancho 1990). *Acacia aneura* was shown to fix nitrogen in Australia, but without an increased buildup of total N within the soil profile under the tree (Anonymous 1980). Catchpoole and Blair (1990) estimated that only 7.6% of the N_2 fixed by *Leucaena leucocephala* shrubs is transferred to the herbage layer. The availability of N for annual plant growth may be increased by applying prunings of *Leucaena* as a green manure, leading to increased crop production (Dommergues 1987). This practice may, however, lead to gradual depletion of soil P (Lal 1991).

The analysis of available data from the SSWA suggests that the presence of nodulating woody legumes does not increase N uptake by the herbage layer. If woody plants significantly increase N availability by N_2 fixation, a correlation should appear between N yield in the herbage biomass and cover of woody legumes. This relationship was tested on a transect from the northern Sahel to the southern Sudan zone by vegetation sampling at 36 distinct sites at the end of the rainy season during 2–4 years. A multiple regression analysis shows that water

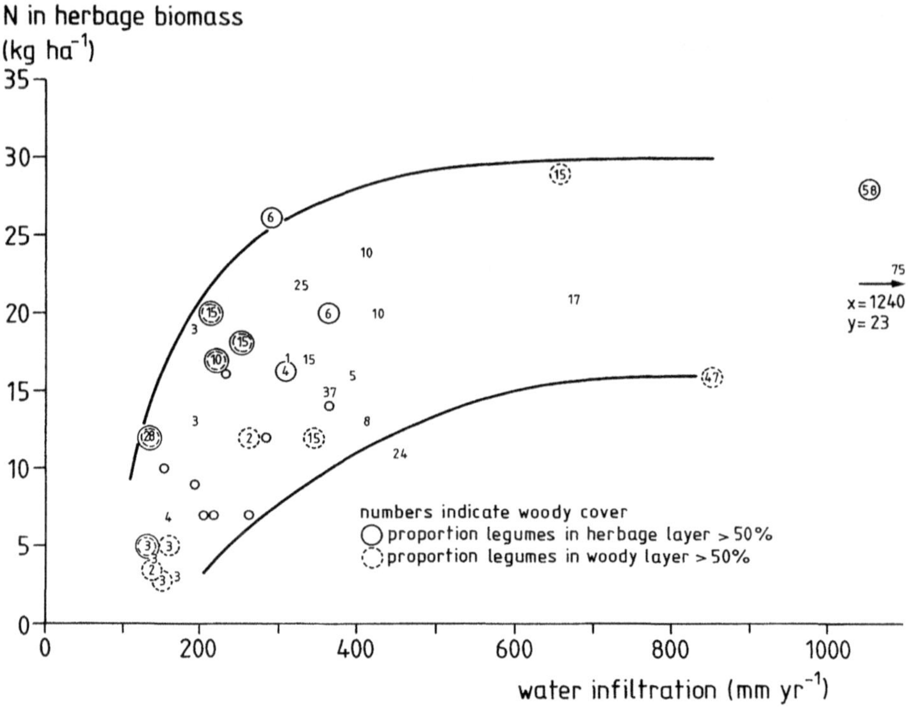

Fig. 4.19. The relationship between annual nitrogen yield in the herbage biomass and water infiltration, for sites with variable woody cover and variable contribution of legumes in the woody and herbage layer. *Lines* indicate outer limits of the relationship

availability at each site accounts for 45% of the variance in N yields of the herbage layer. This can be expected, as more water causes increased depth and period of soil wetting and thus availability of N from deeper soil layers. Total woody cover, or woody cover of leguminous woody plants only, does not show any relationship with N yield and does not improve the explanation of remaining variation (Fig. 4.19). The data presented in Table 4.21 support this conclusion: with increasing woody cover there is no trend of increasing N concentration in the herbage layer, nor of increasing N yields. Where legumes represent more than 50% of the herbs, N yield of the herbage layer is relatively high (Fig. 4.19). This observation supports the hypothesis that N_2 fixed by woody legumes is largely utilised by the woody plants themselves.

Conclusions

It is difficult to make reliable estimates of N_2 actually fixed by woody legumes, or of the potential for N_2 fixation. Roughly, N_2-fixation by woody legumes is estimated at $0.1\,kg\,ha^{-1}\,yr^{-1}$ in the Sahel zones and $0.6\,kg\,ha^{-1}\,yr^{-1}$ in the Sudan zones. This is less than estimates of N_2 fixation from other semi-arid regions. Possible explanations are the prevailing P deficiencies in the SSWA and the long dry season. N_2 fixation by woody plants may be higher at run-on locations. Although the quantities of N_2 fixed by woody legumes can be relatively important, the resulting increase of N available for herbage (crop) production is insignificant. Most N_2 fixed is utilised and translocated by the woody plant itself.

4.5.4.9 Rhizosphere Interactions

Mycorrhizas are specialised rootlike organs, formed as a result of the symbiotic association of certain fungi with fine plant roots. Both ectomycorrhizas and *vesicular-arbuscular* endomycorrhizas occur. Most tropical woody species form *vesicular-arbuscular* mycorrhizas, particularly those with nodules, but only few groups have ectomycorrhiza (Whitmore 1990; Ducousso 1991).

In infertile habitats mycorrhizas are critical for growth and development of woody plants, due to their ability to increase the absorption by roots of nutrients. Particularly poorly mobile nutrients are influenced, in increasing importance of mobility these are $P > NH_4 > K > NO_3^-$ (Stuart Chapin 1980). Several processes may be implicated. Connor (1983) suggests that mycorrhizas absorb P from the same nutrient pool as uninfected roots. Others however found an increased solubilisation of nutrients by those mycorrhizas associated with long-lived roots; which may be related to an increased acidity caused by the mycorrhizas (Stuart Chapin 1980). A correlation was found between low P availability in the soil and the dominance of woody plants with ectomycorrhiza (Högberg 1989).

An additional advantage of mycorrhizas, particularly ectomycorrhiza in seasonal climates, is their large storage capacity of nutrients. Mycorrhizas are also shown to transfer nutrients directly from litter to roots (Vitousek and Sanford 1986).

Mycorrhiza development is favoured by high light intensity (photosynthetic activity) and low to moderate soil fertility (Pritchett 1979). It is suggested that the P/N level is critical for mycorrhizal development (Anderson and Swift 1983). Both a high P/N ratio (sufficient P available) and a low P/N ratio reduce mycorrhiza activity. Mycorrhizas are probably not abundant in regions with a long dry season, as most fine (tree) roots do not survive the dry season and thus suppress mycorrhiza development (Alexander 1989).

In the SSWA, endomycorrhizas are commonly observed in association with herbs. Endomycorrhizas increase the uptake of available P in the soil up to six-fold (Penning de Vries and Djitèye 1991). In spite of this, P availability limits primary production in most semi-arid regions. There is no evidence to assume that the presence of endomycorrhizas increases P availability; they merely facilitate uptake of available P. The potential of the introduction of endomycorrhizas seems limited in view of their already widespread occurrence.

Scarce available data indicate that in the SSWA only woody plants of the genera *Isoberlinia* and *Afzelia* (Caesalpinioideae) are infected with ectomycorrhizas (Alexander 1989; Högberg 1989; Ducousso 1991). These genera occur in the Sudan zones. The occurrence of ectomycorrhizas seems to increase from the dry towards the moist savannas in Africa (Högberg 1989). Endomycorrhizas are widepread among woody plants in the SSWA (Ducousso 1991). Effective mycorrhizas are particularly expected at relatively infertile sites, for long lived woody plants. An eventual increased P uptake via mycorrhizas may lead to increased N_2 fixation by woody legumes (Szott et al. 1991).

4.5.4.10 Conclusions

Table 4.22 summarises the importance of the processes discussed, regarding their potential to redistribute (concentrate) nutrients towards the woody canopy, to reduce nutrient losses, and to increase nutrient availability for the agro-ecosystem. The results are specified for the Sahel (150–600 mm) and Sudan (600–1200 mm) climatic zones. The importance of the processes is quantified in a number of classes, indicating the increased availability of N and P for primary production. The results are expressed on a hectare basis: the improved nutrient availability under woody canopies is considered in relation to their cover in the vegetation as a whole in the pre-drought situation (Fig. 2.4). For each process, Table 4.22 indicates the value corresponding to maximum effectiveness of the process. The effects of the different processes cannot simply be cumulated. For instance, when water erosion is effectively reduced, losses by leaching (following run-on) and by external recycling are less, and so will be the potential for woody plants to reduce such losses. Likewise, the reduction of wind erosion and of fire lead to less capture of wind-blown organic material, dust and ashes.

In the Sahel zones, redistribution processes and processes leading to a reduction of losses seem equally important, whereas in the Sudan zones the latter are most important. The benefits to be expected are related to nutrient availability in the situation without woody plants, which is 0–20 kg N ha^{-1} in the Sahel zones, and 10–25 kg N ha^{-1} in the Sudan zones (variation related to site conditions); P availability is 10% of these values (Penning de Vries and Djitèye 1991). Therefore, in a relative sense, redistribution processes are more important in the Sahel zones

Table 4.22. The maximum effects of various processes associated with woody plants in the Sahel and Sudan zones on nutrient availability for primary production of a mixed vegetation, considering the effects in the canopy area and canopy cover in the vegetation as in the pre-drought situation, in comparison to a vegetation without woody plants

Process		Zone	
		Rainfall: 150–600 mm canopy cover: 2–20% (Sahel zones)	Rainfall: 600–1200 mm canopy cover: 15–30% (Sudan zones)
1. Redistribution/concentration			
atmospheric deposition	– origin dust	+	0
	– origin ocean	0	0
	– origin fire	0	+
Capture of wind blown organic material[a]		+	+
nutrient concentration by lateral roots		+	++
deposition by animals		+	+
2. Reduction of losses			
Decreasing	– wind erosion[a]	+	+
	– water erosion[a,b]	0–+	+–++
	– leaching[c]	+	+++
	– fire (volatilisation)[c]	+	+
Internal recycling		+	++
External recycling		+	++
3. Enrichment			
Nutrient uptake by deep roots[d]		0	+
N_2 fixation		0	+
Rhizosphere interactions[d]		0	+

0 = unimportant; neligible effects.
+, ++ and +++ = N increase by 1–5, 5–10 and $>10\,kg^{-1}\,ha^{-1}\,yr^{-1}$; P increase <10% of these values.
[a] For shrubs highest value.
[b] Depending on nutrient leaving by rivers (lowest value) or internally redistributed (highest value).
[c] For N in particular.
[d] For P in particular.

than in the Sudan zones, leading to spatial concentration of resources under woody plants at the expense of the surrounding areas or regions (depending on the scale of the process involved). The concentrated resources taken up by woody plants enter into a soil-tree-litter cycle which generally involves less losses than the soil-herb-litter cycle so that redistribution leads to a more efficient use of resources by the vegetation as a whole. Processes leading to the reduction of nutrient losses are important in both zones, but different processes are involved. For instance, external recycling is more important in the Sudan zones, as a result of rainfall and leaching from plant tissues, whereas in the Sahel zones internal recycling is more important. Processes that lead to enrichment of the agro-ecosystem have limited potential in both zones, contrary to some humid tropical regions. Enrichment by both N_2 fixation and deep rooting are important at specific locations only. There is no evidence to assume that N_2 fixation by woody legumes significantly increases herbage production or N concentrations in herbs or crops. Nor is there evidence to assume that deep rooting or the presence of mycorrhizas significantly increases availability of P or other immobile nutrients from greater depths. The improvements by woody plants are more important for N than for P, as demonstrated by the greater enrichment factor for N than for P and the stronger gradient between topsoil and subsoil for N than for P.

Based on the available evidence, a hypothetical framework for woody plant development in semi-arid regions is as follows. At early stages of woody plant development, e.g. at the beginning of a new fallow period, the woody plant rapidly takes up available nutrients from the topsoil, and to a limited extent from the subsoil. Here, a laterally extending root system is an important asset for rapid juvenile growth. Woody plant establishment and early growth are best at locations with relatively high nutrient and water availability prior to woody plant establishment. Pre-existing variation in both soil fertility conditions (Scott-Wendt et al. 1988) and water availability (Breman and de Ridder 1991) can be large. Nutrient inputs after the initial phase of woody plant development largely depend on atmospheric inputs (e.g. Richards and Charley 1983), due to the general nutrient poverty of the soils in semi-arid regions. Woody plants are more effective in capturing nutrients from atmospheric origin than are herbs. Additionally, the inputs over large areas are continuously concentrated towards the woody plant-litter-soil cycle, not only by roots, but also by animals.

Once accumulated in the woody biomass, nutrients are effectively recycled, both internally and externally, and are less susceptible to losses than in herbs or in a herbaceous vegetation without woody plants. Biological transport mechanisms leading to spatial concentration of resources (roots, canopy, termites, herbivores), as compared to physical transport mechanisms (wind, water, gravity), are, in a relative sense, more important as conditions become more arid, and are more important for perennial than for annual plants. Concentration of resources by biological mechanisms mainly leads to depletion elsewhere but for the ecosystem as a whole nutrient losses are reduced, and ecological sustainability is enhanced (Sect. 5.2). Redistribution processes and those reducing losses by internal recycling are most effective under low fertility (resource-poor) conditions. Woody plants at relatively infertile locations are more efficient in their use and recycling of nutrients than those at fertile locations, of fertility islands. Although fertility islands form a particular contrast in arid and resource-poor environments, their impact on productivity of the ecosystem as a whole is limited. The rate of increase of canopy cover (Fig. 2.4) is more rapid than the rate of decrease of the enrichment factor (Fig. 4.13). As a result, the contribution of woody plants in nutrient recycling and enrichment increases more than proportionally with increasing rainfall. Since the processes reducing nutrient losses contribute most to fertility improvement under woody canopies (Table 4.22), the potential benefits are greater as initial soil fertility is higher. Since the potential impact by woody plants on reducing losses is greatest for reducing leaching, the benefits of woody plants are greatest on sandy soils and on shallow soil where the subsoil can be reached by roots of woody plants.

4.6 Soil Organic Matter and its Functions in Relation to Woody Plants

4.6.1 Introduction

Increasingly attention is diverted from the role of woody plants as suppliers of nutrients, to their buffering capacities, that lead to reduced nutrient losses. Key issues are nutrient recycling and improvement of soil structure, mainly by increased soil organic matter (SOM) through litter inputs (Beer 1988; Coleman et al. 1989; Young 1989). Organic material inputs to the soil have two major functions:

- maintenance of SOM content, with functions for chemical and physical soil properties, and
- input of nutrients.

Both aspects are essential for plant production: maintenance of good physical and chemical soil characteristics are essential for an efficient use of available nutrients and for an efficient recovery of nutrient inputs originating from natural and/or artificial sources. An increased plant production following organic inputs is attributed to both the supply by the organic manure of nutrients and to the more favourable chemical and physical properties of the soil (de Ridder and van Keulen 1990). However, the two functions are conflicting: a high decomposition rate is most desirable for a rapid release of nutrients, but prevents maintenance of soil chemical and physical conditions (e.g. Coleman et al. 1989).

Under woody canopies SOM content is almost always higher than in open areas (Table 4.10). In this section are investigated the functions and dynamics of SOM in relation to woody plants in semi-arid regions, with an evaluation of the consequences for plant production and stability of the agro-ecosystem.

4.6.2 Effect of SOM on Soil Physical and Chemical Properties

The stability of the aggregates formed from clay and loam particles in the soil is influenced by polysaccharides produced by the soil fauna, hyphens produced by fungi, by plant roots and by high quantities and regular availability of organic material. Aggregate stability improves plant production by better soil aeration and improved resistance to erosion mainly (Janssen 1986; Young 1989; de Ridder and van Keulen 1990). The degree of aggregation, affecting pore volume and pore size distribution, also influences infiltration capacity and soil moisture retention characteristics (Sect. 4.4.2.2).

According to Pieri (1989), the minimal SOM content required for maintenance of soil structure is related to the clay and loam content of the soil as follows: %SOM / %(clay + loam) = 0.05. For sandy, loamy and clay soil in the SSWA, corresponding % SOM would need to be 0.5, 1.0 and 3.5% (i.e. 3.0, 6.0 and 21.0 $g\,C\,kg^{-1}$ soil). Such values are rarely observed (Table 4.17) although in the same sequence increasing SOM contents occur (2.0, 4.0 and 8.0 $g\,C\,kg^{-1}$ respectively). These low SOM contents are mainly related to the high temperatures, increasing decomposition rates. Several other semi-arid regions show higher average SOM contents (Tables 4.10 and 4.11).

The soil chemical property most affected by SOM is the cation exchange capacity (CEC). The CEC is firstly determined by clay content and mineralogical composition of the soil; but in those tropical soils where the clay fraction is dominated by kaolinite and Fe and Al oxides and hydroxides, variation in CEC is determined by differences in SOM mainly (Pichot 1975; Young 1989; Budelman and Zander 1990). In Niger, SOM alone explains 54% of the variation in CEC and shows a significant correlation with total N (ICRISAT 1991). Similar situations occur in other semi-arid regions (Budelman and van der Pol 1992). On the basis of data from semi-arid regions, de Ridder and van Keulen (1990) demonstrated a significant positive correlation between SOM and CEC. A change of 1 $g\,kg^{-1}$ in C concentration resulted in a CEC change of 4.3 $mEq\,kg^{-1}$ (at pH 7). Unexplained variability can be ascribed to differences in origin, nature and quality of the organic material.

The retention capacity of the soil for cations such as K, Mg and Ca is related to CEC and pH. These cations contribute to maintaining base saturation at a high level. Other cations may form complexes with SOM.

Exchange of cations with Al^{3+} and H^{+} counteracts decreasing pH for values lower than 5.5; there is a positive relation between pH and base saturation. CEC thus serves as a buffer for pH changes and against intoxication by free Al^{3+} ions (Janssen 1986). Complexing with inorganic elements however leads to reduced retention of useful cations. Maintenance of a high CEC contributes to agricultural production by both pH stabilisation and cation retention and supply. A minimum CEC for agricultural land-use is 3–4 mEq per 100 g soil (Janssen 1984). For sandy soils in Surinam, a CEC of 3.5 mEq per 100 g was found to be equivalent to 3.5 g C kg^{-1} (Janssen 1986), but in semi-arid regions 8 g C kg^{-1} is required (3.5/0.43; see above). SOM can only effectively increase CEC when pH exceeds 5.5 (Coleman et al. 1989).

4.6.3 Decomposition Processes

Primary sources of organic matter are leaf and wood litter above ground, and root debris in the soil. By decomposition, residues are partly transferred to organic matter pools in the soil. The distinction between litter and SOM is not easy. SOM may be divided into a living component (soil organisms mainly) and a non-living component. The non-living component, making up more than 95% of SOM, can be subdivided into a macro-organic and a humus component. Macro-organic matter constitutes the light fraction of SOM; the humus component is the bulk of SOM and consists of fractions with different degrees of solubility (Coleman et al. 1989).

Different "pools" of SOM are distinguished, characterised by different turn-over rates (Parton et al. 1983; Young 1989; Janssen et al. 1990; Verberne et al. 1990):

1. active, non-humified SOM with a turn-over rate of 0–3 years;
2. slow SOM, or labile humus, decomposing in up to 25 years;
3. resistant SOM, or stable humus, decomposing in about 1000 years.

Active and slow SOM are partly protected against decomposition. Physical protection occurs in fine-textured soils (Sect. 4.6.5); chemical protection includes complex formation with Al and Fe. In general, organic matter inputs rich in lignin and polyphenolics, defined as of poor quality (e.g. wood), lead to a more rapid formation of resistant SOM than plant materials of high quality (Coleman et al. 1989).

The "young" fractions are concentrated in the topsoil; following decomposition, the residues gradually make part of the resistant fraction. The same applies to decay of the living component; faeces of soil fauna is most resistant. Resistant SOM is mostly found in deeper soil layers. The active SOM pool is probably most important with regard to nutrient release (Berendse 1990). Although it has never been demonstrated

(Young 1989), the resistant fraction probably contributes most to physical soil aspects (Coleman et al. 1989). Regarding chemical soil aspects the role of the resistant fraction is unclear; it is possible that this fraction is chemically too inert to improve CEC (Coleman et al. 1989). It is unknown at what stage of decomposition the SOM obtains its various functions.

Organic inputs *in* the soil are expected to lead to a higher proportion of relatively stable SOM fractions than inputs *on* the soil surface, because of the larger contact surface between organics and clay particles with incorporation, and hence the greater aggregation and physical protection (Coleman et al. 1989).

Decomposition rates have been estimated by mass-balance studies, relating litter fall to litter mass, and by direct measurements of weight loss from litter bags.

The first approach defines decomposition as the litterfall divided by the organic matter on the soil under equilibrium conditions (Olson 1963). Values higher than 1, such as prevail in tropical forests, suggest that turnover of SOM occurs within a year. There are, however, disadvantages related to this method:

1. the differentiation between litter input and SOM is difficult and arbitrary for small roots and fine leaf litter;
2. falls of large pieces of wood show a highly variable distribution in space and time;
3. the suggested equilibrium situation of the SOM, with a continuous litter input and a constant decomposition rate does not commonly occur;
4. belowground litter inputs are not considered.

Nevertheless, provided that the same litter components are represented in both litter inputs and SOM, the quotient, referred to hereafter as the litter turnover coefficient, represents the only basis on which geographical patterns of litter decay can be compared throughout different ecosystems (Swift et al. 1981; Anderson and Swift 1983).

The second method measures loss of weight from material enclosed in mesh bags. These bags can include different mesh sizes to exclude different size groups of animals to enable partitioning of the decomposition process into the component processes (1) comminution through the feeding activities of soil fauna and (2) microbial catabolism. Leaching of water-soluble materials may lead to additional losses of nutrients. Shortcomings of this method are reviewed by Swift et al. (1981).

Results of this method show that the decomposition process can be described by a negative exponential function indicating the mass loss (related to carbon loss) from heterogeneous litter mass. Two phases are distinguished: an initial rapid phase, which results in loss of about two-thirds of the residue C, and a second phase in which decomposition follows a simple first order kinetic model. The rapidly decomposable components (sugars, proteins etc.) disappear during the first phase, whereas the decay of the more resistant compounds (lignin, fats etc.) determines the rate of mass loss during the second phase (Janssen 1986; Berendse et al. 1989). Types of litter are characterised by a relative decomposition rate or a half-life value. Decomposition rates obtained in different climates are different but the patterns are similar: the functions can be superimposed using an appropriate conversion factor.

The decomposition of organic material is governed by three groups of variables (Swift et al. 1981): environmental conditions, litter quality, and the decomposer community.

Environmental factors influencing the decomposition process are:

1. Moisture availability. Optimal moisture content of the soil is estimated at 60–80% of maximum water holding capacity.
2. Temperature. Provided sufficient available moisture, decomposition rates double for every increase in temperature of 9 °C (Jenkinson and Ayanaba 1977) until an optimum is reached at about 40 °C. Due to low temperatures, decomposition rates are low at high elevations.
3. Oxygen. Too high soil humidity limits aeration of the soil and thus dissimilation processes. A high soil porosity stimulates aeration.
4. Soil texture. In fine textured soils decomposition rates are low due to poor soil aeration, clay particles that adsorb enzymes responsible for decomposition (Scott 1962) and protection of organic matter particles between soil aggregates (Verberne et al. 1990).
5. Available soil nutrients. A shortage of nutrients may arrest the decomposition process in an early phase.

Regarding *litter quality*, lignin and fats mainly form the stable, relatively inert fraction of SOM, whereas sugars, proteins and (hemi-) cellulose are rapidly decomposed. Polyphenolic substances, including tannins, also increase the resistance towards microbial decomposition (Vallis and Jones 1973). For a range of forest ecosystems, initial lignin, C/N and C/P ratios of leaf litter are correlated with decomposition rate (Coleman et al. 1989). There is evidence to assume that lignin content is a more influential factor than C/N ratio (Laishran and Yadava 1988). In a tropical forest in India, lignin content explained 59% of the variability of weight loss of litter (Upadhyay and Singh 1989). Decomposition in the first year after litter fall is generally correlated with N concentration, but since in tropical regions early decomposition is rapid, overall decomposition rates are correlated with lignin content mainly. Thus, lignin content is a good predictor for decomposition rates in tropical regions.

The same principles probably apply to root litter, although little is known about root turn-over and nutrient release. Nutrient concentrations of root litter are probably relatively high because less translocation of nutrients takes place than with above-ground foliage (Coleman et al. 1989).

The *decomposer community* plays a primary role in comminuting surface litter, incorporating the fragmented products in the soil and converting it into excreta. Fungi predominate in the decomposition of low-quality materials (Swift et al. 1981). Regarding the decomposer community, variation in decay rates is mainly due to variation in the macrofauna (Anderson and Swift 1983). Besides facilitating and accelerating the decomposition of litter by microorganisms, soil meso- and macrofauna contribute to nutrient cycling by grazing on the microflora as well as by mixing soil materials within the soil profile (Coleman et al. 1989). In fresh litter decomposition, Budelman (1988) found that at least phosphorus is lost at a faster rate than carbon, suggesting that soil fauna has a preference for nutrient-rich parts (e.g. cell material).

For forest ecosystems, the soil fauna biomass is often inversely related to the accumulated weight of litter on the soil. Limited evidence from the tropics suggest low soil fauna biomass, so rapid turn-over. It may be assumed that the resident microflora rapidly responds to organic inputs (Verberne et al. 1990).

SOM content of the unprotected active pool is related to litter inputs, i.e. the fraction of plant biomass that dies off annually, and to the decomposition rate (Janssen 1986; Berendse et al. 1989). Both above- and belowground litter inputs should be considered. Variations of the resistant and protected pools of SOM are difficult to predict.

In an equilibrium situation SOM content [SOM] is defined as

$$[SOM] = r_h \times LI/r_d,$$

where LI is the annual litter input, r_h is the humification rate and r_d the decomposition rate (Janssen 1986). The humification rate is defined as the proportion of organic inputs that remains in the soil as humus at 1/n year after application; the decomposition rate is defined as the proportion of soil humus C lost by oxidation during 1/n year (Janssen 1986). $n = 2^{(X-9)/9}$ (X = °C); thus the period 1/n is halved with every 9 °C temperature increase, provided that other site conditions are constant.

A major shortcoming of the above formula is the fact that it is generally unknown which proportion of organic inputs not remaining on the soil is lost from the ecosystem and which proportion is brought into the soil by soil fauna and by leaching (Janssen et al. 1990).

4.6.4 Mineralisation Processes

Mineralisation of litter is related to litter decomposition processes, but the rate of nutrient release is not only related to the rate of decomposition, although for some nutrients such a relation exists for SOM in the topsoil only (Berendse 1990). Three sequential phases can be distinguished in the mineralisation of litter: the initial release phase in which leaching of nutrients from the organic material predominates; the net immobilisation phase; and the net release phase (Berendse et al. 1989; Upadhyay and Singh 1989).

Immobilisation takes place by assimilation of nutrients by soil invertebrates into microbial biomass. The micro-organisms compete with plant growth for limiting nutrients. Ultimately, the nutrients will be mineralised via waste-products and decomposition of dead bodies. Immobilisation during early phases of decomposition occurs for N and P, and mainly for organic inputs with high C/N and C/P ratios. Nutrient release is slowest once sufficient C has been dissimilated to reduce the C/nutrient ratios. Woody litter, characterised by high C/N and C/P ratios, initially acts as a sink of nutrients. In tropical regions, N immobilisation takes place as long as the litter has a C/N ratio higher than 10–30 and P immobilisation is expected at a C/P ratio higher than 100–500 (Janssen 1986; Budelman 1988; Upadhyay and Singh 1989; Janssen et al. 1990). Net release of nutrients takes place when the C/nutrient ratio drops below such "critical levels". During the immobilisation phase resistant components such as lignin accumulate (Schlesinger 1985). Burning of wood (e.g. after clearing) reduces immobilisation (Janssen et al. 1990). The period and degree of immobilisation is also influenced by nutrient availability in the soil: immobilisation of P is generally not observed at relatively nutrient-rich sites, contrary to infertile sites in tropical regions (Vitousek and Sanford 1986).

Polyphenolics (of which tannins are most important) depress decomposition rates. The mineralisation rate of organic material with a polyphenolic/N ratio of less than 0.5 is correlated with N and with lignin concentrations. For materials with a higher polyphenolic/N ratio, mineralisation is not correlated with N or lignin, but is negatively correlated with tannin content (Palm and Sanchez 1991). Polyphenolics thus influence decomposition of high quality organic material, such as litter from legumes. Decomposition of poor quality organic material is correlated with the lignin/N ratio.

4.6.5 SOM Contents in Semi-Arid Regions

Litter inputs required to maintain a certain SOM content can be determined on the basis of the formula $[SOM] = r_h \times LI/r_d$ (Sect. 4.6.3). Since site conditions influence both litter inputs and decomposition rate, the relationship between site conditions and SOM content is complex. Post et al. (1982) showed that SOM accumulation is inversely related to primary production, suggesting that decomposition rates increase faster as a function of temperature than does litter production.

Litter turn-over rates (r_t) determined by the mass-balance method (Table 4.23) show a large variation in seasonal tropical regions. In the SSWA r_t is highest for herbaceous litter, and on loamy and clay soils. Termites, as the main agents removing organic material during the dry season, are less common on sandy soils.

In seasonal climates SOM contents fluctuate throughout the year by several $g\,kg^{-1}$, as do soil nutrient concentrations and CEC (Lambert et al. 1980; Swift et al. 1981). Seasonality of rainfall and of litter inputs is an underlying factor. There is uncertainty as regards the overall effect of seasonality on SOM contents. Post et al. (1982) argue that SOM content is reduced by seasonality, since it reduces litter production more than decomposition rates. Coleman et al. (1989), however, argue the opposite. Fluctuations in both temperature and moisture during the rainy season accelerate decomposition of organic material and mineralisation processes (Birch 1964). It has been suggested that these effects are due to disruption of physically protected organic matter complexes and accelerated turn-over of microbial biomass. Other research, however, suggests increasing stability as a result of seasonality (Duchaufour 1977, cited by Coleman et al. 1989).

In the following are investigated the influences on SOM contents by fire, by termites, by soil conditions, and by agricultural practices. In semi-arid regions, build-up of litter biomass depends on the periodicity of litter fall, including the occurrence of drought periods during which woody plants die, causing massive litter inputs. Apart from grazing and erosion, bush fires have a major impact on litter build-up and thus SOM content. The proportion of carbon lost from the litter layer and the topsoil by burning depends on fire intensity mainly. SOM contents may even increase by fire, particularly light fires, as a result of the addition of (partly ashed) residues, decomposition of litter being stimulated and subsequent increased plant productivity (Raison 1979). However, the major contribution to SOM and nutrient availability in the soil is probably from decay of roots following burning.

Litter biomass available for decomposition (X_n) and the decomposition rate (r_d) are interrelated as follows (Birk and Simpson 1980): $X_n = LI/r_d - (LI/r_d - X_0)\, e^{-r_d n}$. Both LI and r_d may be estimated using non-linear optimisation techniques for values of X collected over several years (n).

In the absence of external disturbances SOM reaches an equilibrium. Birk and Simpson (1980) estimated that under stable conditions *Eucalyptus* forest in semi-arid Australia require 20 to 40 years to reach a steady state of SOM. In chaparral shrubland in semi-arid USA, a stable state of litter mass is reached after 8 to 21 years without fire (Schlesinger

Table 4.23. Litter fall, litter biomass and litter turnover coefficients (r_t) in the natural vegetation in some seasonal tropical regions. Values are on an annual basis and for total small litter (excluding large wood), unless specified otherwise

Country, rainfall and vegetation type	Litterfall ($t\ ha^{-1}\ yr^{-1}$)	Litter biomass ($t\ ha^{-1}$)	r_t	Source
Nigeria, >1500 mm				Hopkins (1965)
evergreen forest	7.2[a]	3.0	2.4[a]	
seasonal forest	4.7[a]	1.0 (wet season) 2.4 (dry season)	2.8[a]	
Nigeria, 1500 mm				Swift et al. (1981)
savanna bush fallow	8.5	8.9 (wet season) 10.9 (dry season)	0.9	
	5.8[a]	3.4 (wet season) 5.2 (dry season)	1.3[a]	
Belize, 1700 mm				
seasonal forest	12.6	7.2	1.8	Lambert et al. (1980)
Australian woodland, 1000 mm				
Eucalyptus marginata	2.9	12.7	0.23	Birk and Simpson (1980)
Eucalyptus diversicolor	6.2	34.4	0.18	
Mali, 200–1000 mm				own observations
woodland savanna, herbaceous litter on				
– sandy soils	2 (0.5–5.0)	1.4 (dry season) 0–1 (wet season)	±2	
– loam and clay soils	1.5 (0.2–6.0)	0.7 (dry season) 0 (wet season)	±4	
woody plant litter on				
– sandy soils	0.1 (0.0–0.3)	0.07 (dry season)	1.4	
– loam and clay soils	0.7 (0.1–1.4)	0.40 (dry season)	1.8	

[a] Leaves only.

1985). The period required is lowest for small shrubs, rapidly reaching maturity, such as also prevail in the Sahel zones.

In semi-arid tropical regions termites play an important role in decomposition processes (Coleman et al. 1989; Lavelle et al. 1990). In semi-arid Texas, transport by termites accounts for 55% of losses of litter from the soil surface, microbial activities accounting for 45% (Bodine and Ueckert 1975). Estimates of the proportion of woody litter consumed by termites in African savannas range from 60% in Nigeria, 86% in Ghana and 91% in Kenya, whereas in South Australia and Arizona only 17% of woody litter is consumed by termites (Buxton 1981). Breakdown by termites of relatively large wood is important to facilitate microbial decomposition (Anderson and Swift 1983). Some termites use fresh plant material to cultivate subterranean fungus combs, which can digest (dry) organic material not utilizable by termites (Whitmore 1990); their importance decreases with decreasing rainfall (Lavelle et al. 1990). In termite mounds and termite channels organic matter can be protected against decomposition by complexes including clay particles (Lavelle et al. 1990; Verberne et al. 1990).

On the basis of 95 soil samples from eastern Africa, Scott (1962) found the variability of SOM content to be most significantly determined by clay + silt content of the soils, and secondly by rainfall. SOM content increases with rainfall until about 1600 mm, ranging from $6.5\,g\,kg^{-1}$ when the soil contains 15% silt and clay, to $15\,g\,kg^{-1}$ for 75% silt + clay. A high clay content of the soil generally stimulates LI and reduces r_d (partly by physical protection of SOM), leading to a higher SOM. In eastern Africa the dual effect of increasing SOM content and more leaching with increasing rainfall leads to an optimal base saturation at about 1100 mm rainfall. Due to the relationship between soil texture and soil fertility, the effect cannot be attributed to soil texture only (Anderson and Swift 1983).

Another illustration of the effects of soil conditions on SOM content is from Sumatra (Hardon 1936; cited by Janssen 1986). Here, SOM content is lowest at pH 5 due to a low LI and a relatively high r_d. A pH lower than 5 suppresses r_d, a pH higher than 5 stimulates LI, both resulting in a higher SOM content.

SOM contents in croplands are usually lower than in the natural vegetation. This is due to crop residues generally being removed (low LI), whereas decomposition rates may be lower (high C/nutrient ratios) or higher (due to opening up of the soil). High quality crops (e.g. green manure) decompose rapidly, whereas poor quality residues decompose slowly. More is needed of a rapidly decomposable organic material to maintain or increase SOM content (Table 4.24).

Natural grasslands in temperate climates commonly have a higher SOM content than woodlands and the rate of nutrient cycling is higher (Janssen 1986). This may be related to a higher turn-over of soil nutrients

Table 4.24. Organic matter inputs required to maintain SOM content in the top 20 cm of a tropical soil, where annual loss of C from existing organic material is $1\,t\,ha^{-1}$, for different sources of organic material

Organic material	C in material ($g\,kg^{-1}$)	C left after humification ($g\,kg^{-1}$)	Organic material required ($t\,ha^{-1}$)
Green manure	250	50	20.0
Compost	300	60	16.6
Animal manure	350	105	9.6
Sorghum straw	450	225	4.4
Tree litter:			
excluding large wood	350–400[a]	70–160[a]	6.3–14.3
including large wood	350–500[a]	70–400[a]	2.5–14.3

[a] Based on Table 4.27, see text.

in grasslands which, in spite of a lower standing biomass, leads to a higher LI. Grasslands also commonly occur at relatively fertile sites where grasses have a competitive advantage to woody plants (Sect. 4.2). Decomposition of grass litter is rapid since the proportion of lignin is low. Simulated SOM dynamics under temperate conditions showed that application of litter with a high lignin content leads to a slightly higher SOM content as compared to similar quantities of litter with a low lignin content (Verberne et al. 1990).

4.6.6 Decomposition of Woody Litter in Semi-Arid Regions

In semi-arid regions, SOM contents in the topsoil under woody plants are generally higher than in the open field (Table 4.10). Differences in SOM content between the woody canopy and the open field vary from $0–10\,g\,C\,kg^{+1}$. In the SSWA, the average SOM content under woody canopies is $7.7\,g\,C\,kg^{-1}$, compared to $3.9\,g\,C\,kg^{-1}$ in the open field (based on data from Table 4.10). Differences are least for parkland trees on agricultural fields.

An analysis of the specific role of woody plants in semi-arid regions on SOM contents requires investigation of factors influencing litter inputs and decomposition rates, in comparison to a herbage layer only (Table 4.25). A greater litter input per unit of area under woody canopies is an expected consequence of the spatial concentration under woody plants of water (Sect. 4.4) and nutrients (Sect. 4.5.4), in combination with a large leaf area surface available for photosynthesis (Sect. 4.7). A comparison of litter inputs by woody plants in the SSWA (Table 3.10) with the herbage production shows that in every climatic zone in the SSWA litter inputs under woody canopies, in an equilibrium situation, are about

Table 4.25. The relative importance of various factors associated with woody plants, regarding effects on decomposition rates and SOM under woody canopies, in comparison to a herbage layer only

Factor		General effect	Effect on decomposition rate	Effect on SOM content under canopy
Litter inputs[a]		About 3x higher	No effect	Increase[c]
		Continuous	Increase?	Decrease?
Environmental conditions:				
	– temperature	Lower	Decrease (30–50%)	Increase (30–50%)
	– humidity	Higher	Increase (20–30%)	Decrease (20–30%)
	– soil conditions	Favourable	Increase	Decrease
Litter quality:[b]				
Foliage	– tannin content	High	Decrease	Increase
	– lignin content	Similar	Similar	Similar
	– C/N, C/P ratios	Low	Increase	Decrease
Wood	– tannin content	Similar	Similar	Similar
	– lignin content	High	Decrease	Increase
	– C/N, C/P ratios	High	Decrease	Increase
Soil fauna activities		More important	Increase	Decrease

[a] Both above- and belowground.
[b] Aboveground only.
[c] Unless litter losses by fire, herbivory and wind are high.

three times higher than those from a herbage layer only. It is assumed that all litter inputs from woody plants are deposited in the canopy zone. This is not very realistic since litter losses occur by herbivory, by fire and by wind. Herbage production under woody canopies, which is highly variable (Sect. 4.2), was not included in the litter inputs under woody canopies, and may be assumed to have the same order of magnitude as litter losses by fire, wind and herbivory.

Litter deposition from a woody plant community is less concentrated seasonally than are litter inputs from herbs only (Sect. 3.5). Since decomposition of litter is concentrated in the wet season, litter from a woody vegetation builds up during the dry season (Table 4.23; Swift et al. 1981). Little is known of patterns of root decay from woody plants.

Regarding litter decomposition under woody canopies the following factors are important (Table 4.25):

1. *Environmental conditions*. Under woody canopies, in comparison to the open field, average temperatures, both of the air and the soil, are

5–10 °C lower (Sect. 4.4.2.3). As result decomposition would be 30–50% lower. On the other hand, since PET is lower under woody canopies (Sect. 4.4.2.3), the soil remains moist and decomposition can continue for a longer period (at most 1 month longer in comparison to the open field). The temperature effect is more important than the humidity effect, so that decomposition rates are 10–20% lower under woody canopies. Other factors influencing decomposition are higher soil nutrient concentrations and a lower soil bulk density under woody canopies (Sect. 4.5.1), both favouring decomposition to an unknown extent. The ultimate effects of environmental conditions are difficult to predict.

2. *Litter quality* (Sect. 3.6.2). Most mature leaves from woody plants have higher N concentrations, comparable lignin concentrations, and higher polyphenolic (including tannin) concentrations than herbaceous litter (Table 3.14). In the SSWA, the polyphenolics/N ratio of leaves is mostly higher than 0.5, considered as the threshold for inhibiting decomposition (Palm and Sanchez 1991). Decomposition of wood is inhibited by high lignin concentrations in the first place. Stuart Chapin (1980) suggests that woody plants adapt to infertile soils by high lignin and high polyphenolics concentrations in their tissues, leading to low decomposition rates. A slowly degrading litter layer also maintains favourable microclimatic conditions (Budelman 1989). Scarce available evidence suggests that small roots have comparatively low lignin concentrations and would be expected to decompose rapidly, whereas decomposition and nutrient release from larger, more lignified root litter is slow (Coleman et al. 1989).

3. *Soil fauna.* Under woody plants soil fauna is more abundant than in the open field (Sect. 4.5.3.4). This is probably the result of both a prolonged availability of litter throughout the year (in terms of inputs and the presence of slowly degradable wood) and a more favourable microclimate under woody canopies (Sect. 4.4.2.3).

Due to the complexity of factors influencing decomposition rates under woody canopies, the overall effect is difficult to predict (Table 4.25). Decomposition rates of woody litter assessed under standard (open field) conditions are biased since environmental conditions and fauna activities under woody canopies influence the decomposition rate, probably in a favourable way.

The available evidence on the relationship between litter decomposition and SOM contents in semi-arid regions is scarce (Table 4.26). In comparison to other trees, Drechsel et al. (1991) show that SOM contents are lowest under *Acacia auriculiformis*, characterised by highest tannin concentrations of litter and highest litter inputs. Here, litter accumulates on the surface and is little incorporated in the soil. Highest SOM contents occur under *Azadirachta indica*, of which the litter has high C/N and C/P ratios. As the study was carried out only 5 years after tree planting, SOM contents in an equilibrium situation may be different.

Table 4.26. Data from semi-arid regions on factors determining litter decomposition and corresponding C concentrations in the topsoil

Country + source	Woody species	Litter quality			Litter production	Soil fauna	g C kg^{-1} in 0–20 cm
		C/N ratio	C/P ratio	tannin (g kg^{-1})			
Togo (Drechsel et al. 1991)	*Azadirachta indical*[a]	40	1036	1.1			19.8
	Albizia lebbek[b]	20	412	1.2			14.2
	Cassia siamea[b]	21	321	5.6			13.7
	Acacia auriculiformis[b]	38	1126	16.7	Highest		12.7
Nigeria (Nwoboshi 1972)	*Tectona grandis*[b]	15		4.9		Absent[c]	12.8
	Cassia siamea[b]	12		5.6		Common[c]	17.1
India (Gill et al. 1987)	*Acacia nilotica*[a]	×			~4 t yr^{-1}		7.5
	Eucalyptus tereticornis[a]	2×			~1 t yr^{-1}		5.5
	Open areas						3.0

[a] Mixed litter.
[b] Leaves only.
[c] Determined on the basis of presence of channels and droppings in the soil.

Also, the influence of decomposition of large wood is not included in this short period. Nwoboshi (1972) investigated tree plantations of 30 years old, but analysed leaf litter inputs only. SOM contents are lowest under *Tectona grandis* (a deciduous species). *Tectona* leaf litter has a higher C/N ratio and shows less soil fauna activities in comparison to *Cassia siamea* (an evergreen species). Leaf litter of *Tectona grandis* accumulates on the soil surface, forming a litter layer of 10 cm, compared to only 1 cm for *Cassia* litter. Gill et al. (1987) investigated 4-year-old plantations of *Acacia nilotica* and *Eucalyptus tereticornis*. *Acacia* gives highest litter production but with lowest C/N ratio and high tannin contents (Sect. 3.6.2). These factors may explain why under *Acacia* SOM contents are highest.

Comparing the results of these studies it appears difficult to conclude which characteristics of woody species stimulate SOM build up. The short term effect of high tannin concentrations may be accumulation on the surface soil (Table 4.26: *Acacia auriculiformis*). Other factors that lead to litter accumulation on the soil are litter deposition concentrated in the dry season, large leaves, and allelopathic effects on the soil fauna. Mott et al. (1982) found, on the basis of litter bag studies, that leaf litter from woody plants in Australia takes 9–11 years to decompose in the absence of fire. This may be related to the large proportion of *Eucalyptus* leaves. Several *Eucalyptus* species are well known for accumulation of litter on the soil surface. Litter accumulation risks losses by fire, wind or herbivores. Poor decomposition due to high C/N or C/P ratio (Table 4.26: *Azadirachta indica*) seems to lead to high SOM contents (incorporation in the soil). The long term effect of poorly decomposable litter (Table 4.26: *Tectona grandis*) is slow turn-over of nutrients and few soil fauna activities and may eventually lead to low plant productivity and low SOM contents in an equilibrium situation. Available data are insufficient to test this hypothesis. Anderson and Swift (1983), on the basis of data from tropical rainforests, are equally inconclusive as to the possibilities of predicting SOM contents on the basis of litter inputs and decomposition rates.

4.6.7 A Quantified Example from the SSWA

In the following, the possibilities of maintaining equilibrium SOM contents by litter inputs from woody plants are worked out for woody plants in the northern Sudan zone (Table 4.27). Several assumptions will have to be made. In the SSWA, the decomposition rate of SOM in the natural vegetation varies between 0.02 (sandy loam) and 0.05 (sandy) (Pieri 1989). On agricultural land it is at least 0.06, due to high exposure of the soil (higher temperatures) and opening up of the soil (de Ridder and van Keulen 1990). Assuming in the natural vegetation an average decom-

Table 4.27. The effect of annual litter inputs from woody plants and the herbage layer on C supply to the top 0–20 cm soil layer, in the northern Sudan zone

Litter component	Annual input ($g\ m^{-2}$)	C in litter ($g\ kg^{-1}$)[c]	proportion of C left after humification (%)[d]	C supply to soil ($g\ m^{-2}$)	proportion (%)
Woody plant:					
leaves	360[a]	350	20	25	7
fruits	75[a]	350	30	8	2
twigs	150[a]	400	40	24	7
wood	585[a]	500	80	234	62
roots	235[b]	400	50	47	13
Herbage layer:					
aboveground	300[e]	300	15	14	4
belowground	200[e]	400	30	24	7

[a] On the basis of data from Table 3.10.
[b] Assuming aboveground production = belowground production and 40% of root mass located in top 0–20 cm of the soil.
[c] On the basis of data from Mott et al. (1982), Young (1989) and de Ridder and van Keulen (1990).
[d] On the basis of data from Janssen (1986) and Janssen et al. (1990).
[e] On the basis of Penning de Vries and Djitèye (1991).

position rate of 0.04 and a soil bulk density of $1.4\,g\,cm^{-3}$, annual C losses by oxidation from the top 0–20 cm of soil are $11.2\,g\,C\,m^{-2}$ for every $g\,C\,kg^{-1}$ soil in the equilibrium situation. C losses by erosion in the natural vegetation are estimated at $1\,g\,C\,m^{-2}$ (Roose 1981).

Most litter inputs from woody plants are according to Table 3.10. Inputs of large wood are estimated at $585\,g\,m^{-2}\,yr^{-1}$ on the basis of the average annual wood increment in an equilibrium situation. Although an equilibrium is doubtful for semi-arid regions, over a long period wood mortality should equal wood increment.

Other sources and assumptions used to calculate annual C supply by a woody plant and associated herbage under the canopy are indicated in Table 4.27. The humification rate is inversely related to the proportion of C left after humification. Total C inputs from the woody plant and the associated herbage layer are $376\,g\,m^{-2}$. These inputs would result, in an equilibrium situation, at an average SOM content in the topsoil corresponding to $34\,g\,C\,kg^{-1}$. This is much higher than most field measurements (Table 4.10). The following adjustments are important in explaining the difference, and are used in subsequent calculations.

1. Of total C return to the soil about 60% originates from inputs of large wood. However, large trees are often either cut or lopped, fallen wood is collected for use as firewood or is lost by fire (in the natural vegetation). It is assumed that 90% of the large wood does not become available for decomposition.
2. Of total C return to soil about 20% originates from herbs, leaves, fruits and small twigs, all of which are susceptible to loss factors such as grazing, fire and erosion. An average of 40% of the herbage biomass is lost, i.e. does not become available for decomposition (Breman and de Ridder 1991). It is assumed that from the litter layer the same proportion of leaves, twigs and fruits are annually lost.
3. Termites transport organic material over large distances, mainly towards their mounds. It is assumed that 25% of the litter available for decomposition is removed from under the canopy towards the open field.
4. Of total C supply to soil about 20% originates from roots. It is assumed that these inputs are not susceptible to losses.

Applying the above adjustments to aboveground litter inputs, total C supply to the soil would be $71 + 50 = 121\,g\,m^{-2}$ (underground and aboveground sources respectively). These inputs are sufficient to maintain SOM under woody canopies at about $10\,g\,C\,kg^{-1}$ soil, corresponding to some field measurements (Table 4.10). With the same litter inputs, equilibrium SOM contents of the topsoil on croplands would be lower since the decomposition rate (0.06) is higher and C losses by erosion are higher (Table 4.15).

Table 4.28. Organic inputs from crops or herbs and woody plants required to maintain SOM contents in the top 0–20 cm soil layer at 6 g C kg^{-1} soil, and the associated canopy cover required, assuming different proportions of total organic litter inputs available for decomposition (see text for explanation and other assumptions), per climatic zone in the SSWA

	Organic inputs ($g\ m^{-2}$)				Required woody cover (%) Proportion available for decomposition		
	Herbage (crop) layer[a]		Woody plants[b]				
	Above-ground	Below-ground[c]	Above-ground	Below-ground[c]	All litter inputs	10% large wood 60% browse + residues	Roots only
Northern Sahel	100	75	600	130	55	>100	>100
Southern Sahel	250	150	790	175	34	>100	>100
Northern Sudan	350	250	1170	235	17	60	>100
Southern Sudan	500	350	1490	300	9	35	>100

[a] Based on data from Breman and de Ridder (1991).
[b] Based on data from Table 3.10.
[c] In 0–20 cm soil layer.

On the basis of the foregoing can be estimated the woody cover required to maintain SOM content on croplands at a level of $6\,g\,C\,kg^{-1}$ soil (considered minimal for maintenance of soil structure with 20% clay for agricultural purposes: Pieri 1989). In the northern Sahel zone crop production is not possible, so the "crop" is pasture. Available litter from woody plants is assumed to be evenly spread over the entire area. Litter inputs from woody plants are based on production estimates per climatic zone (Table 3.10). It is assumed, very optimistically, that crop production is not influenced by woody plants and that woody plant production is independent of woody plant density. For croplands, where the decomposition rate is 0.06, three scenarios are worked out (Table 4.28):

1. no litter losses, so no crop residues removed, nor wood, nor browse material;
2. 90% of large wood and 40% of crop residues removed;
3. only belowground inputs available for decomposition.

The second scenario is most realistic, although crop residues are increasingly removed from cropland. The resulting canopy cover estimates (Table 4.28) show that with scenario (3), belowground organic inputs only cannot maintain SOM contents at $6\,g\,C\,kg^{-1}$ in any climatic zone. With scenario (2) woody plants cannot maintain SOM at minimum levels in the Sahel zones, whereas in the northern and southern Sudan zones woody cover should be 60 and 35% respectively. When no organic inputs are lost, the required woody cover varies from 55% in the northern Sahel zone to 9% in the southern Sudan zone. Note that in the case of high woody plant densities the estimates of woody plant production are increasingly unrealistic.

In the Sahel zones actual woody cover on croplands (Sect. 2.4.4) is even lower than estimates based on scenario (1); in the Sudan zones woody cover in the parklands is similar to estimates from scenario (1). This conclusion explains why SOM contents of croplands are generally lower than the required minimum level, even when some trees are maintained and shows that the role of woody plants in maintaining SOM contents in semi-arid regions should not be overemphasised.

4.6.8 Conclusions

SOM contents in the topsoil are higher under woody canopies than in the open field (4–10 and $2–5\,g\,C\,kg^{-1}$ respectively). Factors that explain the relatively high SOM contents under woody canopies (Table 4.25) are high litter inputs (both above and below ground), a high proportion of litter with poor quality, and litter inputs more evenly distributed throughout the year.

Relatively high litter inputs under woody canopies are due to two factors. Firstly, the woody plant in the natural vegetation occupies another niche than herbs and thus utilises resources for plant production that would otherwise have been lost. Secondly, the woody plant effectively concentrates water and nutrients for its proper use (Sects. 4.4 and 4.5), thus creating a highly productive unit intercepting light by a high leaf area index. The increased nutrient pool associated with the woody plant is partly stored in the soil as part of the woody plant-litter-soil cycle. Due to the relatively high root to shoot ratio of woody plants in semi-arid regions (Sect. 4.2), belowground inputs of organic material and release of nutrients are probably more important than above ground (e.g. Vitousek and Sanford 1986). Seasonal drying of the soil leads to annual mortality of fine roots, increasing their turn-over rate (Davies et al. 1989).

Litter inputs under woody plants are of variable quality. The combination of high quality (easily decomposable) herbs, high and moderate quality foliage and low quality wood is considered a major advantage for ecosystems (Coleman et al. 1989; Palm and Sanchez 1991). The slowly decomposable (woody) component contributes to maintenance of the proportion of the resistant SOM fraction. The optimal mix should consider the two largely conflicting functions of SOM: nutrient release and maintenance of chemical and physical soil conditions. Particularly in semi-arid regions the latter function is crucial. Here CEC is largely determined by SOM. For maintenance of soil structure SOM contents should be at least $3.0–21.0\,g\,C\,kg^{-1}$ (according to soil texture); for minimum CEC ($3–4\,mEq\,g^{-1}$) minimum SOM content should be about $8\,g\,C\,kg^{-1}$. SOM contents of most open field situations are lower (Table 4.10).

It is suggested that woody plants adapt to infertile soils by high lignin and tannin concentrations and high C/nutrient ratios in their tissues, leading to low litter decomposition rates, and by an evergreen status, leading to low litter inputs (Stuart Chapin 1980; Vitousek and Sanford 1986). Under such conditions litter- and nutrient turn-over rates are low. As a consequence, stable SOM components accumulate so that SOM content increases, which is a major advantage for an unfertile site, but plant production is low due to little nutrient release. A comparison between woody plant plantations of different species (Table 4.26) shows that the ultimate effect of variable litter production and -quality on SOM contents is unpredictable.

Above- and belowground biomass of a woody vegetation is higher than of a herbage layer only and, as a consequence, litter production and SOM contents in an equilibrium situation are also higher. On the basis of woody plant and crop production data, the woody cover required to maintain SOM content of a cropland can be estimated, as demonstrated for the SSWA (Table 4.28). The extent to which (woody) plants improve or maintain SOM largely depends on the extent to which litter losses are controlled (Sanchez et al. 1985). In case of average rates of litter losses, a canopy cover of 35–100% would be required (from south to north in the SSWA) to maintain SOM content at $6\,g\,C\,kg^{-1}$. However, competition by woody plants for water, nutrients and/or light will then limit the possibilities of growing a crop. The last aspect (light) is further investigated in the following section.

4.7 Influences of Woody Plants on Light Availability

4.7.1 Introduction

Light is a factor influencing primary production that is often overlooked, since in the semi-arid tropics light rarely seems a limiting factor. However, in agroforestry systems it can be a major factor limiting pasture or crop production.

The amount and spectral distribution of radiation reaching the soil surface depends primarily on "obstructions" that remove or modify the solar beam. As light is considered the photosynthetic active radiation (PAR), i.e. a proportion of incoming radiation. Both climatic and woody plant parameters influence light transmission by woody canopies.

Climatic factors determine light intensity and distribution over the day and the year, and the fraction of diffuse light. Underlying factors involved are the angle of the incoming solar beam, daylength, cloudiness and the amount of dust in the atmosphere, both influencing the amount of diffuse light reaching the soil surface. There are no significant differences between the extinction coefficients of diffuse and direct light. However, the incoming angle of direct incoming light depends on the position of the sun, diffuse light comes from all angles from the sky. Therefore, the overall interception by individual trees is less than by closed canopies; a parallel difference exists between trees-on-a-stem and shrubs.

Interception by a woody canopy of vertically incoming direct light depends on canopy cover mainly; interception of diffuse light depends on canopy volume mainly (Goudriaan 1977). Other factors influencing light transmission the size, density and orientation of the leaves and canopies being broadleaved or coniferous, clumped or dispersed, evergreen or drought-deciduous.

Light transmission by a woody plant follows an exponential relationship. The total amount of light being transmitted in a woody vegetation is given by:

$$TR = IR\,e^{-c(X+Y)},$$

where IR is the amount of incoming light (in Joules $m^{-2}d^{-1}$), c is the light extinction coefficient and X and Y are estimates of leaf and branch surface respectively (in m^2m^{-2}) (Mohren 1987; Jackson 1989). The intercepting surfaces of leaves and branches are estimated on the basis of leaf and branch weight per unit of area and specific leaf and branch area per kg of dry weight.

In the following, two extremes are compared: the situation of isolated woody plants, as occur in the parklands, and the situation of closed canopies, occurring localised or wide-spread in natural vegetations under favourable conditions. Light transmission by the natural woody vegetation in the SSWA is determined on the basis of average canopy cover and foliage production per climatic zone before the drought periods (Sect. 3.4.2). First are discussed general aspects on the relationship between light and plant production.

4.7.2 Light Intensity and Plant Production

4.7.2.1 Light Extinction by Woody Plants

Measurements of light extinction by woody plants in semi-arid regions are presented in Table 4.29 by the proportion of incoming radiation being transmitted to the soil surface. Vetaas (1992) estimates the average light transmission by woody plants in dry savannas at 40–55%; two out of four references in Table 4.29 present lower values.

Table 4.29. The proportion of incoming radiation transmitted by woody canopies at locations in semi-arid regions. Estimates are on a daily basis and for sunny days, unless indicated otherwise

Country + source	Location	Phenology of woody plant	Proportion transmitted (%)
Spain	Under *Acacia caven* trees	No leaves	47
(Ovalle and Avendaño 1988)	id.	Full leaves	35
Kenya (Belsky et al. 1989)	Under savanna trees	Full leaves	55
Rhodesia	Open canopy zone	Full leaves	37
(Kennard and Walker 1973)	Closed canopy zone	id.	25
Botswana	Grassland		85–95[a]
(Veenendaal 1991)	Savanna–edge of canopy	Full leaves	46–50[a]
	Near trunk of tree	id.	12–26[a]

[a] Highest values are for an overcast day.

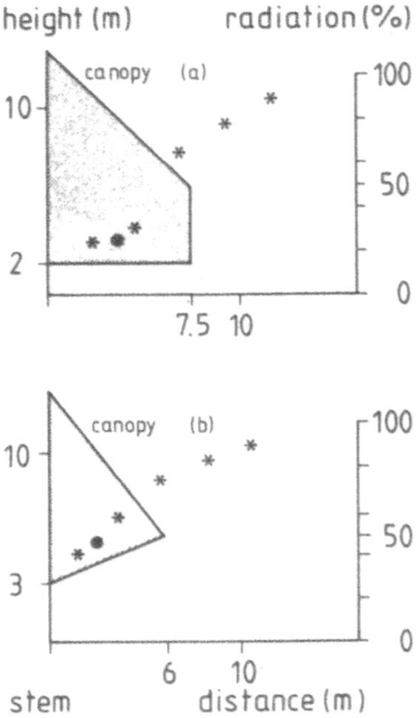

Fig. 4.20a–b. The relationship between canopy structure and tree height of **a** a *Parkia biglobosa* tree, **b** a *Vitellaria paradoxa* tree, both in Burkina Faso, and radiation under the canopy during the rainy season, expressed as the percentage of radiation in the open field, in relation to the distance to the tree trunk. * measured radiation values; ● calculated radiation under the canopy. (After Kessler 1992)

Light extinction by a woody canopy increases with increasing maturity and crown development. This is illustrated by Jackson (1989) for a coconut plantation: 50% transmission for a 3-year old coconut plantation, 70% transmission when 5 years old, and 90% transmission at maturity.

The influence of canopy shape is illustrated by a study on light transmission by two parkland trees (Fig. 4.20). When the lower branches are low and spreading, as is the case for most *Parkia biglobosa* trees, near the stem only 20% of incoming radiation is transmitted; when the canopy has a more erect shape, as is common for *Vitellaria paradoxa* trees, near the stem up to 50% of incoming radiation is transmitted. Woody plants cast long shadows of the early morning and late afternoon sunlight in easterly and westerly directions, depending on its height, and in all directions from diffuse light. Only at a distance of about twice the distance stem-canopy edge transmission nears 100%. When woody plants are randomly distributed, transmitted radiation is more evenly distributed than in case of a row orientation.

Shading under a woody canopy increases with woody plant density, because of the interception of lateral incoming light by other canopies (e.g. Eastham and Rose 1988).

4.7.2.2 Light and Plant Production

In the tropics, the relationship between production and light intensity firstly depends on species being of the C_3 or C_4 photosynthetic pathway. Most tropical grasses are C_4, typical C_4 crops are maize and sorghum, woody plants are all C_3 species. C_3 species use solar energy less efficiently at high light intensity (Fig. 4.21). Of the incoming radiation 45% can be considered as photosynthetic active radiation, the visible light or simply "light" (400–700 nm; Goudriaan 1977), as presented in Fig. 4.21. Individual leaves reach light saturation at about 10 and 20 kJ m^{-2} min^{-1} for C_3 and C_4 species respectively. Plants as a whole do not reach light saturation at the natural levels of light intensity.

Plants may respond to a lower light intensity by increasing leaf area ratio, decreasing leaf-root weight ratio, and by decreasing net assimilation rate and growth rate (Corré 1983). A specific response of plants to the light quality under shade conditions, characterised by a low red/far-red ratio, is one of stem elongation at the expense of leaf production (Corré 1984). Such a response particularly occurs among so-called shade

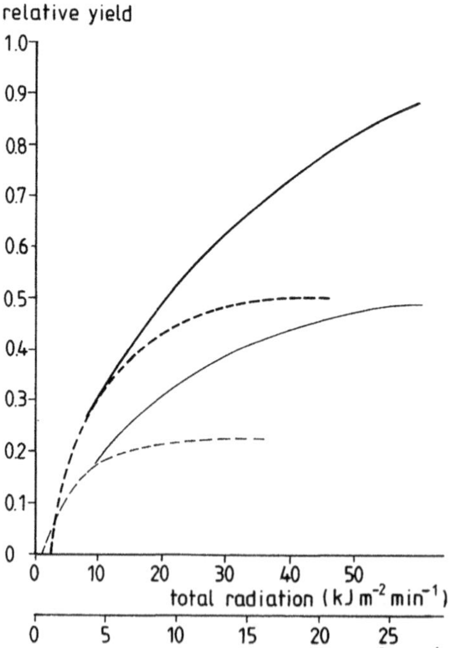

Fig. 4.21. Characteristic net relative yield (CO_2 assimilation) functions for individual leaves and for plants as a whole, of C_3 or C_4 plant species, in relation to total radiation and to photosynthetic active radiation. - - - leaves of C_3 plants; – C_3 crops; - - - leaves of C_4 plants; — C_4 crops. (After Alberda et al. 1977; van Keulen and Wolf 1986)

non-tolerant plants. Under low light intensities, e.g. in the shade of a woody plant, moisture and nutrient availabilities are also different (Sect. 4.8). The absence of a favourable response to high light intensity by some so-called shade plants could be related to factors such as adverse water or nutrient relations. Shading can also influence nutrient availability for plant production via impacts on soil dynamics. Factors governing an increased availability of soil N for uptake by plants under shade (Wilson et al. 1986) may be improved conditions for mineralisation and for soil fauna activities (Table 4.25). N_2 fixation is not enhanced, nor is N translocated within the plant.

4.7.3 Distribution and Quality of Light Under Woody Canopies

Characteristics of daily incoming radiation during the rainy season in the SSWA (Table 4.30) are based on agro-climatological data from the FAO (1984). Figure 4.22 shows global radiation and light availability during the rainy season in relation to average rainfall. The maximum duration of bright sunshine per climatic zone (Table 4.30) also varies as a function of its average latitude (van Keulen and Wolf 1986). Incoming radiation is related to the sinus of the position of the sun (Goudriaan 1977). Direct radiation casts shadows according to the position of the sun and the height of the woody plant. Almost 70% of daily radiation occurs in the period between 9.00 and 15.00 h. At the latitudes of the SSWA, sunshine in this period is almost vertical and lateral incoming light reaching the soil surface under a woody canopy can be neglected. The maximum intensity of daily radiation occurs between 11.00 and 13.00 h (Table 4.30). Comparison with Fig. 4.21 shows that maximum intensity of daily radiation is not even sufficient for maximum photosynthesis of C_4 plants. Every reduction of radiation by woody canopies therefore reduces photosynthesis of both C_4 and C_3 plants in the vicinity.

The fraction of diffuse light (Table 4.30) is estimated on the basis of global radiation in relation to the potential radiation in the rainy season (i.e. with a clear sky). Part of the simulation model for potential and water-limited production (van Laar et al. 1992) was used for this purpose. The fraction of diffuse light during the rainy season increases from about 50% in the northern Sahel zone to more than 60% in the southern Sudan zone (Fig. 4.22).

Apart from diffuse light and the non-intercepted proportion of vertical incoming direct light, lateral incoming direct light reaches the soil surface under a woody canopy. Lateral incoming direct light in the period before 9.00 h and after 15.00 h is 30% of daily radiation. Of lateral incoming direct light during this period, the fraction diffuse light is twice more than of incoming light in the period between 9.00 h and 15.00 h, because lateral incoming direct light has a longer path through the atmosphere.

Table 4.30. Incoming radiation, maximum hours of sunshine, maximum intensity of incoming radiation, and fraction of diffuse light, during the rainy season, per climatic zone. (Based on FAO 1984; van Keulen and Wolf 1986; Penning de Vries and Djitèye 1991; Goudriaan 1977)

Climatic zone	Incoming radiation (MJ m^{-2} d^{-1})		Max. sunshine (h d^{-1})	Max. radiation intensity (kJ m^{-2} min^{-1})	Fraction of diffuse light (%)
	Total	Photosynthetic active			
Northern Sahel	21.3	9.6	12.8	44	52
Southern Sahel	20.2	9.1	12.7	42	55
Northern Sudan	19.1	8.6	12.6	39	58
Southern Sudan	18.2	8.2	12.5	38	62

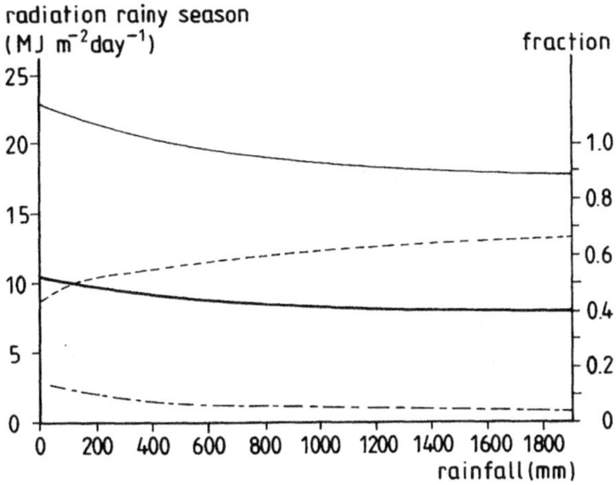

Fig. 4.22. Global radiation during the rainy season (—) and photosynthetic active radiation PAR (**—**) in relation to average annual rainfall in the SSWA; diffuse light (- - - - -) and direct laterla incoming light (– - –), both expressed as a fraction of PAR

Assuming that the presence of clouds and dust in the atmosphere does not vary throughout the day, the fraction of lateral incoming direct light which is not diffused decreases from 10% in the northern Sahel zone to 5% in the southern Sudan zone (Fig. 4.22).

4.7.4 Light Availability Under a Closed Canopy

The formula presented in Section 4.7.1 is used to determine light transmission by a closed woody canopy. In order to estimate the global availability of light for plant growth under a closed woody vegetation, one does not have to differentiate for diffuse and lateral incoming direct light. Both forms of light are intercepted by the closed canopy in an almost similar way.

A distinction has to be made between the extinction coefficient of global radiation and (visible, photosynthetic active) light. For global radiation the extinction coefficient by woody plants varies between 0.5 and 0.7 (Mohren 1987), or between 0.3 and 0.9 (Doyle et al. 1986). Variation depends on leaf geometry mainly, but may be averaged for a mixture of woody plants. An average extinction coefficient of 0.6 is used for global radiation, but for visible light it is 0.7, and for the non-active (infra-red) fraction it is 0.4 (Goudriaan 1977).

To determine branch weight of a woody vegetation in relation to foliage weight, and surface/weight ratios for foliage and branches, data are used from the northern Sahel zone in Senegal (Bille 1977) and from the Sahelian-Sudanian transition zone in Mali (Penning de Vries, unpubl. data). Data from the south African savanna (Rutherford 1979) are used for verification purposes. As foliage to branch weight conversion factors are used a factor 4.0 for the Sahel zones, and 5.0 for the Sudan zones (Sect. 3.3). Surface/weight ratios are determined at $0.02\,ha\,t^{-1}$ for branches, at $0.83\,ha\,t^{-1}$ for foliage in the Sahel zones and at $0.71\,ha\,t^{-1}$ for foliage in the Sudan zones. Leaf and branch surface is determined on the basis of foliage production per climatic zone. It is assumed that all woody plants are in full foliage during the rainy season, the only important exception being *Faidherbia albida* (Table 3.11). The number of leaf layers per climatic zone is extrapolated southwards: 5.5 leaf layers in the northern Guinea zone and 6 leaf layers in the southern Guinea zone.

Transmitted global radiation and visible light under a closed woody canopy are presented in Fig. 4.23. Comparison with Fig. 4.22 shows that transmitted and incoming radiation have different quality, as a result of the difference in extinction of active and non-photosynthetic active light. Transmitted radiation intercepted by green leaves is relatively poor in blue and red light and rich in green and far-red (near-infrared) wavelengths (Reifsnyder 1989). The photosynthetic active fraction of the transmitted radiation depends on the density (total surface) of the foliage and branches, which increases from the northern Sahel to the southern Sudan zone. As a result, the photosynthetic active fraction of the transmitted radiation is less than the normal value of 45%, from 33% in the northern Sahel to 18% in the southern Sudan zone, stabilising at a low 12% in the Guinea zones (Fig. 4.23). The average light intensity under a closed woody canopy is determined by division of total photo-

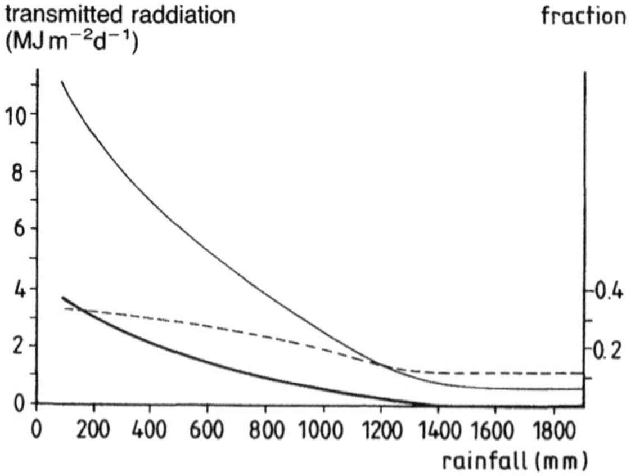

Fig. 4.23. Transmitted global radiation (–) and photosynthetic active radiation PAR (**—**) under a closed woody canopy, during the rainy season, in relation to average annual rainfall in the SSWA; PAR expressed as a fraction of global radiation (- - -)

synthetic active radiation by day length (Table 4.30). For example, in the northern Sahel zone it is $2800/12.8 \times 60 = 3.6\,\mathrm{kJ\,m^{-2}\,min^{-1}}$, in the southern Sudan zone it is only $0.5\,\mathrm{kJ\,m^{-2}\,min^{-1}}$, in the Guinea zone it is almost zero. The consequences for herbaceous production under the woody canopies are determined by applying these data to Fig. 4.21.

4.7.5 Light Availability Under Isolated Woody Plants

Under isolated woody plants diffuse light, lateral incoming direct light and transmitted direct light have to be considered as separate sources of light for plant growth under the woody canopy (Knevel 1993). In the following, diffuse and lateral light are assumed not to be influenced by other trees in the vicinity.

4.7.5.1 Diffuse Light

If the crown has the shape of a compact circular disk, diffuse light, coming from all directions from the sky, is intercepted proportionally to the disk surface (πr^2), similar as vertical incoming direct light. Interception by a spherical shaped crown is proportional to $2\pi r^2$, half of the surface area of the sphere. This is twice as high as the interception of perpendicular direct light, which is proportional to πr^2, the projection of the sphere. In other words, a 100% closed cover is required to intercept all direct light, but a 50% cover is sufficient to intercept all diffuse light. For a cylinder-shaped canopy, interception is proportional to $\pi r^2 + 2\pi r X$ (X is the length of the cylinder, and πr^2 the horizontal section). Thus, even less cover of trees with cylindrical shaped crows is required to intercept all diffuse light.

Because woody canopies are in general not completely closed, interception does not mean that all light is extinguished. The degree of extinction is as described in the previous section, depending on the amount of leaves and branches per unit intercepting surface area. For example, in the case of a spherical crown of an isolated tree, in view of the intercepting surface area being twice as high as the vertical crown projection, the intercepted biomass of leaves and branches should also be twice as high as that of the average tree of the natural vegetation in order to intercept the same amount of diffuse light. This assumption is not unrealistic since isolated woody plants are well developed and have about 1.5 times more foliage biomass than the average woody plant occurring in the land-use type "old fallowlands, rangelands and forests" (Table 3.10). It is assumed that biomass of isolated woody plants increases proportionally with the surface intercepting diffuse light. Therefore, the transmission of light by the crown of isolated trees can be determined using Fig. 4.23 and Table 3.9. For example, a tree in the southern Sahel zone with two leaf layers, and an intercepting surface for diffuse light two times the vertical crown projection, transmits diffuse light like a closed canopy with four leaf layers in the southern Sudan zone (Fig. 4.23). As the upper limit to increasing woody plant biomass is considered a maximum number of six leaf layers. In other words, light transmission in the northern Sudan zone by a tree with three leaf layers is equal to transmission by a tree with four leaf layers, if the intercepting surfaces for diffuse light are twice the vertical crown projection. In that case, the lowest light transmission values of Fig. 4.23 are applicable.

Since diffuse light comes from all directions, light interception casts shadow in a relatively large area around the trunk. Only a fraction of the shadow occurs under the woody canopy. This fraction depends on the ratio between the height of the trunk (i.e. the height of the first main branches) and the crown radius, and on the shape of the crown (Knevel 1993). The average canopy radius for woody plants in the Sahel zones is about 1.5 m; in the Sudan zones it is 3–4 m (Table 2.3). The assumption of 1.0 as the ratio trunk height/crown radius is realistic for both zones, considering that the proportion of foloiage available below 2 m decreases from 25% in the Sahel zones to 15% in the Sudan zones (Sect. 5.4.5.1).

The relationship between the fraction of shadow caused by interception of diffuse light coinciding with the vertical projection of the crown, and the ratio trunk height/crown radius, is calculated for woody plants with circular disk and spherical shaped crowns (Fig. 4.24). When the ratio trunk height/crown radius is 1.0, this fraction is 0.3 for woody plants with spherical shaped crowns, and 0.4 for circular disk-shaped

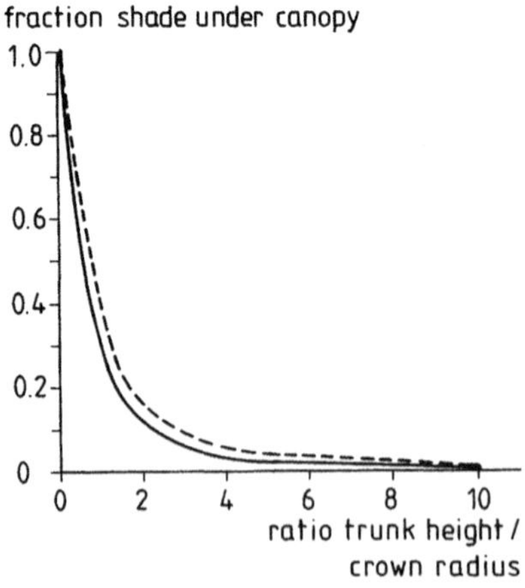

Fig. 4.24. The fraction of shade under a woody canopy caused by interception of diffuse light coinciding with the vertical projection of the crown, in relation to the ratio of trunk height to crown radius, for two different crown shapes. – round-shaped crown, - - - disk-shaped crown

crowns. The fraction increases rapidly if the trunk height becomes less than the crown radius (Fig. 4.24). For rather disk-shaped crowns of *Acacia raddiana*, the fraction of the shadow caused by interception of diffuse light coinciding with the vertical crown projection is almost 0.4; for *Parkia biglobosa* trees, with crown radius up to three times the trunk height, this fraction is estimated at 0.7; for *Bombax costatum* trees, with trunk height exceeding the crown diameter, this fraction is less than 0.1.

On the basis of the above analysis and assumptions, the contribution of diffuse light (DR) to light in the area of the vertical projection of the crown can be calculated as follows:

$$DR = aIR - ab\,(IR - TR),$$

where

IR = incoming radiation (Fig. 4.22);
a = fraction of diffuse radiation (Fig. 4.22);
b = the fraction of shadow caused by intercepted diffuse light coinciding with the vertical crown projection (Fig. 4.24);
TR = transmission of IR by the crown of isolated woody plants (formula Sect. 4.7.1 or Fig. 4.23), using as the basis of the light absorbing biomass the number of leaf layers (Table 3.9), multiplied by the ratio between the intercepting surface of the crown for diffuse light and the surface of the vertical crown projection, up to a maximum of six leaf layers.

The results for a standard tree as defined above are presented in Table 4.31, for both global radiation and photosynthetic active radiation, per climatic zone.

4.7.5.2 Direct Light

Lateral incoming direct light, as the fraction of incoming light, is presented in Fig. 4.22. The average absolute values per climatic zone are

Table 4.31. Total and photosynthetic active radiation (PAR), expressed as $MJ\,m^{-2}\,d^{-1}$, originating from diffuse and from lateral and vertical incoming direct radiation, under isolated woody plants, during the rainy season, per climatic zone

Climatic zone	Diffuse		Lateral		Vertical		Total	
	All	PAR	All	PAR	All	PAR	All	PAR
Northern Sahel	8.5	3.7	1.9	0.9	1.5	0.4	11.9	5.0
Southern Sahel	8.1	3.5	1.6	0.7	0.8	0.1	10.5	4.3
Northern Sudan	8.1	3.3	1.2	0.6	0.3	0.0	9.6	3.9
Southern Sudan	8.0	3.2	1.0	0.5	0.3	0.0	9.3	3.7

given in Table 4.31. Vertical incoming direct light is total light (Fig. 4.22) minus lateral incoming direct light and minus diffuse light. In order to estimate its transmission to the area under the crown of isolated woody plants, the higher biomass of leaves and branches in comparison to canopies of woody species in the natural vegetation has to be taken into account. The approach used to determine the transmission of light (Sect. 4.7.1) is proposed if no field data are available.

4.7.5.3 Total Light

Total light availability under the crown of an isolated tree is calculated by addition of the fractions transmitted diffuse and vertical incoming direct light, and lateral incoming direct light (Table 4.31). Available global radiation under a standard tree as a percentage of total incoming radiation decreases from about 55% in the nortern Sahel to 50% in the southern Sudan zone. Of transmitted radiation a fraction of 0.40, instead of 0.45 of incoming radiation, is photosynthetic active. These results are comparable to field measurements under isolated trees (Table 4.29). Shrubs should be compared with a closed canopy cover, i.e. values presented in Fig. 4.23. Herbaceous yield suppression is highest under well developed shrubs, which is consistent with field observations in semi-arid regions (Sect. 4.2.2.2).

Application of this methodology to *Parkia biglobosa* and *Vitellaria paradoxa* trees with representative dimensions of Kessler (1992) provides estimates comparable to field measurements. The average light availability measured under the *Parkia* tree of Fig. 4.20 is 31% of the incoming light, the calculated value is 28%. These values are respectively 45 and 39% for the *Vitellaria* tree in the same figure. For other trees, reliable estimates of light availability under the crown can be obtained if some dimensions and foliage characteristics are known.

4.7.5.4 Light Availability in the Natural Vegetation

Light availability under crowns of isolated woody plants may be relatively high, but interception of diffuse light and lateral incoming direct light causes shading beyond the canopy zone. The total interception has to be taken into account to determine light availability for the herbaceous layer of a vegetation as a whole, or for crops in a field with trees. Whether woody plants are homogeneously distributed, as in a parkland, or more or less clustered, as in most natural vegetations, does not significantly influence overall light extinction. It is the total canopy cover that counts. The light availability for herbs can be calculated as the weighed average of light availability in the open field and that under a woody canopy. Canopy cover values and the number of leaf layers from the period before the 1970s drought are used (Fig. 2.4). The resulting light availability for the herbaceous layer in the land-use type "old fallowlands, rangelands and forests" is compared with the light availability in the open field without any tree, under isolated trees, and under a closed canopy (Table 4.23). The consequences for herbaceous production can be derived from Fig. 4.26.

For crops, light availability can be determined in the same way. On agricultural fields canopy cover varies from 0–10% in the Sahel zones and from 5–20% in the Sudan zones (Sect. 2.4.4). For a 10% canopy cover with six leaf layers in the Sudan zones, average light availability for crops, determined as a weighed average of 10% closed canopies and 90% open field, is about 8 MJ $m^{-2} d^{-1}$. However, light availability varies between 8.6 MJ $m^{-2} d^{-1}$ in the open field far from any tree, to 3.8 MJ $m^{-2} d^{-1}$ under the canopy of an isolated tree, to less than 3.8 MJ $m^{-2} d^{-1}$ under the canopy of a tree with other trees in the vicinity. Light interception by *Faidherbia albida* trees, without foliage in the rainy season, is by branches only. This characteristic is accounted for by modifying input data to the formula in Section 4.7.1. The result shows that in a *Faidherbia albida* parkland, with a 40% canopy cover (Sect. 2.2.1.2), average light availability for crops is 7 MJ $m^{-2} d^{-1}$. With leaves, during the dry season, it is 5 MJ $m^{-2} d^{-1}$.

4.7.6 Conclusions

A methodology was developed to estimate light availability under woody plants. Three components of light should be distinguished: diffuse light, lateral incoming direct light and vertical incoming direct light. Woody canopies have variable influence on light transmission of these three light components. The results appear comparable with field measurements.

Comparison of light saturation values for C_3 and C_4 plants (Fig. 4.21) with average daily incoming photosynthetic active radiation in the open

Table 4.32. Average (transmitted) photosynthetic active radiation (MJ $m^{-2} d^{-1}$) available for herbaceous production in the open field, under closed canopies, under isolated woody plants, and in the natural vegetation with canopy cover as before the 1970s drought, during the rainy season, per climatic zone

Climatic zone	Open field, no trees	Closed canopies	Isolated trees, canopy zone	Natural vegetation
Northern Sahel	9.6	2.8	5.0	9.2
Southern Sahel	9.1	1.8	4.3	8.1
Northern Sudan	8.6	1.2	3.9	6.7
Southern Sudan	8.2	0.4	3.8	5.7

field, under closed canopies, under isolated woody plants and in the natural vegetation (Table 4.32), shows that even in the open field C_4 plants do not reach maximum productivity. Under woody canopies productivity of C_4 plants is further depressed. C_3 plants do not reach maximum productivity under closed canopies or under isolated woody plants in any climatic zone, nor in the natural vegetation in the Sudan zones. There is, however, considerable variation around these average values, depending mainly on the ratio between trunk height and crown radius, plant phenology and leaf density. The methodology allows estimates of light availability for herbs under the woody plant canopy with known canopy dimensions, leaf density and phenology.

4.8 Potential Influences of Woody Plants on Plant Production

4.8.1 Introduction

Two aspects are dealt with in the analysis of the influence of woody plants on plant production: the influence on total plant production, and the influence on herbaceous production only. To do so for a particular region, the plant production factors of that region are analysed, as well as the actual and potential influence of woody plants on those factors, in general, and on herbaceous production in particular.

Section 4.2 presented the apparently contradictory evidence concerning the interaction between woody and herbaceous species: both suppression and stimulation of herbs and crops under woody canopies are observed. In Sections 4.3 to 4.6 it was discussed how woody plants may improve the availability of water, nutrients and SOM. The woody plants will firstly benefit from these improvements, but, in principle, herbaceous plants can also benefit, at least locally.

Analysing plant production in the SSWA, Penning de Vries and Djitèye (1991) identified N, P and water as the main determining factors. Their analysis mainly deals with herbaceous production in the absence of a strong interaction with woody species. Here, light availability is not a limiting factor for herbaceous production. However, light can easily become limiting for herbaceous plants in the vicinity of woody species (Sect. 4.7). Water is the main limiting factor for primary production in the northern Sahel zone, whereas in the southern Sahel and the Sudan zones nutrients are the main limiting factor. The importance of nutrients as a limiting factor is related to the extremely low fertility of the majority of West African soils. For the average low fertility status of the soils, the transition from water to nutrients as the main limiting factor corresponds to the availability for plant production of about 250 mm yr^{-1} of infiltrated water. For water availability less than 250 mm yr^{-1}, nutrients are less limiting than water, but do influence plant production. Where water infiltration exceeds this value, plant production can be improved most by adding nutrients. Where infiltrated water is 250 mm yr^{-1} or more, the average N availability for the aboveground production of the herbaceous layer varies between 10 and 25 kg ha^{-1}; that of phosphorus varies between 1.0 and 2.5 kg ha^{-1}.

Important variation of the "average" situation per climatic zone is caused by local variation in water and nutrient availability. Variation in water availability is mainly due to redistribution of rainwater by run-off and run-on patterns. Nutrient limited situations can occur at run-on sites in the northern Sahel zone, whereas water-limited situations can even occur at run-off sites in the southern Sudan zone. Averaged for the SSWA as a whole, annual rainfall is 300 mm in order to obtain 250 mm infiltrated water. Where soil fertility is higher than "average" (e.g. pockets of volcanic outcrops), the transition from water to nutrient limiting growth occurs at more than 250 mm yr^{-1} of infiltrated water. The inverse is true for very poor or exhausted soils.

In the SSWA, 16 soil-climate combinations can be made of the 4 climate zones and 4 geomorphological units (Sects. 1.2 and 1.3). The consequences of the characteristic limiting factors of these soil-climate combinations on the influence of woody plants on plant production, are determined by the following approach:

1. Average net production is calculated for a herbage layer only and for a mixed vegetation (with woody plants), using theories and models on primary production developed by Penning de Vries and Djitèye (1991) and Breman and de Ridder (1991). If possible, the proportions of net production by herbs and by woody plants in the mixed vegetation are specified.
2. Consecutively are calculated:
   - net production limited by the natural light availability only;
   - net production limited by water availability only;
   - net production limited by the availability of water and nutrients.

By comparing production levels limited by light, water or nutrients, the relative importance of each factor is determined.

3. Calculations concerning the mixed vegetation are based on canopy cover estimates from the pre-drought period in the land-use type "old fallowlands, rangelands and forests" (Fig. 2.4). This average cover, being the highest in our era, can be considered as the potential woody cover (Sect. 2.2.1.1). An equilibrium situation of canopy cover and standing biomass is assumed. By using this approach the maximum effect of natural densities of woody plants can be determined.

4. The results are used to identify those combinations of ecological conditions and woody plant properties (-species) that are beneficial for certain land-use system objectives (Chap. 5). The focus is on possibilities to optimise the relationship between herbs and woody plants.

4.8.2 Net Production Limited by Light

Under conditions of sufficient water and nutrient availability, the relationship between plant production and radiation depends on the plant being of the C_3 or C_4 photosynthetic type, and whether active regulation of stomata takes place. Variations in incoming radiation, cloudiness and temperature cause important variations in photosynthetic activity of plants throughout the year. On the basis of average daily radiation values net plant production (above- and belowground) in the rainy season is calculated for the West African north-south transect with the SUCROS simulation model (Spitters et al. 1989).[1] The resulting potential production of C_3 and C_4 plants is presented in Fig. 4.25. Light availability being highest in the north, the potential C_4 production decreases from 390 kg $ha^{-1} d^{-1}$ of dry matter in the northern Sahel to 370 kg $ha^{-1} d^{-1}$ in the southern Sudan zone. The C_3 production is about 90 kg $ha^{-1} d^{-1}$ lower. The production of woody species limited by light could be represented by the C_3 curve. Net annual production with full radiation available (in the open field) would accordingly be about 135–140 t ha^{-1} of dry matter for C_4 species and 105–110 t ha^{-1} for C_3 species.

The estimated light transmission by woody plants (Fig. 4.23 and Table 4.32) allows the analysis of their impact on herbaceous production, such as crops on arable fields. Simulation with the SUCROS model shows net annual production in relation to daily available PAR during the rainy season on the north-south transect (Fig. 4.26). The PAR values ranging from 6 to 11 MJ $m^{-2} d^{-1}$ only (= 15 to 30 MJ m^{-2} global radiation), are extrapolated to 0 MJ on the basis of values from Fig. 4.21. Data from Fig. 4.26 in combination with data from Figs. 4.22 and 4.23 allow estimates of potential herbaceous production on the transect (a) in the open field, (b) in the mixed vegetation with woody cover as before the drought periods, (c) under individual standard woody canopies (with spherical crowns with a radius equal to the length of the stem); and (d) under a closed canopy (Fig. 4.27). In cases (a) and (b) the relative importance of C_3 and C_4 species in the vegetation has been taken into account, in cases (c) and (d) only C_3 species are concerned.

---

[1] Without the availability of SUCROS and a computer, the potential production can be calculated using 4.3 and 3.3 g $m^{-2} d^{-1}$ per MJ of PAR.

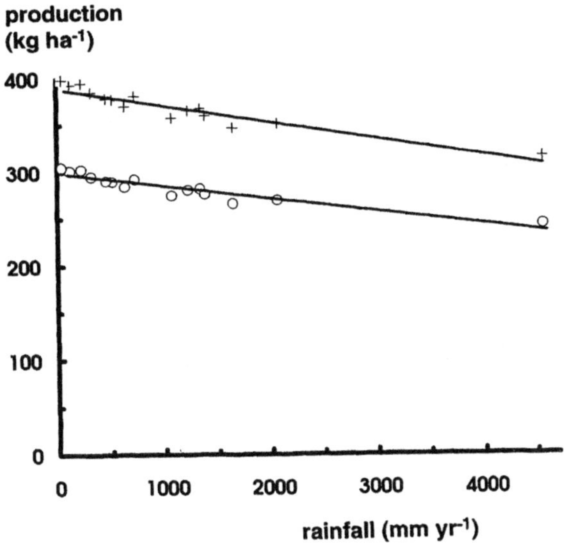

Fig. 4.25. Simulated daily net dry matter production limited by light in relation to average annual rainfall in West Africa, for C_3 and C_4 crops. ○ C_3 crops; + C_4 crops

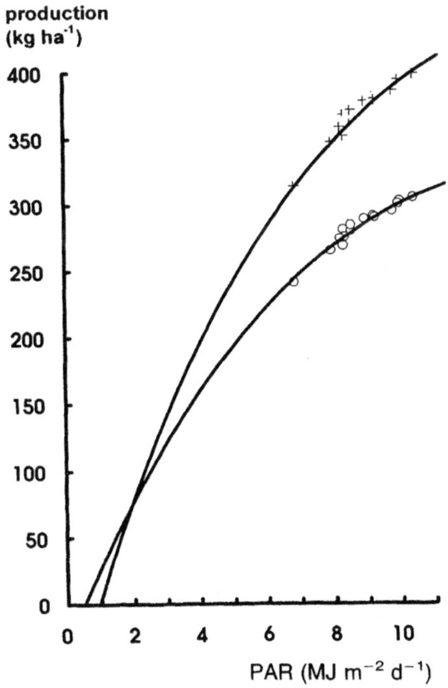

Fig. 4.26. Simulated daily net dry matter production limited by light in relation to photosynthetic active radiation, for C_3 and C_4 crops. ○ C_3 crops; + C_4 crops

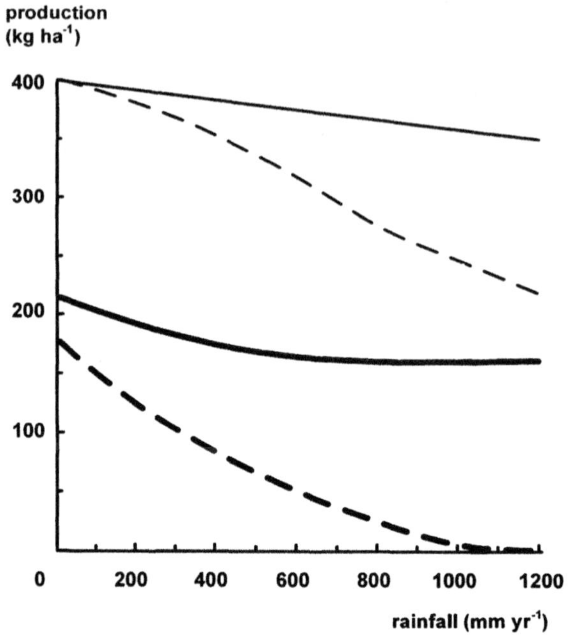

Fig. 4.27. Simulated daily net dry matter production during the rainy season of herbaceous plants in the SSWA, in relation to annual average rainfall, for four different situations. – open field without woody plants; ---- vegetation with woody canopy cover as found before the drought periods; — under individual woody canopies; --- under closed woody canopies

Averaging radiation available for the herbage layer in a mixed vegetation (weighed average of curves a and d = curve b) can mask local variability, for example the reduction of herbage production where dense woody plants do not allow lateral light penetration (curve d). Such situations of low light availability for herbs are rare in the Sahel zones but become the general situation south of the southern Sudan zone, as the woody layer gradually closes with increasing rainfall (Sect. 2.4). In the Sahel zones situations of low light availability for herbs are encountered only on loamy and shallow soils where run-on occurs (Sect. 2.2.1.1). Variation in the herbage production of curve (d) is associated with the number of leaf layers of the woody canopy (Table 3.9). For instance, the average herbage production under closed canopies in the southern Sudan zone (3 kg ha^{-1} d^{-1}) is found in clay depressions (La, M) in the southern Sahel zone, whereas herbage production on shallow soils (Sq) is 30 kg ha^{-1} d^{-1} in the southern Sudan zone. Curve (c) is applicable to isolated trees on rangelands and croplands. However, production will be higher than presented (Fig. 4.26) if the crop is a C_4 species. In case of woody plants with trunk height much less than the crown radius (i.e.

wider than the assumed standard), the production of associated herbs is better described by curve (d).

4.8.3 Net Production Limited by Water

Net production is limited by water where the natural soil fertility is high, or has been substantially improved, and at locations with low rainfall or strong run-off (Sect. 4.8.1). Net production limited by water is calculated on the basis of water availability and water-use efficiency of plants.

4.8.3.1 Water Availability

Water availability in relation to soil depth is determined as follows (Breman and de Ridder 1991):

1. Average annual rainfall is corrected for evaporation losses of water intercepted by the herbage and woody layer; interception is estimated at 1 and 2 mm per rainfall event respectively (Sect. 4.4.2.1). Knowing the number of rainfall events per climatic zone and the cover of woody species (before the drought periods), the amount of rainfall reaching the soil surface can be calculated as rainfall minus evaporated interception losses. In a year with average rainfall, interception losses are 20 mm yr^{-1} in the northern Sahel, 45 mm yr^{-1} in the southern Sahel, 75 mm yr^{-1} in the northern Sudan and 115 mm yr^{-1} in the southern Sudan zone.
2. The presence of woody plants being largely associated with water availability beyond 2 m soil depth (Sect. 2.5), redistribution of non-intercepted rainfall by run-off is important. Redistribution depends on soil texture and slope. Run-off values are:
   - geomorphological unit Os+Ps 15%;
   - Pl 45% in the northern Sahel, 50% elsewhere;
   - Sq 15%, 40% and 50% for the northern and southern Sahel and the Sudan zones respectively;
   - La −25% (i.e. infiltrated water 125% of gross rainfall).

The proportions of land showing run-off, showing run-on (where run-off water concentrates), and without water redistribution are as follows:

- Os+Ps 35% run-off area, 15% run-on, 50% without significant redistribution (i.e. with homogeneous infiltration);
- Pl 75% run-off, 25% run-on;
- Sq 65% run-off and 35% run-on.
- La receives water from other units, internal redistribution is neglected.

3. The run-off fractions leaving the geomorphological unit are: Os+Ps loose 3% of the water reaching the soil surface; Pl 11%; Sq 3%, 7% and 9% in the northern and southern Sahel and the Sudan zones respectively. The unit La receives part of these losses, the rest will leave the region. The values chosen for the fraction of run-off not captured in depressions and leaving the units can be too low where the run-on area is (very) small in relation to the run-off area.
4. The depth of water infiltration depends on the amount of infiltrated water, the field capacity of the soil and its depth. The following values of field capacity ($cm^3 cm^{-3}$) are used:
   - Os+Ps: 0.10, but run-on depressions 0.125;
   - Pl: 0.15, but run-on depressions 0.20;
   - Sq: 0.15, but run-on depressions 0.20;
   - La: 0.30.
5. Soils are assumed to have depths exceeding 2 m except in unit Sq, where the average soil depth is 40 cm only. Here, roots from woody plants may penetrate deeper if rock fissures are present.
6. Water beyond 2 m soil depth, or beyond 40 cm in unit Sq, is considered unavailable for roots of annual herbaceous plants. Water percolation beyond 2 m depth is calculated according to a simple model:

$$X = 0.69I - 45Y - 0.22Z - 7.8(Y \times Z) - 74$$

in which X = amount of percolated water ($mm\,yr^{-1}$); I = amount of infiltrated water ($mm\,yr^{-1}$); Y = field capacity ($cm^3 cm^{-3}$); Z = depth of the profile (cm).

Total water availability and the amount of water within the reach of roots of annual herbaceous plants (Table 4.33) are the weighed mean of the water availability per sub-unit (i.e. the areas with run-off, with homogeneous infiltration and with run-on). The results show that in geomorphological unit Sq in the northern Sahel zone water can already be available beyond the reach of herbaceous roots. For water to infiltrate beyond 2 m soil depth in the unit La, rainfall exceeding 600 $mm\,yr^{-1}$ is required (northern Sudan zone).

The following discusses the factors related to water availability that influence the interaction between the herbaceous and the woody layers.

Run-Off

Run-off, causing water redistribution within and between geomorphological units, is highest on loamy soils, with variably sized particles and heterogeneous soil texture (units Pl and Sq). Such soils are sensitive for the formation of surface crusts, limiting water infiltration and causing

Table 4.33. The total amount of infiltrated water (all), the availability of water for herbaceous roots in the topsoil (top)[a] and beyond the reach of herbaceous roots (sub)[b], in a year with average rainfall, expressed in mm yr^{-1}, per geomorphological unit and per climatic zone

Geomorpho-logical unit	Soil layer	Northern Sahel	Southern Sahel	Northern Sudan	Southern Sudan
Os+Ps	All	222	418	654	906
	Top	222	404	483	565
	Sub	0	14	171	341
Pl	All	206	382	619	858
	Top	206	339	476	618
	Sub	0	43	143	240
Sq	All	<223	<400	<614	<851
	Top	206	280	346	419
	Sub	<17	<120	<268	<432
La	All	312	594	937	1312
	Top	312	594	889	1005
	Sub	0	0	48	307

[a] Water in first 200 cm (or 40 cm for unit Sq).
[b] Water below 200 cm (or below 40 cm for unit Sq).

run-off, even on slopes of 1%. As the soil becomes coarser run-off decreases. In the unit Os+Ps the maximum value of 15% runoff is applicable to soils with a loamy sand texture. On coarse sand a moderate slope is required to cause run-off. Reduced vegetation cover causes increased soil crusting and run-off. Woody plants play a secondary role in reducing run-off by means of vegetative cover because for most combinations of geomorphological unit and climatic zone, run-off is required in order to have sufficiently deep water infiltration for woody plants to establish. Reduced run-off due to the presence of woody plants would ultimately be at their own disadvantage. The validity for this reasoning is provided by the fact that furthest north woody plants are most commonly found on soil types showing water redistribution by run-off. Here, on the run-on sites woody cover reaches up to 30% in the northern Sahel zone. On the contrary, in this zone woody plant survival is most difficult on sandy soils.

Slope

The ratio of the run-off and the run-on area determines the penetration of water beyond the reach of herbaceous roots in run-on areas. The average values of this ratio are a factor 2–3 for most geomorphological units. However, for specific sites these average values can be highly variable. Extremes found can be as low as a factor 0.5 (35:65 run-off to run-on area). A factor 0.5 in unit Sq, where run-off from slopes and plains is 50%, leads to run-on areas receiving additional water equal to less than a quarter of the annual rainfall (35/65 × 0.5 × non-intercepted rainfall). On the other hand, extremes can be as high as a factor 9, e.g. in unit Os+Ps, where run-off from slopes is 15%, leading to run-on areas receiving additional run-off water equal to 90/10 × 0.15 = 1.35 times the annual rainfall.

Water Holding Capacity

The water holding capacity of the soil is determined by its field capacity and its depth. The field capacity increases as the soil is more finely textured (Fig. 2.5). Clay soils can store several times more water than sandy soils: heavy clays have a water holding capacity of about $0.3\,cm^3\,cm^{-3}$; sandy soils $0.1\,cm^3\,cm^{-3}$. On sandy soils, redistribution of water is little because run-off is negligible; on the other hand, because of the low water holding capacity, $250\,mm\,yr^{-1}$ of rainfall infiltrating homogeneously is already sufficient for water to reach beyond 2 m soil

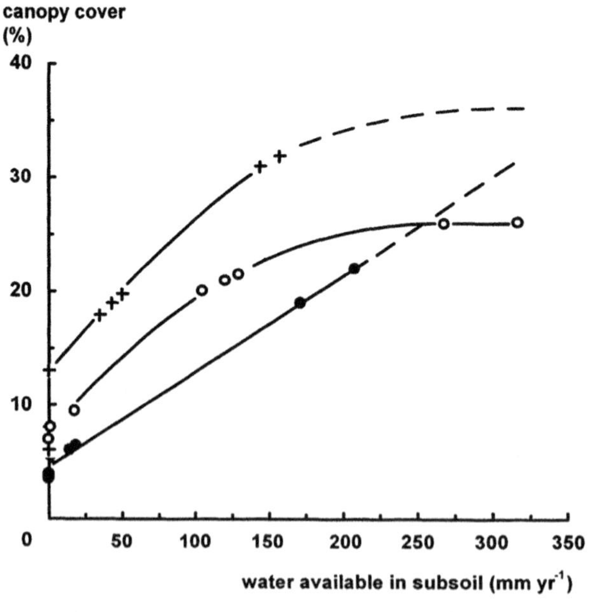

Fig. 4.28. Woody cover before the drought periods in the land-use type "old fallowlands, rangelands and forests" in the SSWA, in relation to water availability beyond the reach of roots of herbaceous plants, for three different geomorphological units (Sect. 2.1.3). • sandy soils (Os+Ps); + sandy-loam plains (Pl); ○ shallow loamy soils (Sq)

depth on coarse sandy soils. The water holding capacity of run-on areas is usually higher than that of the run-off area, since depressions are usually more finely textured. Where such is not the case, deep water infiltration in run-on areas is more important.

Table 4.33 indicates the amount of water not available for annual herbaceous roots, by percolation beyond 2 m soil depth or into rock fissures beyond 40 cm soil depth in unit Sq. These values are presented in Fig. 4.28 in relation to the woody cover of the land-use type "old fallowlands, rangelands and forests". Geomorphological unit La has been left out, since firm data to make the necessary calculations are not available. With increasing rainfall in this unit, stagnating surface water increasingly limits woody plant survival. The highest values of woody cover for unit Os+Ps should be obtained by linear extrapolation, because sandy soils are rare in the Sudan zones and so are data on woody cover from the period before the drought periods.

In units Sq and Pl, with increasing water beyond 2 m soil depth, the increase of woody cover is not as rapid as on sandy soils. This is due to the fact that other factors than water limit woody cover, fire being the most important (Sect. 2.4). For the three geomorphological units investigated, Fig. 4.28 shows similar curves of the relationship between water beyond the reach of herbaceous roots and canopy cover before the droughts periods. If, instead of 200 cm, 170 cm would have been chosen as the limit of herbaceous roots, water availability in the subsoil would increase and the curves would overlap. The overlapping slopes in the first part of the curve suggest that 50 mm yr^{-1} beyond the reach of herbaceous roots can support 10% canopy cover. In unit Sq maximum woody cover is about 25%. Hence, the average amount of water available in fissures under the 40 cm of soil profile probably does not exceed $2.5 \times 50 = 125$ mm yr^{-1}; excess water is lost by subsoil run-off.

However, if post-drought canopy cover is taken as a basis for the above reasoning, 10% canopy cover corresponds to 100 mm yr^{-1} beyond the reach of herbaceous roots.

4.8.3.2 Water-Limited Plant Production

Water-limited total plant production (above- and belowground) is determined on the basis of a plant growth rate limited by water availability, assuming a sufficient seed availability at the start of the growing season. From the start of growth it takes 24 days for an annual crop to reach a closed cover, corresponding to a net production of 2.9 t ha^{-1} and a water utilisation of 132 mm (including 50–60 mm for transpiration). Subsequent growth is linear and depends on the plants having a C_3 or C_4 photosynthetic pathway: water utilisation is 240 and 170 l kg^{-1} dry matter (above- and belowground) for C_3 and C_4 plants respectively (Penning de

Vries and Djitèye 1991). For woody plants, water utilisation of 240 l kg^{-1} is only realistic for young and rapidly growing woody plants. For other woody plants a water-use efficiency up to 480 l kg^{-1} is used (Sect. 4.4.2.3).

Water losses by evaporation from the soil occur before or after the main growing season mainly. Following the methodology described in Penning de Vries and Djitèye (1991) total evaporation losses are 80, 125, 160 and 180 mm yr^{-1} for an average rainfall year in the four zones from the northern Sahel to the southern Sudan zone respectively. These amounts are substracted from infiltrated water presented in Table 4.33. Net production based on the remaining water gives potential water limited production. The estimates neglect percolation beyond the reach of woody roots as well as losses by subsoil run-off through shallow soils. The average proportion of C_3 and C_4 plants in the vegetation (Fig. 4.5) is used to determine the weighed means of transpiration coefficients.

For the vegetation with woody plants two levels of water limited production are determined, corresponding to the high and the low transpiration coefficient of woody plants. The transpiration coefficient of the total vegetation (herbs and woody plants) increases from 180 l kg^{-1} in the northern Sahel to 240 l kg^{-1} in the southern Sudan zone, if only efficiently transpiring woody plants are involved; for inefficiently transpiring woody plants these values are 200 and 320 l kg^{-1} respectively.

In the absence of woody plants, herbs obtain moisture from the first 2 m of the soil profile only (or 40 cm in case of unit Sq). In that case, the transpiration coefficient of the total vegetation varies from 170 l kg^{-1} in the northern Sahel to 180 l kg^{-1} in the southern Sudan zone, as a result of the slight increase of C_3 species towards the south.

If water availability is limited, the methodology described above is very sensitive for small variations in the parameters used. Therefore, in case of water availability less than 250 mm yr^{-1}, a methodology is used which was particularly developed for water-limited plant production (Breman and de Ridder 1991). Here plant production is determined by the formula $X = aI - b$, in which X is the aboveground plant production (kg dry matter ha^{-1}) and I the amount of infiltrated water (Table 4.33). For geomorphological unit Os+Ps parameters a and b are respectively 8.49 and 676, for units Pl and Sq respectively 7.74 and 611. Belowground production estimates are elaborated in Section 4.8.4.1.

Plant production limited by water availability only is determined following the described methodology for a vegetation of woody and herbaceous species and for herbaceous species only, per climatic zone and per geomorphological unit (Fig. 4.29). It is impossible to determine the proportion of the production of a mixed vegetation to be attributed to woody plants. Most of woody roots are found in the topsoil, competing for water with herbaceous roots. With increasing soil fertility, the cover of woody plants probably becomes lower than in the pre-drought situation. The water use of the increased herbage production is re-

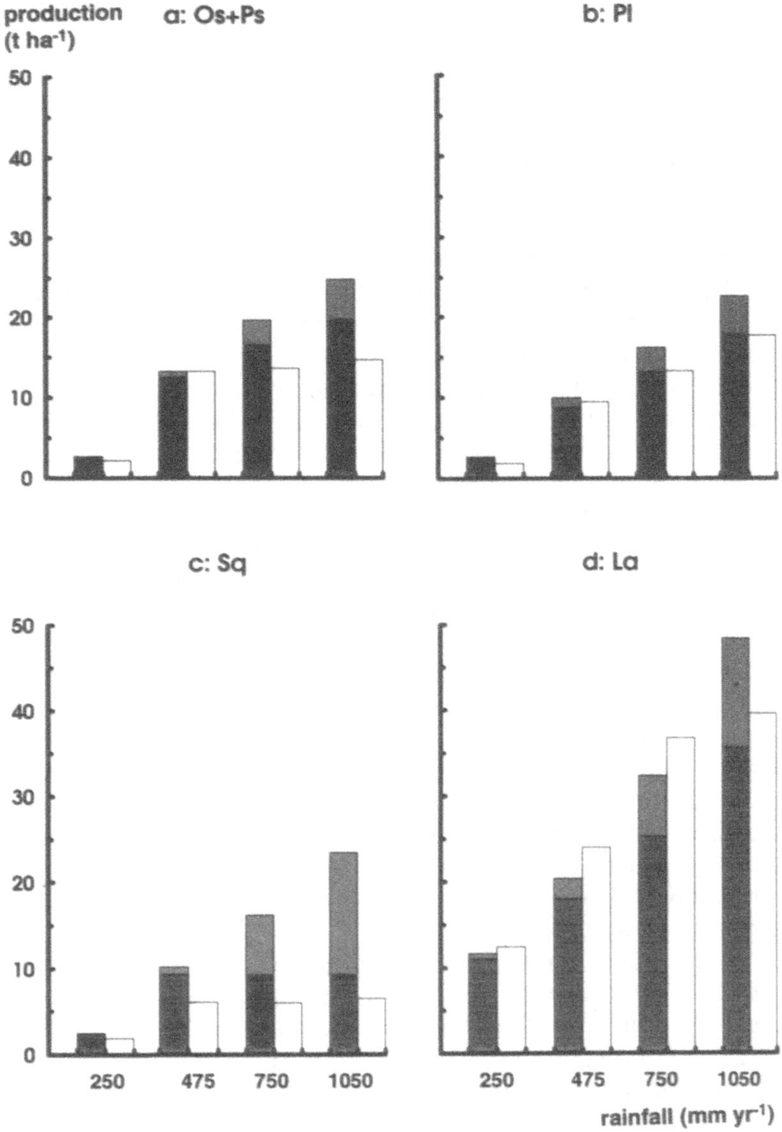

Fig. 4.29a–d. Simulated net dry matter production limited by water availability of a vegetation without woody plants (□), and of vegetations with woody cover as before the drought periods, in case of woody plants with low (■) and high (■ + □) water-use efficiency, in the SSWA, in relation to average annual rainfall, per geomorphological unit(s) **(a–d)**

sponsible for the decrease of deep infiltration and of dry season water reserves, thus threatening survival of perennials.

Net production determined by water availability varies per geomorphological unit and per climatic zone, according to the values presented in Table 4.33: unit La is most productive; units Pl and Sq are least productive. Plant production of the mixed vegetation in unit Sq is probably overestimated because all water beyond 40 cm soil depth was considered available for plant production. If only 100 mm yr^{-1} would be accessible (Fig. 4.28), plant production does not exceed 10 t ha^{-1}.

The higher water availability for plant production in the mixed vegetation, in comparison to herbaceous species only (Table 4.33), does not lead to a proportional increase of plant production because of the relatively low water use efficiency of woody species. This can be illustrated by plant production in unit La: with similar water availability, production is usually lower for a mixed vegetation than for a herbaceous vegetation only.

If water is the plant production-limiting factor, the potential of woody plants to increase plant production is inversely proportional to the water holding capacity of the fraction of the soil profile accessible for herbaceous roots. This potential is high for shallow soils, but only if the rocky subsoil has a high water holding capacity. For deep soils this potential increases with decreasing water holding capacity. Therefore, the potential for woody plants to increase plant production is lowest for unit La, intermediate for unit Pl and highest for unit Os+Ps. Figure 4.29 shows that total (units Pl and La) or most (units Sq and Os+Ps) of the increase in plant production related to the presence of woody plants is realised only if the woody plants have an efficient water use (i.e. young, rapidly growing woody plants). In most cases of water limited plant production, herbaceous production in the mixed vegetation is lower than without woody plants.

If the upper limit of the soil profile accessible for herbaceous roots is set at 170 cm instead of 200 cm, as suggested in the previous section, plant production of herbaceous species is lower than indicated in Fig. 4.29: in a vegetation with herbaceous plants only decreasing about 3 t ha^{-1} in unit Os+Ps, 4 t ha^{-1} in unit Pl and 7 t ha^{-1} in unit La (the variation being related to variable water holding capacity).

4.8.3.3 Specific Influences of Woody Plants on Water Availability

Specific influences on water availability by processes associated with woody plants are summarised in Table 4.8. In the following are analysed the influence of these processes on plant production in the mixed vegetation, and on the contribution by herbs in particular.

Stemflow

Deep water infiltration by stemflow reduces evaporation losses. In the following, the influence of stemflow is elaborated for sandy soils where its potential benefit on plant production is highest, for two reasons. In the first place, once infiltrated into sandy soils, water does not easily remount to the surface. In the second place, the beneficial effect of woody plants on water limited plant production is highest on sandy soils. On the other hand, woody plant survival in unit Os+Ps is difficult where rainfall is low (Sect. 2.3.3).

On sandy soils (field capacity $0.1\,cm^3\,cm^{-3}$), more than $400\,mm\,yr^{-1}$ infiltrated water is required for water to percolate below 2 m of the soil profile (Table 4.4). The maximum value observed for stemflow is 20% of the rainfall intercepted by the woody canopy on an annual basis, with 10% interception losses and 70% throughfall (Sect. 4.4.2.1). It is assumed that the soil under the woody canopy is improved in such a way that run-off has become negligible. If stemflow water infiltrates vertically around the stem in an area equal to 3% of the total canopy area (hereafter called stemflow area), infiltration in the stemflow area can be described by $I = 0.7R + 0.2R \times 30$ (R = rainfall) (Table 4.4).

Based on the pre-drought woody cover, what is the additional water availability if woody species show the maximum stemflow as described above, as compared to woody species without significant stemflow? In the northern Sahel zone, the infiltration in the stemflow area is at most $1525\,mm\,yr^{-1}$, from which 1125 mm percolates beyond 2 m depth. Considering the pre-drought woody cover in the northern Sahel zone of 4%, the additional amount of water in the subsoil would be $0.03 \times 0.04 \times 1125 = 1.3\,mm\,ha^{-1}\,yr^{-1}$ only. For the southern Sahel zone this value is 4 mm, for the northern Sudan zone it is 21 mm. Considering that every 50 mm beyond 2 m soil depth can support 10% of woody cover (Sect. 4.8.3.1), maximum stemflow in the southern Sahel zone can increase woody cover from 6 to 7%, and in the northern Sudan zone from 19 to 23%. This increase corresponds to woody plant production of 200 and $900\,kg\,ha^{-1}$ respectively, if water is the limiting factor. Herbaceous species can also benefit in an indirect way due to the overall beneficial effects of woody plants on water availability (Table 4.8). Stemflow may enable woody species to reach the ground water, otherwise beyond the reach of woody roots, thus reducing competition for water in the topsoil (e.g. in fossil river beds in the Sahel zone).

Infiltration

Table 4.5 shows to what extent woody plants can improve water infiltration under their canopy. As a consequence run-off decreases. Reduced run-off and a higher moisture retention capacity of the soil under woody plants are due to improved SOM contents, improved soil structure and a direct protection of the soil by herbs and/or litter. Thus, positive effects can be expected only when the exploitation rate of both the herbage layer and the woody plants are relatively low. This characteristic is of particular interest on loamy soils (units Pl and Sq). With decreasing run-off the redistribution of water decreases and so will the amount of water percolating to the subsoil. Where homogeneous infiltration of rainfall is not sufficient to moisten the first 2 m of the soil profile, woody plants reducing run-off cause less water percolation to the subsoil, and less possibilities for woody plants to survive. In a general way, improved water infiltration is beneficial for woody plants where it exceeds 500 mm ha^{-1} yr^{-1}.

Of total run-off, the fraction is lost that cannot be stored in depressions or valleys. For relatively undisturbed vegetations of a first order basin in the Sudan zones, run-off losses are at average 11% of total rainfall in unit Pl and 3% in unit Os+Ps (Sect. 4.8.3.1). In unit Sq losses are more due to low storage capacity in the subsoil. In unit Pl these losses equal a water limited plant production of 2 to 3 t ha^{-1} (Fig. 4.29), involving woody plants mainly since water availability in the subsoil is reduced mainly.

Evapotranspiration

The improved water availability under woody canopies in comparison to open areas is partly the result of reduced evaporation in favour of transpiration. The average reduction of PET under canopies of isolated trees is 30–40%, as a result of the reduced radiation and the drying power of the air (Sect. 4.4.2.3). Maximum effects of woody plants on PET are expected where much moisture is available and when PET is high in the open field: for instance, valleys in the Sahel zones during the dry season. As a result of the improved soil fertility under woody canopies herbage transpiration is favoured in comparison to evaporation under woody canopies, limiting the reduction of the evapotranspiration rate under woody canopies to at most 20% (Sect. 4.4.3.2).

On the basis of the processes indicated in Table 4.8, the increased water availability in the soil up to 2 m depth under woody canopies can be estimated at 10–50 mm yr^{-1} in the Sahel zones and 50–200 mm yr^{-1} in the Sudan zones, corresponding to at most a 7% increase in the Sahel zones and a 30% increase in the Sudan zones (comparison with Table 4.33). This increase is due to the processes stemflow, improved soil structure and improved SOM content mainly. The highest values refer to fine-textured soils; the lowest values refer to coarse-textured (sandy) soils, where woody cover is low and soil moisture retention (in the topsoil) is also low.

A simulation model is used to investigate to what extent herbs (crops) profit from this additional water availability in the topsoil. The following assumptions are made, corresponding to a situation where a maximum profit can be obtained by the herbage layer from the increased water availability:

- under woody canopies the reduction of radiation and corresponding evaporative demand is at least 50% (Sect. 4.7);
- WUE of C_3 herbs under the woody canopy is at most twice as high as in the open field (Sect. 4.4.2.3);
- soil fertility under woody canopies is so much higher that water is always the production limiting factor;
- the water use by herbs and woody plants is proportional with the evaporative demand under and above the woody canopy respectively, and inversely proportional with their water use efficiency.

Under the woody canopy, the shift from evaporation to transpiration is insufficient to compensate for the competition with the woody plant. If the woody plant transpires during the whole day and obtains almost all water from the topsoil, as in most parts of the Sahel zones, water availability for herbs is 50% lower than without woody plants. If the woody plant transpires during 2/3 of the day only, or obtains 1/3 of total water requirements from the subsoil, water availability for the herb layer is reduced by 30%. (In the Sudan zones at most 1/3 of rainwater infiltrates beyond 2 m soil depth; Table 4.33).

The combined effect of the maximum increase of water availability in the topsoil due to woody plants (7% in the Sahel zone, 30% in the Sudan zone), and the optimum shift from evaporation to transpiration under the woody canopy, results in a reduced water availability for herbs under woody canopies equal to at least 25% in the Sahel zones and 10% in the Sudan zones.

Although the increased water availability due to woody plants is less than increased transpiration by woody plants, plant production of the mixed vegetation (herbs and woody plants) can be higher than of herbs only. This is valid in spite of the assumed shift from C_4 herbs in the open field to C_3 species under the woody canopy, but only as long as the woody plant transpires at the lowest rate ($240 \, l \, kg^{-1}$). This is only possible for young, rapidly growing plants.

The outcome of the competitive relationship for topsoil water between woody plants and herbs depends on their WUE, and on the density and spatial distribution of their root systems. Herbs can particularly benefit from the increased water availability under woody plants if:

- soil moisture remains in the topsoil;
- radiation interception does not exceed 50%;
- there is a strong shift from evaporation in favour of transpiration;
- soil fertility under the woody plant is high;
- roots from woody plants are in the subsoil mainly, or those in the topsoil have a small absorbing surface (Sect. 4.3).

It can be concluded that it is difficult to select or manipulate woody plant characteristics in order to obtain maximum benefits from an increased water availability for herbs (crops). Whereas processes improving water availability are most important for mature and large trees, young, small and rapidly growing woody plants have highest WUE, but have predominantly shallow roots, and as a result compete most severely for limited moisture in the topsoil (Singh et al. 1989; Malik and Sharma 1990; Lal 1991). Pruning of large trees may maintain a high WUE but will reduce other processes such as improved soil structure and improved soil fertility under woody canopies (Sect. 5.3.4).

4.8.4 Actual Net Production, Limited by Water and Nutrients

The analysis in this section is mainly on net production limited by nutrients, corresponding to regions with more than 250 mm yr^{-1} of infiltrated water. For situations with less than 250 mm yr^{-1} of infiltrated water, the methodology to estimate above-ground herbage production was discussed in Section 4.8.3.2. The analysis focusses on nitrogen (N), which is usually most limiting. Situations where phosphorus (P) is more limiting than N are considered in the second place. As regards the influence of woody species on the availability of nutrients, both N and P are considered because the availability of N is determined, among other factors, by the P-dependant biological N_2 fixation.

4.8.4.1 Methodology

Herbaceous Production Limited by Nitrogen, Without Influence by Woody Species

For situations with nutrient limited production, Breman and de Ridder (1991) developed a formula explaining herbage production on the basis of N availability: $N_h = 0.0083\ I/(f_h - 0.13)$. Nitrogen annually available for the aboveground production of herbs in an equilibrium situation between N inputs and -outputs is represented by N_h. N_h is calculated on the basis of infiltrated water (I) in the first 2 m of the soil profile (or less in case of shallow soils), and of the average fraction of nitrogen in the aboveground herbaceous biomass that is lost annually (f_h). The values of I are presented in Table 4.33. The average value of f_h increases from 0.3 in the northern Sahel zone to 0.5 in the southern Sudan zone, mainly as a result of increasing losses by fire (Table 4.14).

Net production is determined on the basis of N_h, using average N concentration of herbaceous peak biomass at the end of the growing season. The average N concentration of the aboveground herbage biomass decreases from north to the south from 14 to 7, 6 and 5 $g\,kg^{-1}$ in the four climatic zones respectively (decrease due to increasing dilution). In the absence of woody plants, the average N concentration of herbs is up to 40% lower than in the mixed vegetation (increasingly towards the south) due to a higher proportion of C_4 plants (characterised by a lower N concentration) in the open field. C_4 plants produce more biomass with the same amount of nutrients as a result of a more efficient water utilisation and dilution of nutrients in plant tissues. Belowground production of herbs is determined in relation to the aboveground production. The ratio belowground to aboveground production decreases from 4.0 when aboveground production is less than 200 $kg\,ha^{-1}$ to 1.0 at 1 $t\,ha^{-1}$ aboveground production, with a maximum of 1.5 $t\,ha^{-1}$ of roots when aboveground production exceeds 2.5 $t\,ha^{-1}$.

Herbaceous Production Limited by Nitrogen, with Presence of Woody Species

Herbage production can be adequately explained by the above method, in the Sahel zones where woody cover does not exceed 15% with two leaf layers. It appears unnecessary to account for the variable presence of woody plants in the vegetation, which supports earlier conclusions (Sect. 4.2.5) that up to a woody cover of 15% with two leaf layers, beneficial effects by woody plants balance competition with herbs. Benefits by woody plants can be interpreted as the potential for niche differentiation. If woody cover exceeds 15% and there are more than two leaf layers, competition with herbs risks to exceed benefits. Figure 4.2 shows that in the Sudan zones herbage production decreases when woody cover exceeds 15–20%. Towards the south, possibilities for niche differentiation do not increase much, in spite of increasing rainfall and associated N inputs. This is caused by the increasing proportion of shallow soils with low water storage capacity in the rocky subsoil (Sect. 4.8.3.1), and by the increasing proportion of fine-textured soils with nutrients concentrated in the topsoil, resulting from poor soil fertility and a high water holding capacity of the topsoil. The potential for niche differentiation is highest on well-drained, deep and fertile soils, but such conditions are rare in the Sudan zones.

Where herbaceous and woody species compete, Breman and de Ridder (1991) adjusted the N availability for aboveground plant production, by the formula $N_{hw} = 0.0083\ I/(f_{hw} - 0.13)$ in which I now represents the total amount of infiltrated water (not only the fraction in the topsoil available for herbaceous production) and f_{hw} is the average fraction of N in the above-ground biomass lost by herbs ànd woody plants.

The value of f_{hw} is a weighed average of the f_h value for herbs and the f_w value for woody plants, based on canopy cover in the pre-drought situation. For woody plants N losses are limited, because an important fraction is translocated to the woody structures before leaf fall. As a result, N losses from woody plants are also less susceptible to fire. Additionally, translocation as well as low digestibility and high tannin content of foliage make woody plants less susceptible to herbivory than herbs. For woody plants the average fraction of N in the aboveground biomass that is annually lost (f_w) is estimated at 0.2, representing volatilisation losses only. Hence, it is assumed that the woody plants are not exploited.

Woody Plant Production

Woody plant production is directly related to canopy cover (Chap. 3). In order to determine net production and competition with herbs the following methodology is followed, per climatic zone and per geomorphological unit.

1. Canopy cover from the pre-drought period as a starting point (Fig. 2.4);
2. Foliage production determined on the basis of canopy cover and leaf layers (Sect. 3.4.3) using the average number of leaf layers as indicated in Table 3.9.
3. An equilibrium situation is assumed, in which N requirements are based on the annual net production of foliage and fruits, and their average N concentrations only (Table 3.16).
4. Only a fraction of N requirements are obtained from the soil, in competition with herbs (N_{hw}). As a result of translocation and internal recycling by woody plants, 40% of annual N requirements are assumed to originate from the woody plant itself, whereas the woody plants obtain $6\,kg\,N\,ha^{-1}$ by niche differentiation, by increased rooting depth and prolonged growing season in relation to a vegetation with herbs only. The value of $6\,kg\,N\,ha^{-1}$ corresponds to the N uptake required for foliage and fruit production of 15% woody cover with two leaf layers, the maximum assumed not to compete with herbs. Additionally, woody plants increase N availability by N_2 fixation, but this effect is of limited importance (Table 4.20).

5. Relationships between leaf production and (i) wood production, (ii) total aboveground production and (iii) belowground production (Sect. 3.4), are used to estimate total (above- and belowground) production and its components.

Where water is the factor limiting plant production, N requirements by woody plants are not relevant. Here, production by woody plants is simply added to that of the herbage layer (calculated by the formula with N_h), because there is no evidence for competition exceeding niche differentiation. Where N is the production limiting factor, herbaceous production is based on (i) N remaining available after subtraction of the proportion used by woody plants from the total N_{hw}, and (ii) N concentrations in herbaceous biomass produced.

4.8.4.2 Nitrogen Availability

Following the above methodology, N availability is estimated for the vegetation without and with woody plants (Table 4.34). N availability is largely related to water availability (I in the above formula), but varia-

Table 4.34. Nitrogen annually available for plant production ($kg^{-1} ha^{-1}$), per climatic zone and per geomorphological unit, for two vegetations: a herbaceous vegetation without woody plants, and the mixed vegetation with woody cover as in the pre-drought situation in the land-use type "old fallowlands, rangelands and forests". In the last case, total N availability and that for the herbaceous layer only are specified. Total N does not include the fraction of internally recycled N of the woody plants

Geomorphological unit	Vegetation	Northern Sahel	Southern Sahel	Northern Sudan	Southern Sudan
Os+Ps	Herbaceous	12[a]	14	13	≥13
	Total mixed	14[a]	17	28	≥36
	Herbs of mixed	12[a]	14	18	≥14
Pl	Herbaceous	13[a]	12	13	≥14
	Total mixed	17[a]	21	28	≥33
	Herbs of mixed	13[a]	13	16	≥10
Sq	Herbaceous	13[a]	10[a]	10[a]	10[a]
	Total mixed	16[b]	22[b]	21[b]	19[b]
	Herbs of mixed	13[b]	15[b]	19[b]	13[b]
La	Herbaceous	≥12	≥20	≥25	≥23
	Total mixed	≥16	≥29	≥40	≥42
	Herbs of mixed	≥13	≥17	≥8	≥18

[a] Production limited by water.
[b] Assuming 125 mm yr^{-1} stored in the subsoil within the reach of roots from woody plants (Sect. 4.8.3.1).

tion occurs in relation to variable soil fertility. Table 4.34 shows that in the northern Sahel zone the differences in N availability between geomorphological units are little, as here water is most limiting.

For the vegetation without woody plants, there is little variation in N availability between climatic zones and between geomorphological units. This is due to the fact that water availability for herbs increases less than proportional to rainfall (Table 4.33), whereas towards the south N losses (f_h) also increase (due to fire mainly). An exception is unit La, where N availability is higher than in the other units, because of N inputs with run-on and clay soils with a high water holding capacity.

For the vegetation with woody plants, N availability is higher. The increase is related to increasing canopy cover towards the south, and the much lower N losses from woody plants (f_w) than from herbs (f_h). The beneficial effect on N availability is highest in unit La, where both water availability and canopy cover are high. The beneficial effect is lowest in unit Sq, where no increase of N availability with increasing rainfall is detected, because of the water holding capacity in the subsoil limited to 125 mm (Sect. 4.8.3.1).

The proportion of N in the woody vegetation that is utilised by herbs is indicated in Table 4.34. Herbs can only benefit from the increased N availability in a vegetation with woody plants as long as this improvement is more important than the competition for nutrients with woody plants. Table 4.34 shows that for most geomorphological units the beneficial effect of woody plants on herbaceous production is largest in the northern Sudan zone. An exception is unit La, where an opposite pattern is observed. Here woody cover increases rapidly with increasing rainfall (Fig. 2.4), and so does competition with herbs. However, in the southern Sudan zone, woody cover in unit La is lower, because of water logging and flooding (Sect. 2.3). It is doubtful whether herbs can utilise all N remaining available in unit La.

All above estimates are based on a herbage layer of annual plants only. However, perennial grasses and herbs appear in the natural vegetation from the northern Sudan zone southwards, and occur in unit La throughout the SSWA. Proportional to the importance of perennial grasses and herbs in the herbaceous layer, N availability will be (slightly) higher than indicated in Table 4.34, as a result of N translocation and internal recycling, similar to woody plants.

The equilibrium between N inputs and N losses assumes a SOM content correlated with the f value and inherently average constant organic matter inputs. Penning de Vries and Djitèye (1991) showed that mineralisation from organic matter is the major source of N available for plant production. Other N inputs are by rainfall and N_2 fixation. Major N losses from the soil-plant-litter system are by "normal" volatilisation, by leaching, by fire and by exploitation through herbivory. The f values used in the above analysis are from Penning de Vries and Djitèye (1991), who based their data on exploitation levels such as prevailed during the 1970s.

The f_h values (N losses from herbaceous biomass) are assumed to vary per climatic zone, but not per geomorphological unit. This is a simplification: bush fires occur more frequently on sandy soils, whereas losses by herbivory are highest on loamy soils (in view of herbage quality). However, where losses by herbivory and grazing are high, losses by fire are less. The N losses by volatilisation are estimated at 20% ($f = 0.2$). This is considered also as the loss factor of N from woody biomass ($f_w = 0.2$) assuming that no losses take place by fire and no fruits are exploited, or leaves browsed before N translocation from foliage has been completed. Clearly this is not according to reality: for instance, foliage is browsed before translocation because otherwise it has become useless. Therefore, in reality f_{hw} is higher than 0.2, and so N availability in the mixed vegetation is lower. The values indicated in Table 4.34 can be considered as hypothetical maxima.

Table 4.35 demonstrates the influence of variable N losses, and of variable canopy cover, on N availability in a wooded vegetation on sandy soils in the northern Sudan zone. The number of leaf layers is 3.0 (Table 3.9), irrespective of canopy cover. For f_w two extremes are taken: 0.2 assuming no other N losses than volatilisation, and 0.43 assuming N losses similar to the herbage layer, as recorded for exploitation levels prevailing in the 1970s. In reality, N losses from woody plant biomass are probably closer to 0.2 than to 0.43 because:

- fires destroy only parts of the leaves (a limited proportion being on the soil or within the reach of fires) and the burnt leaves have completed N translocation;
- when expressed in terms of live weight, only a limited fraction of the herbivores are browsers, exploiting leaves and fruits during the growing season; the majority are cattle (grazers) exploiting leaves and fruits during the dry season only, when their availability and quality are low (Sects. 3.5, 3.6 and 5.4); for cattle, on an annual basis, woody plant browse represents only about 15% of intakes (Breman and de Ridder 1991).

Table 4.35 shows that decreasing f values lead to increasing N availability for plant production in the mixed vegetation. N availability also increases with increasing canopy cover, which is related to lower f_{hw} values as a result of the greater importance of woody plants. In the northern Sudan zone the greatest positive impact of woody plants on N availability for herbs occurs when canopy cover is 10% and f_w is 0.2. Here, additional N due to the presence of woody plants is 15 kg ha^{-1} yr^{-1}, of which 8 kg is used by the woody plants. The 8 kg utilised by the woody plant originates from translocation (3 kg) and niche differentiation (5 kg), without competition with herbs. With increasing canopy cover

Table 4.35. The influence of canopy cover and annual N losses from above-ground woody plant biomass (f_w values) on annual N losses from the mixed vegetation (f_{wh}), on N availability for plant production (kg ha^{-1} yr^{-1}), on the origin of N inputs from woody plant production, and on some production and biomass parameters (t ha^{-1}), on sandy soils in the northern Sudan zone

	Canopy cover (%)						
	0	10[a]		20[b]		40[c]	
f value woody plants (f_w)	–	0.2	0.43	0.2	0.43	0.2	0.43
f value mixed vegetation (f_{wh})[d]	0.43	0.40	0.43	0.38	0.43	0.34	0.43
N availability (kg ha^{-1} yr^{-1}):							
for herbs	13	20	18	18	14	12	4
for woody plants (leaves and fruits)	–	8	8	17	17	34	34
Origin N for woody plants (kg ha^{-1} yr^{-1}):							
translocation		3	3	7	7	13	13
niche differentiation		5	5	6	6	6	6
N competed for with herbs		–	–	4	4	15	15
Biomass (t ha^{-1}):							
herbaceous biomass	2.2	3.3	3.0	3.0	2.3	2.0	0.7
biomass woody plants	–	5.5	5.5	10.5	10.5	22.1	22.1
N in biomass woody plants (kg ha^{-1})	–	27	27	52	52	110	110

[a] Present canopy cover.
[b] Pre-drought canopy cover.
[c] Maximum canopy cover recorded.
[d] f_h in all cases 0.43.

inputs from niche differentiation do not further increase, so that N utilised by woody plants increasingly competes with herbs. With increasing canopy cover, there is an increase of the amount of N translocated (internally recycled), in relation to N stored in woody biomass (Table 4.35). Higher proportions of N stored in aboveground woody biomass, in stead of in the soil, increases the susceptibility of the ecosystem to losses by uncontrolled intensive exploitation. Table 4.35 also shows the increase of standing woody biomass with increasing canopy cover, which is more than proportional to the corresponding decrease of herbaceous biomass.

4.8.4.3 Water- and Nutrient-Limited Production

Water- and nutrient-limited production of a herbaceous and of a woody (mixed) vegetation (Fig. 4.30a–d) show a relationship with N availability (Table 4.34), since production in most climatic zones and geomorphological units is limited by N availability. It is assumed that no organic inputs nor chemical fertilisers are applied. The production values thus correspond to an equilibrium between natural N inputs and N losses, in a situation where woody plants are not exploited. As production of woody plants, only foliage and fruits are considered. In order to determine an equilibrium situation, only the fraction produced with N taken from the soil is considered (Sect. 4.8.4.1) The actual production is higher (Fig. 4.31) due to the additional availability of internally and externally recycled N (Sect. 4.5).

The northern Sahel zone shows little or no variation in plant production between geomorphological units. In the other climatic zones, plant production increases in the sequence of unit Sq, unit Pl, unit Os+Ps and unit La. The contribution by woody plants to plant production in the mixed vegetation is highest in the Sudan zones, particularly in units Pl and La (25–50%). Plant production in the mixed vegetation is always higher than of the herbaceous vegetation without woody plants, except for unit La. Where canopy cover is low (northern Sahel zone, and unit Os+Ps in southern Sahel zone), woody plant production is additional to herbaceous production without woody plants. Synergistic effects appear in the northern Sudan zone: here the herbaceous production in the mixed vegetation is higher than the herbaceous production without woody plants. Table 4.35 demonstrates how this synergistic effect depends upon the fraction of N losses (i.e. the exploitation intensity of woody plants).

Unit La is particular because of considerable water and nutrient inputs by run-on, irrespective of the presence of woody plants (Table 4.34). Here, differences in plant production between the herbaceous vegetation with and without woody plants are related to the increasing

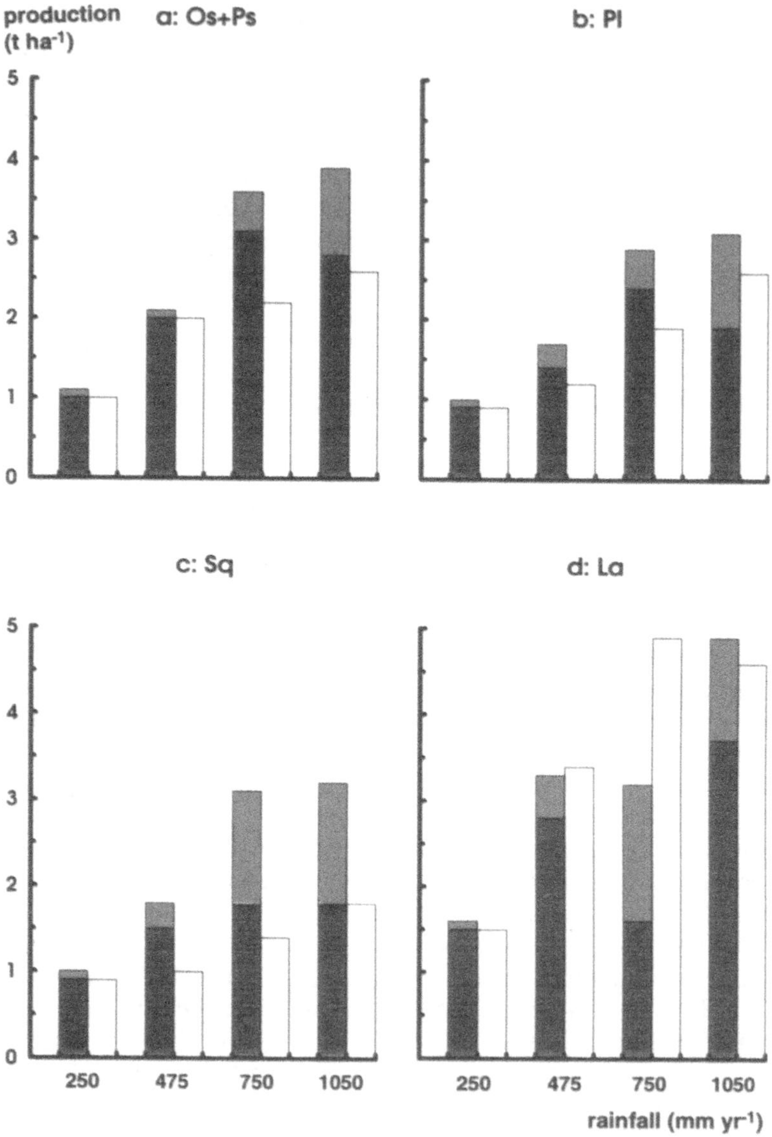

Fig. 4.30a–d. Simulated above ground dry matter production limited by water and nutrient availability of a vegetation without woody plants (□), and of the vegetation with woody cover as before the drought periods, differentiated for herbaceous production (▩) and foliage + fruit production of woody plants (□), in the SSWA, in relation to average annual rainfally, per geomorphological unit(s) (**a–d**). Foliage + fruit production related to internal recycling of N in woody plants is neglected

proportion of C_3 plants in the mixed vegetation. C_3 plants are characterised by higher N concentrations in plant tissues (Sect. 4.8.4.1). Although this is a disadvantage in terms of dry matter produced, it is an advantage for animal husbandry, because C_3 species provide more nutritious forage.

The analysis so far concentrated on N as the limiting nutrient. Although in the SSWA N appears most limiting for plant production (Penning de Vries and Djitèye 1991), phosphorus is most limiting in various tropical regions, including the semi-arid and sub-humid tropics (Allen 1985). P deficiencies limit N_2 fixation by leguminous plants.

Following the spatial concentration of nutrients under woody canopies (Sects. 4.5.3 and 4.5.4), woody plants are less nutrient stressed than herbs, particularly when they have reached maturity (build up of soil-tree-litter cycle). There are several arguments to assume that, in comparison to N, for woody plants P is more limiting than for herbs.

- The enrichment factor under woody canopies is lower for P than for N (Fig. 4.13).
- The P/N ratio in woody plant foliage is at average 0.08 (Table 3.15); for herbaceous biomass an average of 0.10 is found (Sect. 4.5.4.8).
- Translocation of P from foliage is estimated at 70% of annual uptake (Sect. 4.5.4.6) which is higher than for N and for most other nutrients.
- In view of the low P concentrations in foliage (Table 3.14), P returns to the soil by litter are low. For a wooded vegetation on sandy soil in the northern Sudan zone, annual P returns by woody litter are estimated at $0.2\,kg\,ha^{-1}$. The major P losses are by fruits, from which little translocation takes place.

4.8.4.4 Specific Influences of Woody Plants on Nutrient Availability

Specific influences on nutrient availability by processes associated with woody plants are summarised in Table 4.22. In the following are analysed the influence of these processes on plant production of the wooded vegetation, and the contribution of herbs in particular.

Spatial Concentration

The effective spatial concentration of nutrients under woody canopies by biological and physical processes can lead to an enrichment of the agro-ecosystem as a whole, because the soil-woody plant-litter cycle is relatively closed, and because concentration of scarce resources can be beneficial in terms of average production levels (Sect. 5.3.5). The woody plants themselves are most likely to utilise the additional nutrients concentrated under their canopy. Data presented in Table 4.35 and in Fig. 4.30 indicate the situations where herbs can also benefit from the additional nutrient availability. However, this is possible only when:

- N losses by exploitation of woody plants are negligible (Table 4.35);
- light is not a limiting factor (Sect. 4.8.2);
- herbaceous production in the open field adjacent to the woody plant does not decrease more than proportional to production increase under the woody canopy (as a response to reduced nutrient availability in the open field).

The last condition is a difficult one because a reduced vegetation cover easily leads to soil crusting on loamy soils, with subsequent increased run-off and erosion. In the Sahel zones the undisturbed herbaceous production is already insufficient for full protection of loamy soils (Sect. 5.4.4). Thus, even if the reduction of herbaceous production in the adjacent open field is less than the increase under woody plants, it easily leads to increased run-off and limited water availability. The ultimate result may be the tiger bush pattern, occurring naturally in the Sahel zones, but nowadays extending into zones with higher rainfall as a result of land degradation (Le Houérou 1989). To become intensive, wind erosion on sandy soils requires a more dramatic reduction of vegetation cover (Sect. 5.4.4).

In order to compensate for increased susceptibility to soil degradation by reduced herbaceous production in the open field of woody vegetations, the woody plants themselves should provide soil protection. This is possible with a high density of small shrubs, such as occur on fallowlands (Sect. 2.4.4), contrary to few large trees. However, small shrubs compete most severely with herbs, reducing herbaceous productivity (Sects. 4.2.2.2 and 4.5.3). Hence, protective (sustainability) and short-term productive criteria are conflicting (Sect. 5.4.4).

Enrichment

N_2 fixation by woody plants enriches the agro-ecosystem. Calculating plant production limited by N availability (Sect. 4.8.4.1), an average of 5% of the herbaceous production is assumed to provide N inputs by N_2 fixation. In the vegetation with woody plants the contribution by leguminous plants to total plant production is higher than 5% (Sect. 4.5.4.8), although under woody canopies the proportion of leguminous herbs is often less than in the open field (Sect. 4.2.4). However, whereas N uptake by herbaceous leguminous plants originates from N_2 fixation mainly (up to 75% of N uptake, corresponding to $5\,kg\,ha^{-1}\,yr^{-1}$), N_2 fixation contributes only 0.2–$0.8\,kg\,ha^{-1}\,yr^{-1}$ to N uptake by nodulating woody leguminous plants (Table 4.20).

Assuming that all woody plants in the vegetation are nodulating leguminous plants, and assuming that P availability remains limited as it has always been, the additional N availability for the agro-ecosystem is 0.2, 0.6, 2.4 and 5.0 kg N $ha^{-1} yr^{-1}$ from the northern Sahel to the southern Sudan zone respectively. The additional plant production to be expected from the increased N availability is 15, 60, 260 and 580 kg ha^{-1} yr^{-1} from the northern Sahel to the southern Sudan zone respectively, mainly by increased productivity of the woody plants.

The combination of maximum N_2 fixation (all woody plants nodulating legumes) and maximum P uptake from the subsoil corresponds to the maximum enrichment of the agro-ecosystem by woody plants. Maximum P uptake from the subsoil can be expected where woody plants root deep and show maximum stemflow. Predominantly shallow rooting patterns, without stemflow, cannot be expected to lead to significant P uptake from the subsoil (Sect. 4.5.4.4). Assuming deep rooting and maximum (20%) stemflow, a maximum of 5% of P requirements originates from the subsoil (Sect. 4.5.4.4). Penning de Vries and Djitèye (1991) found increased P availability to lead to increased N_2 fixation by a ratio 1:20. As a result, additional N_2 fixation resulting from maximum P uptake from the subsoil may amount to 0.2, 0.6, 1.1 and 1.5 kg N $ha^{-1} yr^{-1}$, corresponding to plant production of 20, 60, 120 and 190 kg $ha^{-1} yr^{-1}$, for the northern Sahel to the southern Sudan zone respectively. The application of P fertilisers (such as rock-phosphates) seems more effective to stimulate N_2 fixation (Hingston et al. 1982) than the selection and planting of deep rooting woody legumes with much stemflow.

Reduction of Losses

The reduction of losses by woody plants is most effective in increasing nutrient availability for plant production (Table 4.22). The processes involved may be direct (e.g. internal recycling) or indirect (reduced losses due to high SOM content under woody canopies). Maximum impact of the processes involved occurs when exploitation of woody plants is negligible. Nutrient losses by fire and exploitation ($f_w > 0.2$) lead to lower plant production than indicated in Fig. 4.30. Multi-purpose woody plants, producing desirable products, are therefore undesirable if production levels of herbs (crops) in the agroforestry system should be maintained.

In view of the predominantly importance of processes reducing nutrient losses, the importance of woody plants increases with increasing nutrient availability in the soil (by natural or artificial causes). Although processes such as internal recycling and N_2 fixation become less effective as nutrient availability increases, the improved SOM content is most important as long as increased soil fertility leads to increased dry matter being produced and recycled (Sect. 4.6).

4.8.5 Woody Plants and Plant Production Factors

4.8.5.1 The Sahel and Sudan Zones of West Africa

Net Dry Matter Production

Estimates of plant production limited by light, water and nutrients in previous sections allow an analysis of the impact of woody plants on plant production in an agro-ecosystem, and on herbaceous (crop) production in particular, by their impact on these plant production limiting factors. In the following, the results of previous sections are summarised for the geomorphological unit Os + Ps and unit Pl. For unit Sq too much variation occurs regarding soil depth and the possibilities for the subsoil to store water; for unit La too much variation occurs regarding the amount of water and nutrients entering the unit from surrounding areas.

To allow comparison between light, water and nutrient limited production, the results of previous sections are modified as follows.

1. Total annual plant production limited by light is based on total daily plant production limited by light (Fig. 4.25) and the length of the growing season on the basis of water availability, estimated at 25, 70, 125 and 165 days for the northern Sahel to the southern Sudan zone (FAO 1984). For the two units investigated total herbaceous production is calculated per climatic zone and for three distinct situations: under a closed canopy, under an isolated standard tree and in the vegetation with pre-drought canopy cover.
2. Production limited by nutrients, as determined for above-ground biomass (Fig. 4.30), is transformed into total above- and belowground production using for belowground production by woody plants the relationships presented in Section 3.2, and using for herbaceous plants ratios provided by Bille (1977) and by Penning de Vries and Djitèye (1991). Firstly the above-ground biomass of woody plants is extended with the production related to the recycled N, wood and a fraction of leaves and fruits (Sect. 4.8.4.3). Production levels (Fig. 4.31) show that in the Sahel zones herbage production is highest on sandy soils, but production of herbs and woody plants is highest on unit Pl. On both units herbaceous production does not further increase from the northern to the southern Sudan zone as a result of increasing competition with woody plants. Woody plant production continues to increase from the northern to the southern Sudan zone.

Figure 4.32 compares three situations of light-limited production of the herbaceous component of vegetations, the water-limited production

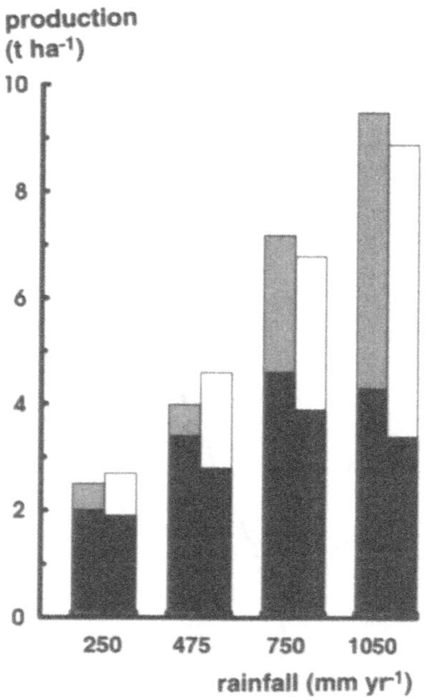

Fig. 4.31. Simulated net dry matter production limited by water and nutrient availability of the vegetation without woody plants and of the mixed vegetation, with woody cover as before the drought periods, for two different geomorphological units, in the SSWA, in relation to average annual rainfall: Os+Ps, herbaceous production only (▨); unit Os+Ps, additional plant production in the mixed vegetation (□); unit Pl: herbaceous production only (▨); unit Pl, additional plant production in the mixed vegetation (□)

by vegetations as a whole, with woody plants with low and high WUE, and the nutrient-limited production by the natural vegetation with pre-drought canopy cover. The differences between units Os+Ps and Pl, appear to be small. Care should be taken to extrapolate dry matter production levels, as presented in Fig. 4.32, to production of useful products such as seeds (cereals), fruits or forage, because both harvesting ratios and plant nutritive values may change in relation to variable availability of plant production factors and plant species composition.

It appears from Fig. 4.32 that in the natural vegetation (land-use type "old fallowlands, rangelands and forests") dry matter production limited by light exceeds water-limited production only by a factor 1.5 to 2.0. Water-limited production is next best, except for the northern Sahel zone, where water-limited production is lower than nutrient-limited production (Sect. 4.8.1). Under woody canopies nutrients are less limiting than in the open field, as a result of the spatial concentration of nutrients. This is visualised in Fig. 4.32 by arrows: the production increase under woody canopies is proportional to the enrichment factor (Fig. 4.13), and

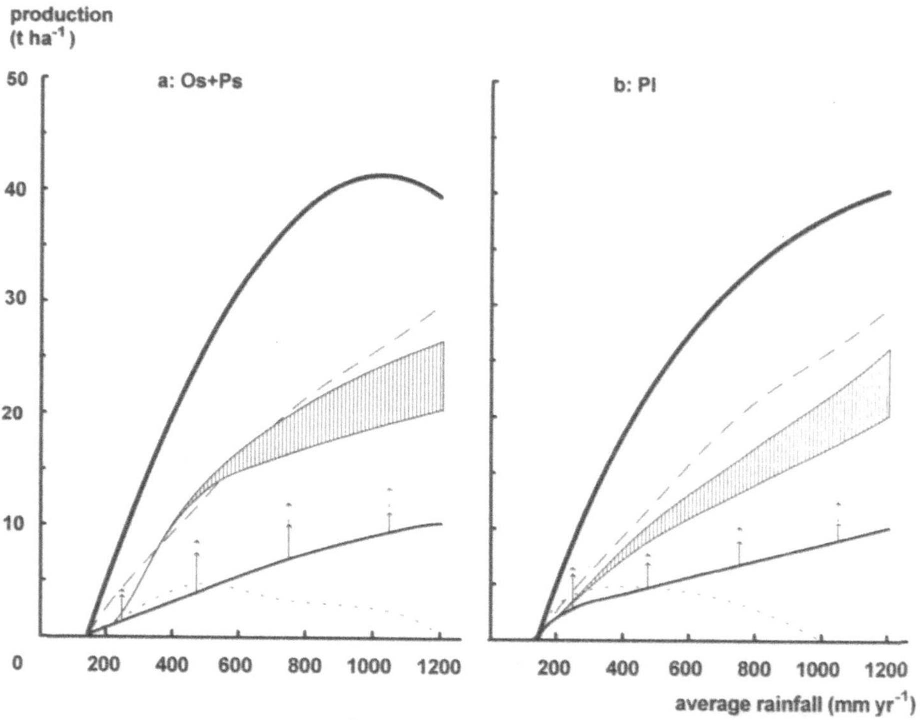

Fig. 4.32a,b. Simulated net dry matter production during the rainy season in the SSWA, in relation to average annual rainfall, for geomorphological units Os+Ps (**a**) and Pl (**b**), for situations with different production limiting factors. —— herbage production limited by light availability in the vegetation with woody cover as before the drought periods; – – · herbage production limited by light availability under a standard tree (Sect. 4.7.5); · · · herbage production limited by light availability under a closed woody canopy; ▭ net production of the mixed vegetation limited by water availability (woody plants with low water use efficiency, lower limit; woody plants with high water use efficiency, upper limit); ┴┴ net production of the mixed vegetation limited by nutrient and water availability; → expected local increase due to P enrichment under woody canopies; →› expected local increase due to N enrichment under woody canopies, in case P does not become limiting

is higher for N than for P. If nutrient enrichment under the canopy is accounted for, production of the woody plant and of herbs under the woody canopy becomes more limited by water. Now the shift from water to nutrients limiting plant production corresponds to an annual rainfall of about 350 mm yr^{-1}.

Another deviation from the average picture presented in Fig. 4.32 deals with the length of the growing season. It was argued in Section 4.4.3.2 that the growing season under woody canopies can be prolonged by 25% as a result of reduced PET. Herbs with a longer growth cycle than in the open field may benefit from the longer growing season under woody canopies. Thus can be explained why herbs with a longer growth cycle as well as perennial grasses (Sect. 4.2.4), normally found in the Sudan zones, occur furthest into drier (Sahel) zones at locations which are relatively densely wooded (e.g. the vegetated parts of the tiger bush pattern) (Breman and de Ridder 1991).

The Proportion of Herbage (Crop) Production

To what extent can herbage production be expected to increase in comparison to a vegetation without woody plants?

In situations of water limited production, it is unlikely to expect an increase of herbaceous production under woody canopies (Sect. 4.4.3.2). Herbaceous production levels under woody canopies similar to those in the open field, can only be expected where woody plants are selected in such a way that processes improving water availability are optimal (Table 4.8). An exception should be made for those situations where a specific rainfall distribution in the Sahel zone causes seed availability in the open field to be more limited than under woody canopies (Sect. 4.2.4). In most situations with higher herbaceous biomass under woody canopies than in the open field, plant production is nutrient-limited.

In situations of nutrient-limited production, woody plants may increase nutrient availability for herbaceous production, as indicated in Table 4.34 (N availability) and Fig. 4.30 (dry matter production). In the southern Sudan zone the production increase of the herbage layer could be even higher than indicated in Fig. 4.30 if competition with woody plants is reduced. Competition can be reduced by limiting canopy cover to at most 35% and two leaf layers, whereas at the same time the fraction of nutrient losses should be limited to 20% ($f_w = 0.2$). If the number of leaf layers is three, or if f_w is 0.3, canopy cover should not exceed 20%, etc. Present canopy cover and leaf layers in the southern Sudan zone are higher (Fig. 2.4; Table 3.9).

Herbaceous production levels limited by nutrients in the vegetation with woody plants (Figs. 4.30 and 4.32) are average values in a large area. Considerable variation occurs between the canopy area and the open field. On the one hand, the increased nutrient availability under woody canopies (indicated by arrows in Fig. 4.32) stimulates herbaceous production; herbs will usually be able to utilise 80–90% of additional nutrients (according to the relations indicated in Fig. 4.30). On the other hand, in the open field adjacent to the woody plant, herbaceous production may be reduced. Herbaceous production under woody canopies is intermediate between production limited by light under closed canopies and production limited by light under isolated standard trees. Under the woody canopy light is not a factor more limiting than nutrients as long as light availability does not decrease with more than 30% of light availability under the standard tree. Hence, woody plants can have a slightly larger crown radius (in relation to trunk height) or densities and distributions can be such that shadow patterns overlap to some extent (Sect. 4.7.5; Fig. 4.24). In reality, many shrubs occur in the natural vegetation of the land-use type "old fallowlands, rangelands and forests". About 50% of the woody plants have an unfavourable ratio between trunk height and crown radius (less than 1.0). The proportion of shrubs increases as a result of drought and exploitation levels (Sect. 2.2.2.3). In most geomorphological units, woody plants occur in clusters mainly (Sect. 2.2), resembling a closed canopy, so that reduction of light availability is more than 30%. Only under tall and isolated woody plants can herbs benefit from the higher nutrient availability. This is the common situation in unit Os+Ps, where trees usually occur scattered and are well developed (Sect. 2.3).

Selection of woody species and suitable site conditions to maximise potential benefits (Tables 4.8 and 4.22) can further improve potential production values indicated in Fig. 4.32. This is not likely for nutrient limited growth, where the assumed N losses ($f_w = 0.2$) are already very low. Regarding water limited production, by improved management and plant species selection run-off might be reduced by 10% in the Sahel zones and by 20% in the Sudan zones (Sect. 4.8.3.1). However, since in the majority of the SSWA nutrients are most limiting (Fig. 4.32), the possible increase in water availability does not easily lead to increased plant production. Higher-than-average soil fertility conditions, or nutrient inputs are required. Where both water and nutrient availability are improved, light becomes the limiting factor (as indicated by light-limited production under a standard tree; Zoungrana et al., in press). It is therefore more probable that the woody plant itself, the production of which is not limited by light availability, profits from an increased water and nutrient availability.

4.8.5.2 Conclusions

Woody plants have the potential to increase the availability of water and nutrients, and thus of plant production, in comparison to a vegetation without woody plants. Potential benefits are mainly due to a reduction of nutrient and water losses, i.e. a more efficient use of their availability. The potential varies according to woody plant characteristics and site conditions; it is inversely correlated with the exploitation intensity of woody plants, and is positively correlated with rainfall and soil fertility. The relationships are more than proportional: regarding rainfall, the more than proportional relationship with canopy cover is shown in Fig. 2.4. Regarding soil fertility and exploitation rate, the interaction with protective biomass, litter and SOM content leads to an exponential relationship: increasing litter quantities and SOM content further improve water and nutrient availability by reducing losses, further enhancing plant production (Sect. 4.6). An exception are situations of excessive water availability, stagnating in depressions with fine-textured soils, unfortunately becoming more widespread as soil degradation proceeds.

Due to the presence of woody species, the shift from water to nutrient limited production occurs with higher water availability than the 250 mm yr^{-1} for a herbaceous vegetation only. In situations of water-limited production, increased water availability by woody plants is at most a 25% increase of the period of plant growth under the canopy. The woody plants do not lead to increased herbaceous production in the vegetation as a whole. In situations of nutrient-limited production, increased nutrient availability by woody plants is more important for N than for P, so that P increasingly becomes the major production limiting factor. N availability for herbaceous production may be increased with at most 3 kg ha^{-1} yr^{-1} in the southern Sahel and 6 kg ha^{-1} yr^{-1} in the northern Sudan zone. Similar estimates are obtained (i) based on the assumption that woody plants can avoid N losses by leaching or run-off (Table 4.15), (ii) based on additional N requirements for increased millet growth in *Faidherbia albida* parklands (Shaikh et al. 1988). In the southern Sudan zone woody biomass and production increase so much that the additional N availability is largely utilised by the woody plants themselves. Here, herbs can make maximum use of the beneficial effect of woody plants on N availability, only if N losses from woody plant biomass do not exceed 30%, as long as canopy cover does not exceed 20% and the number of leaf layers is not more than two (other possible combinations are 20%–35%-2 or 20%–20%-3), and woody plants are evenly distributed with a favourable ratio of trunk height to crown diameter (isolated large trees).

5 Synthesis, Generalisation and Recommendations

5.1 Introduction

This chapter provides a synthesis of the benefits to be expected from woody plants and the conditions determining this potential, a generalisation of the results obtained from the SSWA to other semi-arid regions, and recommendations for the use of woody plants in agrosylvopastoral land-use systems in semi-arid regions.

In Section 5.2 the relationship between woody cover and climatic and physical site conditions is summarised, followed by an extrapolation of the results obtained in the SSWA to other semi-arid regions. In Section 5.3 the potential production of a vegetation with woody plants is assessed, maximising all positive influences by selection of specific woody species and site conditions and by appropriate management. The role of woody plants in land-use systems in semi-arid regions is dealt with in Section 5.4. The analysis focuses on the impact of woody plants on agricultural production, security and sustainability objectives, for sylvopastoral and agrosylvicultural land-use systems. As throughout this book, the focus is on ecological factors: climatic and ecological site conditions are worked out to optimise the beneficial role by woody plants, as well as the most desirable characteristics of the woody species involved.

5.2 Woody Plants and Plant Production Factors

In order to extrapolate the results from previous chapters on the role of woody plants in the SSWA to other semi-arid regions, the influences of woody plants on the plant production factors light, water and nutrients are summarised, and the SSWA is compared to other semi-arid regions.

The climatic and ecological site conditions influencing the plant production factors light, water and nutrients, summarised in Tables 5.1 and 5.2, are both simple (e.g. wind, radiation, soil texture) and complex (e.g. PET, soil fertility). Between semi-arid regions, these conditions show considerable variation. The influence of these conditions on plant

Table 5.1. Climatic conditions and their influence on plant production factors, directly and via their influence on cover of woody plants

Climatic factor	Influence
Radiation – intensity and duration	As radiation intensity and day length increase, light interception by woody plants less easily leads to plant production under woody canopies being limited by light availability
– radiation angle'	A negative correlation between (a) radiation angle and (b) fraction direct radiation under woody canopies, and between (a) and (b) the percentage diffuse light; a positive correlation between (a) and (b) light intensity
Rainfall – intensity and size of showers	A positive correlation between (a) rainfall intensity and shower size and (b) stemflow, water leaching, run-off, and water erosion intensity; therefore a positive correlation with (b) (horizontal and vertical) water redistribution and woody cover; a negative correlation with (b) interception losses
Rainfall : PET relation	A positive correlation between (a) the ratio rainfall : PET during the rainy season and the length of the rainy season and (b) woody cover
Dry season – aridity (i.e. length and PET)	A negative correlation between (a) the length of the dry season and PET and (b) woody cover; as rainfall is increasingly two-peaked, the persistence of the dry season decreases, improving conditions for woody plants
PET	A positive correlation between (a) PET and (b) the potential of woody plants to improve microclimate under woody canopies; a negative correlation between (a) PET and (b) the isohyet at which growth becomes water limited
Wind	A positive correlation between (a) wind intensity and (b) wind erosion intensity and PET

Table 5.2. Ecological site conditions and their influence on plant production factors, directly and via their influence on woody cover

Ecological site factor	Influence
Soil texture	A positive correlation between (a) particle size and (b) infiltration rate; a negative correlation with (b) moisture retention capacity and run-off, and therefore a positive correlation with (b) water beyond the reach of herbs, and woody cover; a negative correlation between (a) particle size and (b) organic matter contents, but a positive correlation with (b) decomposition rate; a negative correlation between (a) particle size and (b) wind erosion risk, but a negative correlation with (b) water erosion risk
Soil texture – cation exchange capacity	Where the clay fraction is kaolinitic with Fe and Al oxides and hydroxides, soil organic matter largely determines CEC
Soil depth	A positive correlation between (a) soil depth and (b) potential water stocks, and therefore with (b) water available for roots from woody plants, and woody cover; impenetrable soil layers limit root development by woody plants
Slope	A positive correlation between (a) slope and (b) run-off, and therefore a positive correlation with water redistribution and woody cover
Texture and slope	A negative correlation between (a) fine textured depressions receiving run-off and (b) woody cover (due to water stagnation); a positive correlation between (a) coarse textured depressions receiving run-off and (b) woody cover
Soil fertility – nutrients	A positive correlation between (a) soil fertility and (b) woody cover, unless water is more limiting or competition with herbs is severe (e.g. shallow soil); a negative correlation between (a) and (b) processes leading to nutrient enrichment and internal recycling (because of negative feed back); a positive correlation between (a) and (b) processes leading to reduced nutrient losses by improved soil structure and SOM; a positive correlation between (a) the availability of P and (b) the potential for N fixation
Groundwater	A positive correlation between (a) the availability of ground water within the reach of roots of woody plants and (b) woody cover

production factors is both direct and indirect, the latter via their influence on woody plants.

Direct Influences

Regarding radiation, extrapolation from the SSWA to other semi-arid regions is not problematic. The variation of total available radiation in the SSWA does not significantly differ from that in other semi-arid and sub-humid climatic zones: reduced radiation further from the equator is compensated for by increased day length during the growing season, whereas cloudiness is more or less similar for each rainfall regime. For total plant production in semi-arid regions, nutrients and water are always more limiting than radiation. As radiation intensity and day length increase, light interception by woody plants less easily leads to plant production under woody canopies being limited by light availability. Increased day length is associated with a smaller radiation angle and therefore leads to a higher fraction of lateral direct radiation. However, a smaller angle also leads to decreased radiation intensity and to direct light as a fraction of total radiation decreasing in favour of diffuse light.

Regarding water availability for plant production, the SSWA is unique because of its extremely high PET. Northern Australia most closely resembles the SSWA (Papadakis 1975), but as it lies further from the equator and has a smaller landmass, it has a lower annual PET rate in the rainy season and a higher air humidity in the dry season (Breman 1992). Rainfall intensity and rainfall variability in the SSWA are not extreme when compared to other semi-arid regions (Nix 1983). Lower PET values in other semi-arid regions lead to a larger fraction of rainfall being available for plant production. Hence, the shift from water- to nutrient- limited plant production is likely to occur with lower rainfall in other semi-arid regions. For instance, in the mediterranean region, water availability for plant production is twice as high as in the SSWA with similar rainfall (van Keulen 1975). As a result, the shift from water to nutrient-limited production occurs at about 125 mm water infiltration, in stead of 250 mm in the SSWA (Sect. 4.8.1). Factors that may lead to a greater proportion of rainfall being available for plant production in other semi-arid regions are higher elevation (e.g. several areas in East and South Africa), greater distance from the equator (e.g. semi-arid Australia and India), and the coincidence of rainfall with low (winter) temperatures (mediterranean climate). As compared to one rainy period only (in the majority of semi-arid regions), a rainy season with two rainfall periods (e.g. in parts of East Africa and Australia) has advantages for perennial plants because of reduced rainfall concentration and therefore reduced water losses and reduced length of the dry season, by better spreading of rain throughout the year.

Regarding nutrient availability, inherent soil fertility in the SSWA is extremely low due to the predominance of old, weathered rocks. Less weathered soils are more common in other semi-arid regions (e.g. China, India), as well as young, volcanic soils and alluvial valleys (East Africa, California, China). Local variability between soil types (texture, fertility, depth, etc.) influences water and nutrient availability (Sect. 4.8.3; Table 5.2). For instance, a higher soil fertility causes the shift from water- to nutrient-limited plant production to take place with higher rainfall. At some locations, extreme P fixation by high aluminium contents and low pH levels limits plant production. Altogether, as compared to the SSWA, the lower PET values in most other semi-arid regions predominate, and causes the shift from water to nutrient-limited production to occur at lower rainfall levels.

The shift from water to nutrient limited plant production is visualised in Fig. 5.1 for the SSWA. Important variation is caused by water redistribution on finely textured soils, causing water infiltration to vary between run-on and run-off sites. In extremely arid conditions, and where land degradation is severe, the available vegetation is an even more limiting factor than water or nutrients. Figure 5.1 shows three situations, differentiated according to underlying dynamics:

I Production limited by insufficient vegetation (i.e. starting biomass):
  Ia by natural origin;
  Ib due to land degradation.
II Production limited by water availability:
  IIa due to relatively low rainfall;
  IIb due to water losses by run-off.
III Production limited by nutrient availability:
  IIIa due to low soil fertility;
  IIIb due to run-on.

The areas covered by the different situations are determined by the importance of climatic zones (isohyets), soil types and the proportion of run-off and run-on areas. Table 5.3 shows estimates for the SSWA as a whole, but several data are missing. Arrows in Fig. 5.1 indicate how the boundaries between the different situations change as a result of higher effective rainfall or higher soil fertility in other semi-arid regions.

Whereas Fig. 5.1 shows the variation occurring as regards the production limiting factors, in relation to annual rainfall and infiltration rate, Fig. 4.32 shows net production limited by light, water or nutrients, for one geomorphological unit, in relation to rainfall. Here, one geomorphological unit includes both run-off and run-on areas, whereas these situations are differentiated in Fig. 5.1.

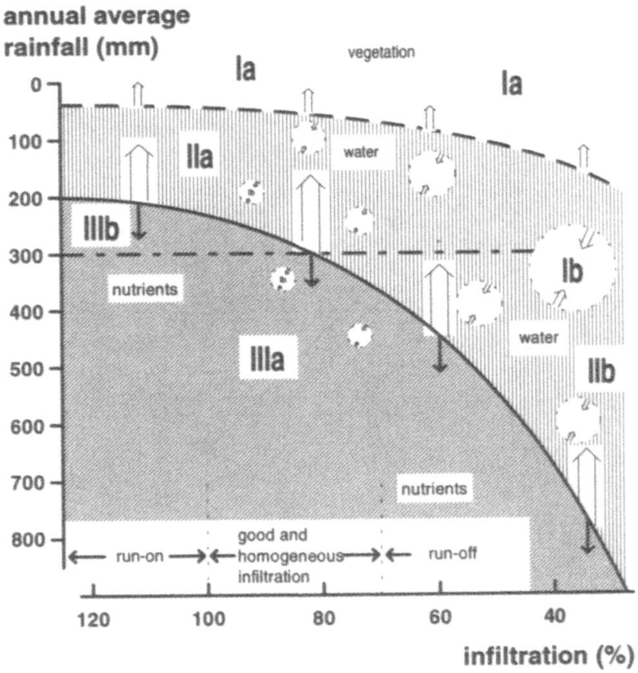

Fig. 5.1. The boundaries of plant production in the SSWA limited by vegetation (seeds), water, or nutrients, in relation to annual average rainfall and redistribution of water (% infiltration). *Arrows* indicate how the boundaries shift as a result of increasing effective rainfall (⇒) or increasing soil fertility (→). For explanation of symbols see text

Indirect Influences

Climatic and ecological site conditions influence woody cover and productivity of woody plants (Chaps. 2 and 3). As woody cover increases the influence of woody plants on water and nutrient availability increases (Chap. 4), thus influencing plant production factors in an indirect way (Tables 5.1 and 5.2).

Based on canopy cover data in the SSWA, in the land-use type "old fallowlands, rangelands and forests" (Fig. 2.4), the different situations indicated in Fig. 5.1 are characterised by woody cover of different periods (Table 5.3). Woody cover is lowest in situation I (where a starting biomass is insufficient), in situations IIa and IIIa, and in situation IIIb on sandy soils in the southern Sahel. In these situations the influence of the scarce woody cover on plant production factors cannot be significant.

In the SSWA woody plants establish and survive where water is available beyond the reach of herbaceous plants (Sect. 2.3.5; Fig. 4.28).

Table 5.3. The relative importance of areas characterised by different plant production limiting factors (Fig. 5.1), per climatic zone and geomorphological unit, expressed as the proportion of the area covered per climatic zone, and their potential woody cover (pre-drought) and their actual cover (1990, post-drought)

Production-limiting factor	Climatic zone and geomorphological unit	Relative importance (%)	Canopy cover	
			1970	1990
Insufficient vegetation				
Ia by natural origin	Sahara	100	~0	~0
Ib by land degradation	Sanddunes	[a]	~0	~0
	Large crusted areas and erosion gullies	[a]	~0	~0
Water shortage				
IIa due to relatively low rainfall	Northern Sahel – Os+Ps	65 – a	4	2
	– Pl	5	13	1
	– Sq	25	10	2
	– La	5 – b	6	2
IIb due to run-off	Southern Sahel – Pl	10 – c	19	3
	– Sq	35	21	4
	Northern Sudan – Pl	20 – d	31	<20
	– Sq	40 – e	26	<20
	Southern Sudan – Sq	f	28	28

Limited by nutrient availability				
IIIa due to low soil fertility	Southern Sahel – Os+Ps	40	6	3
	– Pl	c	19	3
	– La	15	15	10
	Northern Sudan – Os+Ps	25	19	<10
	– Pl	d	31	<20
	– Sq	e	26	<20
	– La	15	32	<30
	Southern Sudan – Os+Ps	15	33	33
	– Pl	30	35	35
	– Sq	40 – f	28	28
	– La	15	24	24
IIIb due to run-on	Northern Sahel – Os+Ps	a	>4	>2
	– La	b	>6	>2

[a] Very limited areas involved.
a,b etc.: areas of unknown importance.

Here, niche differentiation is essential for woody plant survival. More water becomes available in the subsoil as rainfall increases, as rainstorm intensity increases and as soil texture is more coarse (Sect. 2.3). In other semi-arid regions, lower aridity during the dry season, a shorter dry season as a result of a two peaked rainy season, and/or less frequent dry years, make woody plants less dependent on subsoil water reserves for survival during the dry season. One consequence is that in other semi-arid regions woody cover and biomass for a specific rainfall are often higher than in the SSWA (Sect. 3.2.2.2). Bush encroachment processes are also more important in other semi-arid regions (Sect. 2.4.3.2), supporting the above conclusion. A higher inherent soil fertility is also beneficial for woody plants, unless water is more limiting, and reduces competitive relationships between woody plants and herbs (Sect. 4.2.2.1). As a result, the curves presented in Fig. 2.4 may be more steep in other semi-arid regions than in the SSWA, and consequently the positive effects of woody plants on water and nutrient availability can be greater. However, it is a dilemma that an increasing density and cover of woody plants is, on the one hand, desirable to optimise processes associated with woody plants leading to increased nutrient and water availability, but, on the other hand, causes increased competition with herbs for limited resources.

5.3 Productive Benefits by Woody Plants

5.3.1 Introduction

In the SSWA, standing biomass and net production of woody vegetations are higher than of an herbaceous vegetation only (Fig. 4.31). A number of factors are involved:

- the higher water and nutrient availability resulting from niche differentiation between woody plants and herbs, but up to a certain canopy cover, depending on climatic and site conditions (Sects. 4.4 and 4.5);
- less losses of nutrients from mixed vegetations due to efficient external recycling of nutrients and build up of soil organic matter, efficient internal recycling by woody plants, and relatively little exploitation of woody plants in relation to herbaceous plants;
- low nutrient concentrations of woody plant tissues;
- the benefits related to the spatial concentration of scarce resources (Sect. 5.3.5).

Where ecological conditions for woody plant survival are more favourable than in the SSWA, the canopy cover representing maximum niche differentiation and potential plant production increases. As a result the potential benefits of woody vegetations are higher (Sect. 5.3.4). Benefits may be realised by selection of woody species with specific characteristics for selected sites, and by management of woody vegetations (Sects. 5.3.2 and 5.3.3).

5.3.2 Woody Plant Characteristics

Table 5.4 summarises the characteristics of woody plants determining the influences of woody plants on water and nutrient availability, wind and water erosion, and shading (Sects. 4.8.3.3 and 4.8.4.3). Indicated are those aspects that provide more-than-average benefits. Every increased plant production resulting from increased water and/or nutrient availability can contribute to reducing wind and/or water erosion; specific beneficial characteristics of woody plants are indicated in Table 5.4. The potential to provide protection against water erosion is greatest for species with shallow roots, a low trunk height/crown radius ratio (shrubs), perennity of leafs (evergreen) and maximum production of above- and belowground living biomass and soil organic matter. Competition with the herbage layer should be limited, in order not to reduce direct soil protection provided by the herbage layer (Sect. 5.4.2.4). Unfortunately, these characteristics are not compatible.

In most cases, the opposite characteristics, as those indicated in Table 5.4, have opposite effects (e.g. high water use efficiency improving water availability; low WUE has an opposite effect), unless it is indicated otherwise. Some beneficial characteristics are compatible (e.g. deep and shallow rooting), others are not (e.g. deep rooting and rapid growth, or low palatability and high decomposition rate).

Specific adaptations for woody plant establishment and survival are fire resistance and a high regeneration (resprouting) capacity. There are remarkable adaptations of woody species in arid and seasonally dry climates to drought, grazing, cutting and fire (Sect. 2.4). Vegetative regeneration from trunks (after cutting) and regrowth from root stocks is more important than reproduction from seeds (Nyerges 1989). Characteristics indicated in Table 5.4 are beneficial in association with specific climatic and site conditions. Whereas deep rooting is beneficial if moisture is available in deep soil layers, shallow rooting is effective to concentrate scarce nutrients. A rapid growth rate is beneficial to effectively compete with herbs, but leads to a high uptake of (scarce) water and nutrient resources (Sect. 4.5.3). A slow growth rate coincides with the development of a deep and widespread root system, contributing to survival in drought years, and to build up of above- and belowground biomass (vegetative biomass and soil organic matter), optimising the positive influences on nutrient and water availability (Sect. 4.6). As a result, a long life span prevails in semi-arid regions (Mai Moussa

Table 5.4. The potential of woody plant characteristics to improve aspects of agrosylvopastoral land use

Woody plant characteristic	Water availability[a]	Nutrient availability	Wind erosion	Water erosion	Shade
High regeneration capacity	–[b]	–	+	+	+
Rapid growth rate	0	0	+	+	+
Slow growth rate/long life span	0	+	+	+	+
High water use efficiency	+	0	+	+	+
Effective internal nutrient cycling	0	+	0	0	0
Deep rooting	+	0	0	0	0
Shallow rooting	–[b]	0	+[c]	+[c]	0
Presence of mycorrhiza	0	+	0	0	0
Biological N_2 fixation	0	+	0	0	0
Canopy characteristics:					
– low ratio trunk height/crown radius	0	0	+	+	0
– open canopy	–	0	0	0	0
– tall closed canopy	+	+[d]	0	+[c]	+
– stimulating stemflow	+[d]	0	0	0	0
Foliage characteristics:					
– allelopathic properties	0	0	0	–[f]	0
– high decomposition rate	0	+[e]	0	0	0
– low palatability	0	+	+	+	0
– evergreen phenology	+[d]	+[d]	+	+	+
– shedding (early) rainy season	0	+	+	–	0

+: positive influence; 0: no significant influence; –: negative influence.
[a] Increased water availability generally corresponding to prolonged crop growing season.
[b] Short-term disadvantages due to high water and nutrient use by woody plant.
[c] Unless severe competition with herbs.
[d] Improved availability largely/totally utilised by the woody plant.
[e] Rapid decomposition is probably most positive, provided that poorly decomposable litter (woody material) is also available.
[f] Through suppression of herbs, most likely under arid conditions.

et al., in press). Different phenological patterns reflect different survival strategies (Sect. 3.5). Evergreen species are more adapted to stress, reducing nutrient losses and effectively suppressing herbaceous growth under their canopies, but show low productivity. Species with photosynthetic activity in stems can reduce transpiring leaf surface, reducing water stress, and are less susceptible to grazing (Sect. 2.4.3.1). Woody plants that shed leaves early in the dry season provide less soil protection against wind erosion during the dry season; woody plants shedding leaves in the early rainy season provide less soil protection against water erosion during the rainy season. Leaf shedding during the dry season risks loss of litter by fire. Woody plants remaining in foliage during the dry season are important in terms of reducing high PET values under their canopies, and may show more important hydraulic lift, whereas the foliage is less susceptible to losses by fire.

The presence or absence of relevant characteristics is listed in Table 5.5 for the woody species most commonly occurring in the SSWA (distribution indicated in Tables 2.4 to 2.7). Systematic data are not available of all the characteristics discussed above.

5.3.3 Management of Woody Plants

Little is known of existing management techniques of woody plants in agrosylvopastoral land-use systems in semi-arid regions (Savenije 1993). Research has so far hardly addressed management issues, although these are crucial to obtain maximum benefits of woody plants. In Section 5.4 a number of specific management techniques are mentioned when discussing the role of woody plants in land-use systems. More generally, the following can be mentioned as important management techniques that enable optimal use of woody plants in agrosylvopastoral systems (Savenije 1993): protection and stimulation of natural regeneration by affecting woody plants, herbs or site conditions; woody species selection, planting configurations (densities and orientations) and inputs required for artificial regeneration; pruning, pollarding, lopping, thinning, cutting and controlled harvesting/exploitation of woody plants and weeding of associated herbs to optimise the role of woody plants.

For most semi-arid regions, present woody cover is lower than considered desirable from a sustainability point of view. Whether this is also the case from a productive point of view depends upon the situation. For reforestation to take place, natural (passive) regeneration should be stimulated wherever possible and requires protection against grazing and fire mainly. However, where woody species have disappeared, artificial (active) regeneration is required, preferably with local species adapted to the climatic zones (Tables 2.4 to 2.7). Whereas some species resprout easily and predominate on young fallowlands (Sects. 2.2.1.2 and 2.4.4),

Table 5.5. Woody species commonly occurring as native or introduced in the SSWA, common French and English names, indication of the presence or absence of characteristics mentioned in Table 5.4 and utilisation of products

Botanical name	French/English name	Characteristics									Utilisation				
		N fixation	Deep roots	Decom-position	Pala-table	Foliage shedd.	Growth rate	Ration	Canopy structure	Height canopy	Fire wood	Construc-tion	Forage	Fruit	Medi-cinal
Native															
Acacia nilotica		+	0	0	0	0	+	0	–	0	+	+	0	–	–
Acacia raddiana		+	0	0	+	0	0	–	0	+	+	0	+	–	–
Acacia senegal	Gommier	+	0	+	+	–	0	0	0	0	+	0	+	–	0
Acacia seyal	Mimosa épineux	+	0	+	+	0	0	+	+	0	+	0	+	–	–
Adansonia digitata	Baobab	–	–	+	0	0	–	–	0	+	–	–	0	+	+
Afzelia africana	Mahogany bean	–	0	?	+	–	–	0	0	+	?	+	+	–	?
Anogeissus leiocarpus		–	?	0	0	–	–	–	0	+	0	+	0	–	0
Balanites aegyptiaca	Desert date	–	0	+	+	0	0	0	+	0	+	+	+	+	+
Bombax costatum	Kapokier rouge/Red silk cotton	–	–	?	0	–	?	0	0	+	+	0	0	+	0
Borassus aethipium	Rônier/fan palm	–	?	?	–	0	0	+	+	+	0	+	–	+	0
Boscia senegalensis		–	–	+	+	0	+	–	0	–	0	–	+	–	+
Burkea africana		–	0	?	0	–	?	0	0	+	?	?	0	–	?
Combretum spp.		–	0	?	0	0	0	0	0	0	0	–	0	–	0
Commiphora africana		–	–	?	+	–	?	0	+	0	0	0	+	–	+
Detarium microcarpum		–	?	?	0	–	?	0	0	0	0	?	0	+	?
Euphorbia balsamifera		–	?	–	–	–	0	+	–	–	–	–	–	–	0
Faidherbia albida	Winterthorn	+	+	+	+	+	–	+	0	+	0	0	+	–	+
Ficus spp.		–	?	–	0	0	–	0	0	+	0	0	0	+	0
Grewia spp.		–	–	?	+	–	0	0	0	0	0	0	+	–	+
Guiera senegalensis		–	0	0	0	0	+	–	0	–	0	0	0	–	+
Hyphaene thebaica	Doum/dum palm	–	?	?	–	0	–	+	+	+	0	+	–	+	0
Isoberlinia doka		–	?	?	0	0	?	0	0	+	?	?	0	–	?
Khaya senegalensis	Caïlcedrat/mahogany	–	?	0	+	0	0	0	0	+	0	+	+	–	+
Lannea acida	Raisinier	–	?	0	0	–	?	0	–	0	0	0	0	+	0
Parkia biglobosa	Néré/locust bean	–	0	0	0	–	–	–	0	+	0	0	0	+	+

Piliostigma spp.		–	0	0	0	0	+	–	0	–	0	0	0	0	0
Prosopis africana		0	?	?	0	0	?	0	+	0	+	0	+	0	+
Pterocarpus spp.		+	0	+	+	–	0	0	0	0	+	0	+	–	0
Sclerocarya birrea	Prunier	–	–	0	+	–	–	+	+	0	+	0	+	+	0
Tamarindus indica	Tamarinier/tamarind	–	?	0	+	0	–	–	0	+	0	0	+	+	+
Terminalia avicennioides		–	?	?	0	–	0	0	–	0	0	0	0	–	0
Vitellaria paradoxa	Karité/shea butter	–	0	–	0	0	–	0	0	+	0	0	0	+	0
Ziziphus mauritiana	Jujubier	–	–	+	+	0	+	–	0	0	0	0	+	+	0
Introduced															
Albizia lebbeck		+	0	+	+	0	+	0	0	0	0	–	+	–	–
Anacardium occidentale	Cashew	–	0	0	–	–	–	–	–	+	0	0	–	+	–
Azadirachta indica	Neem	–	0	0	0	0	+	0	0	0	+	0	0	–	+
Bauhinia rufescens		–	–	0	+	0	+	0	0	–	0	–	+	–	0
Cassia siamea		–	0	0	0	0	+	0	0	0	+	+	0	–	–
Eucalyptus camaldulensis		–	–	–	–	0	+	+	0	+	+	+	–	–	+
Jatropha curcas		–	?	–	–	–	0	+	0	–	–	–	–	–	0
Leucaena leucocephala		+	0	+	+	0	0	0	0	0	0	0	+	–	–
Mangifera indica	Mangue/mango	–	0	–	0	0	–	–	–	+	0	0	0	+	+
Prosopis juliflora		+	+	0	0	0	0	–	0	0	+	0	0	0	–

Characteristics:
- N fixation: capable of N_2 fixation
- Deep roots: deep roots (+ = type 1; 0 = type 2; – = type 2; – = type 3; Section 4.3.1)
- Decomposition: foliage decomposition rate (+ = rapid; – = low)
- Palatable: palatability of foliage (+ = high; – = low)
- Foliage shedd.: foliage shedding pattern (+ = rainy season; – = dry season)
- Growth rate: growth rate (+ = rapid; – = slow)
- Ratio: ratio trunk height/crown radius (0 = 1.0; + = higher; – = lower)
- Canopy structure: canopy structure (+ = open; – = dense)
- Height canopy: height of canopy (+ = tall tree; – = small shrub)

(where not specifically indicated, 0 indicates the situation between + and –)

Utilisation:
- Fire wood: firewood
- construction: construction wood
- Forage: livestock forage
- Fruit: fruits for human consumption
- Medicinal: Medicinal products

+ positive for characteristic. – negative for characteristic. 0 variable for characteristic. ? unknown.

Important sources: von Maydell (1986); Guinaudeau (1987); Kessler and Boni (1991); pers. observ.

others establish in the shade of colonising species, show slow growth and a long life span. For rapid reforestation in situation III (Fig. 5.1), woody species may be selected with the characteristics of rapid juvenile growth, effective internal nutrient recycling, shallow rooting, capacity for N_2 fixation, presence of mycorrhiza and a closed canopy (Table 5.4). In situation II (water-limited plant production), a high water use efficiency, deep rooting and a high potential for stemflow are important characteristics. In situation Ib reseeding with species adapted to severe erosion conditions is indispensable for restoring woody cover.

Artificial regeneration usually requires watering during the dry season, since the infiltration capacity of deforested soils is low, and protection against grazing and fire (Kessler and Boni 1991). Additional P supply in planting holes, by organic manure or (rock) phosphates, is very effective. Woody plants more effectively concentrate N than P (Fig. 4.13) and P stimulates growth of young plants (Penning de Vries and Djitèye 1991) and N_2 fixation.

5.3.4 Maximum Productive Benefits by Woody Plants

Where plant production is limited by water (Fig. 5.1: situation II), an increased water availability resulting from the presence of woody plants (Table 4.8) can lead to an increased net production (Fig. 4.29). Water availability can be improved by selection for woody species and management techniques that reduce water losses and maximise the efficiency of water utilisation (Table 5.4). What is the maximum benefit that can be expected?

Water losses are 3% for unit Os+Ps (sandy soils) and 11% for unit Pl (loamy) (Sect. 4.8.3.1). Where such losses can be avoided, net production on loamy soils can increase with 0.3 to 2.6 $t\,ha^{-1}$, for annual rainfall increasing from 250 to 1000 mm respectively, as long as water remains the production limiting factor.

Water-use efficiency increases proportionally to a reduction of PET, which leads to increased net production with similar rainfall. In most other semi-arid regions PET is already lower than in the SSWA, so that the additional influences by woody plants on PET are limited.

Woody cover increases with 10% for every additional 100 $mm\,yr^{-1}$ beyond the reach of herbaceous roots (Fig. 4.28), corresponding to at most 2.1 to 4.2 $t\,ha^{-1}$, for low and high water-use efficiency of woody plants respectively (Sect. 4.8.3.2). In the SSWA, 100 $mm\,yr^{-1}$ additional water beyond the reach of herbaceous plants leads to a production increase of only 0.8 $t\,ha^{-1}$ in the northern Sahel and 1.8 $t\,ha^{-1}$ in the southern Sudan zone (Figs. 2.4, 4.28 and 4.31). The difference with the theoretical maximum is due to (i) low woody cover in arid zones, not enabling the use of all water beyond the reach of herbaceous plants, (ii) the predominance of woody plants with low water-use efficiency, and (iii) low soil fertility leading to nutrient-limiteded production. In other semi-arid regions, situations of water-limited production are likely to occur with less than 250 mm yr^{-1} water infiltration (Sect. 5.2). Thus, estimating the maximum productive benefits at 250 mm indicates the highest possible deviation of water-limited production between the SSWA and other semi-arid regions. In view of the amount of water available beyond the roots of herbaceous plants (Table 4.33), at most 1 t ha^{-1} additional production can be expected, if the woody plants have a high WUE.

Where plant production is limited by nutrient availability (situation III in Fig. 5.1), several processes associated with woody plants can increase nutrient availability (Table 4.22). The enrichment processes of N_2 fixation and rhizosphere interactions, and the reduction of losses by internal recycling, become less efficient as soil fertility increases, as a result of negative feed-back. Efficient internal recycling is an adaptation to resource-poor environments. Other processes show a positive correlation with soil fertility: processes leading to reduced losses by run-off and leaching are more effective as soil fertility increases, and so are processes of reducing nutrient losses associated with the improvement of SOM contents (Sect. 4.6).

In a vegetation with woody plants with efficient water use (240 l kg^{-1}) the maximum production increase due to increasing nutrient availability compared to plant production in the open field can be threefold, before water becomes the production-limiting factor. But under woody plants soil fertility is already higher than in the open field; the nutrient limited production, indicated in Fig. 4.32 represents the weighed average of both the open areas and the areas occupied by woody plants of mixed vegetations in the SSWA. Elsewhere, due to higher nutrient availability, the production can be higher in the part of the vegetation occupied by woody plants, and because of larger areas occupied by this wooded part. The average nutrient enrichment under woody canopies in semi-arid Africa is indicated by arrows in Fig. 4.32. The highest enrichment values are 30–50% higher than these average values (Fig. 4.13), and are probably associated with situations where exploitation levels of the vegetation are low and where inherent soil fertility is high. If nutrient enrichment by woody plants is according to these high values, net production of the vegetation part occupied by woody plants can increase by about 1.5 t ha^{-1} yr^{-1}. Reinforcement of the effect is expected where inherent soil fertility is higher, filling the gap between the maximum enrichment to be expected by woody plants and the level of water-limited production, but only as long as the nutrient enrichment by woody plants is not limited by the feedback mechanisms outlined above. Since the enrichment factor by woody plants is higher for N than for P, higher inherent P availability is more likely to increase production in a vegetation with woody plants. In view of the difference between the levels of production limited by P, as indicated by the arrows in Fig. 4.32 and the production levels limited by water, the production in the part of the vegetation occupied by woody plants can at most double for water availability exceeding 400 mm. Hence, net production may increase at most with 5 t ha^{-1} at 500 mm rainfall, and with 10 t ha^{-1} at 1000 mm rainfall by the effect of both higher soil fertility and higher nutrient enrichment (Fig. 4.32). Since woody cover in these rainfall zones is about 10 and 30% respectively, for the vegetation as a whole production should increase with 0.5 and 3 t ha^{-1} at 500 and 1000 mm rainfall respectively.

The production increase may be greater if woody cover is higher than in the SSWA. Of course, increasing woody cover will lead to reduction of the enrichment factor. Extrapolation of Fig. 4.13 to rainfall where the enrichment factor is 1.0 (1200–1400 mm yr^{-1}), and of Fig. 2.4 to woody cover at 1200–1400 mm yr^{-1}, gives 50% as the canopy cover where no further enrichment under woody canopies takes place. The theoretical maximum increase of nutrient limited production due to increased woody cover is equal to the difference between the level of water limited production of a mixed vegetation and the nutrient limited production of a herbaceous vegetation. Figure 4.32 shows this maximum to be 0.02, 0.10 and 0.16t ha^{-1} for each additional percentage woody cover at 250, 500 and 1000 mm yr^{-1}. Doubling the canopy cover in comparison to the SSWA, will lead to a maximum production increase of 0.1, 1.6 and 3.2t ha^{-1}, at 250, 500 and 1000 mm yr^{-1} respectively. At 900 mm yr^{-1} a canopy cover of 50% is attained. Higher rainfall may lead to higher woody cover, but without additional increase of the nutrient limited production. These values are applicable for woody plants with a high water-use efficiency; where WUE is low (480 l kg^{-1}) additional plant production would be 0.1, 1.4 and 2.3t ha^{-1} yr^{-1} for the three rainfall regimes respectively.

Integration of the theoretical upper limit resulting from the increased availability of water and nutrients and from the increased woody cover, indicates the maximum increase of plant production due to woody plants. Annual sustainable exploitation is only a fraction of this increased production. The theoretical potential corresponds to a shift from nutrient to water limited plant production in the vicinity of woody plants. The chances are limited to reach this theoretical potential, but in other semi-arid regions the shift from water- to nutrient-limited production is likely to occur at >350 mm yr^{-1} in the SSWA (Sect. 4.8.5.1).

5.3.5 Woody Plants and Spatial Concentration of Resources

A specific effect of woody plants in resource-poor environments is the redistribution of resources leading to areas of concentration and areas

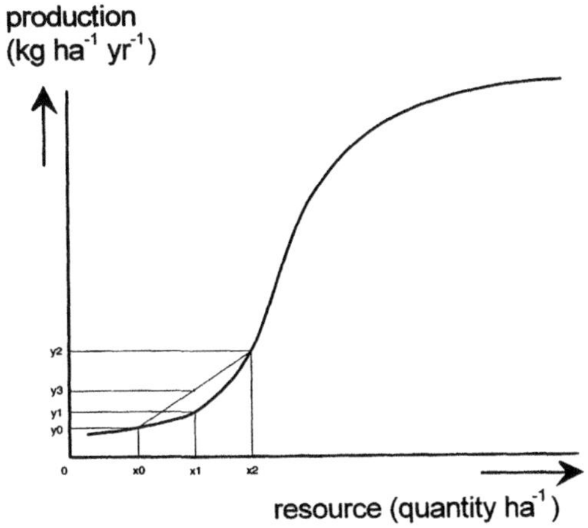

Fig. 5.2. The effect of resource concentration (e.g. by woody plants) on productivity in resource-poor environments: x1 = resource availability in situation of homogeneous distribution, leading to production level y1; following concentration resource availability is x0 in areas depleted and x2 in enriched areas, where x1 − x0 = x2 − x1

of depletion (Fig. 5.2). In resource-poor environments patchiness of resources is a natural phenomenon, due to biological and physical processes, of which woody plants are one, but termites, herbivores, gravity and wind are others. The spatial concentration of water and nutrients under woody canopies (Sects. 4.4 and 4.5) is achieved partly at the expense of the availability of these resources in the open field. The higher SOM contents under woody canopies result from higher plant production associated with the higher nutrient and water concentrations. Hence, "fertility islands" are created. Note that spatial concentration of resources by woody plants often builds onto (i.e. enhances) resource patchiness already occurring prior to woody plant establishment (Sects. 4.4.2.2 and 4.5.2.1).

Where plant production is severely limited by low water or nutrient availability, spatial concentration of resources, such as by woody plants, is an advantage for the (agro-)ecosystem as a whole. Shmida and Burgess (1988) postulated that in (semi-)arid regions biological fluctuations will be lower and total productivity will be higher with strong spatial variability than with an even distribution. This is due to the fact that as long as the relationship between production and resource availability is exponential (the lower part of the sigmoid curve), the average production of a system where scarce resources have been concentrated is higher than in the original situation, characterised by a homogeneous distribution of scarce resources (Fig. 5.2). Whether spatial concentration leads to such benefits depends on (i) the slope of the sigmoid curve, (ii) the relationship between impoverished areas and enriched areas and (iii) the resource losses occurring during the redistribution process.

5.4 The Role of Woody Plants in Land-Use-Systems

5.4.1 Introduction

Agrosylvopastoral land use systems can make use of the beneficial role of woody plants according to the objectives of maximising productivity, security or sustainability (Conway 1987; Martin 1988). In most farming systems these objectives are not focussed on overall plant productivity, but on productivity of crops considered useful for consumption or for marketing. Therefore, first the influences of woody plants on crop or pasture production are considered, as they are the major production goals of agrosylvopastoral land use (Sect. 5.4.2). Woody plant products are considered in Section 5.4.3. Besides the influence on production levels of crops or pastures, maintenance of woody plants may be beneficial in terms of reducing production variation (security or stability), and for enhancing sustainability (i.e. long-term security or stability). These aspects are dealt with in Section 5.4.4. The conditions for maximum benefits for woody plants are elaborated for sylvopastoral and agroforestry systems (Sects. 5.4.5 and 5.4.6).

Obviously, socio-economic conditions also determine the value of woody plants and their products, in relation to other crops, but the variability of these conditions between sites (even between villages) and in time, has not been object of systematic studies. In Section 5.4.7 some socio-economic aspects related to woody plants in land use systems are discussed.

5.4.2 Woody Plant Influences on Crops and Pastures

In the majority of semi-arid regions, there is potential for livestock keeping, agricultural cropping and forestry land use, but land scarcity nowadays increasingly leads to competition between these land-use systems (Sect. 2.4). In areas where plant production is limited by water availability (Fig. 5.1: situation II), vegetative biomass is characterised by high nutrient concentrations and pastoral land use is the only productive land-use system, with no competition with other land-use systems.

Table 5.6 summarises the potential of woody plants to improve productivity, security or sustainability per land-use system, and per situation as characterised by its plant production-limiting factor (Fig. 5.1). For instance, where drought risks occur, particularly in the drier parts of

Table 5.6. Land-use systems, their occurrence in situations characterised by plant production limiting factors (Fig. 5.1), and the potential role of woody plants that may be realised to improve crop or pasture productivity, security or sustainability

Land-use system	Situation (cf. Fig. 5.1)	Potential role of woody plants that may be realised		
		Objective	Improvement	Specification of situations
Arable farming (cropping and horticulture)	IIIa and b, II[a]	Production/security	Prolonged crop growing season	Driest parts
			Increased nutrient availability	Everywhere
			Improved germination and crop establishment	Wind erosion risk
			Increased water availability	Everywhere
		Sustainability	Reduced wind erosion	Sandy soils
			Reduced water erosion and physical degradation	Slopes, loamy soils
			Improved buffering capacity	Slopes, shallow soils, plateaux etc.
Livestock keeping (meat and milk)	IIa and b, III[b]	Production/security	Increased pasture production	During dry years in II
			Improved pasture quality	During dry season in III
			Shade sustainability: see above	Everywhere in the tropics
Livestock keeping (draught and manure)	IIIa and b	Production/security Sustainability: see above	Increased pasture production and better quality	Everywhere

[a] In situation II where soil fertility is so hight that water becomes limiting in spite of water availability exceeding 250 mm yr^{-1}.
[b] In situation III, only in case of low stocking rates or the availability of supplementary feeds.

situation III, woody plants can be beneficial by increasing both productivity and security. Where in situation IIb woody plants reduce water losses by run-off, the situation shifts to III, and pastoral land use loses its specific advantage. Note that situation IIIb (high run-on in low rainfall areas leading to nutrient-limited growth) occurs rarely and represents conditions where agriculture is difficult because of flood risks.

The potential of woody plant characteristics to improve aspects of agricultural land use can be obtained from Table 5.4. Aspects of agrosylvopastoral land use, as indicated in Table 5.6, are increased water availability, increased nutrient availability, reduced wind erosion, reduced water erosion and shade. The first two aspects are beneficial for crop and pasture production and security objectives (including production levels and/or quality aspects), in situations of water or nutrient limited plant production. The reduction of wind and water erosion enhances sustainability. Shade is beneficial for human beings and animals. The assessments in Table 5.4 provide qualitative indications only. To quantify the effects requires data on canopy cover, woody plant productivity, the level of exploitation, and tree-crop competition relations. The latter depend on the woody plant and on the crop or pasture characteristics: C_3 or C_4 phenological pattern, rooting system, germination characteristics, growth cycle, perennity and capacity to fix nitrogen. Table 5.4 provides a framework for a first assessment of the potential that can be expected from woody plants in specific situations. In the following sections quantitative estimates are made on the basis of the data collected in the SSWA.

Three principles can be formulated to optimise the influences from woody plants on crop and pasture production.

1. Maximum positive effects on water and nutrient availability, implying selection for the characteristics water use efficiency, internal recycling of nutrients, deep rooting and long life span.
2. Maximum niche differentiation between woody plants and herbs, both in space and in time, implying selection for the characteristics deep rooting (minimum uptake of nutrients and water by shallow roots) and leaf shedding during early growing season (to reduce light competition by minimum overlap of growth cycles). Niche differentiation is positively correlated with several climatic and site conditions (Tables 5.1 and 5.2), such as deep and fertile soils, annual rainfall, run-on and the presence of ground water within the reach of woody plant roots. Characteristics of herbs are preferably most different from those of woody plants: short lived, shallow rooting, short growth cycle.
3. Minimum light interception, implying selection of species with open canopies and a high ratio of trunk height/crown radius. Species showing photosynthetic activities by their stems meet the first con-

dition, but this characteristic has little impact on nutrient availability, and these species can only improve pasture production in situation II.

In tropical climates shade is an important factor improving conditions for animal survival. Heat stress due to high temperatures causes disruption of physiological processes, causing reduced animal and human productivity. For animals, these effects are more important as average animal production levels are higher.

It is difficult to select woody species with multi-purpose effects enhancing all the above three objectives at the same time. For instance, woody species should be as open as possible to reduce light interception, but open canopies have little biomass, so provide little shade and soil protection, and contribute little to enhancing nutrient or water availability. Desirable characteristics vary according to land-use system and site conditions (Table 5.6).

5.4.3 Production by Woody Plants

Products of woody plants that may be beneficial in agrosylvopastoral land use are wood, fruits and browse material; less important products that are not considered are medicinal products, honey (indirectly), and fibres. The methodology presented in Section 3.3 enables estimates of the production of these products for specific situations, based on canopy cover, the number of leaf layers, and production relations between wood, fruits and browse (Fig. 2.4, Table 3.9 and Sect. 3.3 for the SSWA). The proportions of net production (above- and belowground) represented by wood, browse (foliage, fruits and twigs) and by roots show condiderable variation between sites, e.g. climatic zones in the SSWA. The proportion of wood production varies from 30 to 45% with increasing rainfall from the northern Sahel to the southern Sudan zone; wood production divided by 0.85 gives wood production volume in m^3.

Production estimates of the pre-drought situation in the SSWA are presented in Fig. 3.8 and Tables 2.15 and 3.10 (aboveground) and Figs. 4.30 and 4.31 (total production). Compared to the SSWA, production levels in other semi-arid regions can be higher, as estimated in Section 5.3. Extrapolation to other semi-arid regions should consider the fact that as woody plants grow bigger and older, the proportion of total photosynthesis taken by net production decreases because of increasing expenditures for maintenance (Sect. 3.2). Also the proportion of net production taken by browse decreases. On the contrary, shrub vegetations show relatively high production levels of browse material (Sect. 3.3). The variation in these parameters is not further included in following production estimates, as available evidence from semi-arid regions, apart from the general insight, is not available.

Production by woody plants is generally highest where climatic and site conditions are also best for arable farming, corresponding to loamy and sandy soils in situation IIIa of the southern Sahel and Sudan zones. Here, annual production levels of woody plants are as indicated in Fig. 4.31: in the southern Sahel zone 0.6 and 1.8 $t\,ha^{-1}$ on sandy and loamy soils respectively, 2.6 and 2.9 $t\,ha^{-1}$ in the northern Sudan zone (for sandy and loamy soils respectively) and more than 5 $t\,ha^{-1}$ for both sandy and loamy soils in the southern Sudan zone (based on pre-drought canopy cover).

Apart from woody plants on (potential) croplands, most woody plants are found in areas with marginal site conditions: shallow soils (including slopes), dry areas and moist valleys (in situations II and IIIb respectively). Here, land use can be characterised as sylvopastoral. However, in view of the high susceptibility of these soils for land degradation (water erosion and physical soil degradation) woody covers should be maintained at levels of the pre-drought situations (Fig. 2.4; Table 5.3). On shallow soils and in dry areas, woody plant production levels are low, as can be derivated for the SSWA from Fig. 4.30 for unit Sq (shallow soils): less than 2.5 and 3.6 $t\,ha^{-1}$ in the northern and southern Sudan zone respectively. Where woody plants establish at run-on sites on loamy, shallow soils in the Sahel zones, high woody plant densities reach annual production levels of 1.3–4.2 $t\,ha^{-1}$, whereas average values in the entire unit, including the run-off sites, are a factor 3 lower.

In general, contrary to annual herbaceous plants, woody plant production continues to increase with increasing water availability: whereas in the SSWA herbs reach maximum production levels in the northern Sudan zone and additional water causes increased leaching, woody plants by their deep roots and increasing SOM contents under their canopies can continue to benefit from increased water and associated nutrient inputs. Moist valleys and depressions are an exception, where woody plants reach maximum production levels in the northern Sudan zone, as a result of water stagnation and anaerobic growth conditions.

If considered desirable from a production point of view (wood, fruits, or browse), or to optimise environmental protection goals, woody cover can be enhanced beyond pre-drought levels, e.g. by protection of woody plants against fire and grazing. Figure 2.8 shows the maximum canopy cover that can be reached in the SSWA in relation to rainfall; net production by woody plants can reach 3, 7 and 12 $t\,ha^{-1}$ for the southern Sahel, northern and southern Sudan zone respectively. Of these trees, wood production levels can be doubled, if considered desirable from a forestry point of view, by weeding and the selection of exotic tree species showing rapid growth (Table 3.2). In that case, nutrient inputs (organic or inorganic) are indispensable to sustain such production levels: N outputs by wood only are 4 $kg\,m^{-3}$, as compared to annual N inputs of 3–9 $kg\,ha^{-1}$ in the SSWA (Table 4.13). Fruits and foliage have higher nutrient concentrations (Sect. 3.6), requiring higher nutrient inputs in order to balance their exploitation. For further technical data on tree plantations in semi-arid regions, the reader is referred to the recent summary edited by Felker (1986).

5.4.4 Woody Plants and Production Security and Sustainability

Apart from productive benefits of woody plants in agrosylvopastoral systems, maintaining woody plants has non-productive benefits (e.g. Robins, in press):

1. socio-cultural factors: property rights of woody plants and their products, woody plants used for demarcation, cultural values attached to certain woody species;
2. economic factors: production variation of crops or pastures in relation to that of woody plants, and associated price relationships between products from woody plants (fruits, wood, forage), crops and livestock products;
3. ecological sustainability: (i) "hard" sustainability, i.e. the desirability to maintain natural ecosystems and biodiversity, usually reflecting national or international concerns, maybe aimed at touristic potentials; and (ii) "soft" sustainability, i.e. maintaining long-term ecosystem productivity in order to sustain agricultural livelihood systems.

Security

The potential of woody plants to enhance (economic) production security refers to reducing production variation of the agro-ecosystem as a whole. Improved security due to woody plants can result from reduced variation of production levels and from reduced variation in product quality (nutrient concentrations). Plant production in a mixed vegetation is less susceptible to drought than herbaceous production in a direct and in an indirect way. Direct effects are related to the woody plant production, which is less drought-prone and better distributed throughout the year (woody plants greening earlier and remaining green within the dry season: Sect. 3.5). Indirect effects are related to the beneficial effects by woody plants on plant production factors, and therefore on herbage plant production. Increased security to drought is obvious where woody plants can tap ground water reserves (contrary to herbs). Whereas woody plants appear relatively productive in dry years (e.g. Table 3.5), long-term survival and resilience are lower than in herbs. Indirect effects are due to the increasingly beneficial effects of woody plants on water availability as climatic conditions tend to become more extreme (Table

5.1), and are due to the beneficial effects of concentrating resources under resource-poor conditions (Sect. 5.3.5).

Sustainability

Protective (environmental) benefits of woody plants, oriented at maintaining and enhancing ecological sustainability, can be interpreted as long-term security. The flexibility of an (agro-)ecosystem is one component of ecological sustainability, the reduction of water and nutrient losses by land degradation is another component.

Flexibility is related to maintenance of the regeneration capacity and a diversity of plant and animal species; in short, maintenance of biodiversity. In comparison to an open grassland, the presence of a variety of woody species creates diversity, corresponding to new niches for plant species and animals. For instance, under woody species the growing season is longer, soil fertility is higher, light availability is lower, and the microclimate shows less seasonal variation (Sect. 4.4.2.3). Some herbage species occur almost exclusively under woody canopies (Sect. 4.2.4), whereas the shade conditions under woody canopies offer more suitable conditions for germination and establishment of other woody species. Animals attracted by woody plants are birds and insects in particular, but also specific pests and parasites. These animals may have undesirable effects on crop or livestock production, but may also include predators of important crop pests. Honey-making bees are one other positive feature associated with woody plants. The regeneration capacity of woody plants varies according to species (Sect. 5.3.3).

Spatial concentration of resources in "fertility islands" improves both short-term security and long-term sustainability, particularly under resource-poor conditions (Sect. 5.3.5, Fig. 5.2). Under conditions of water stress during dry years (definition see Sect. 2.1.2), survival is highest where water is concentrated (valleys, depressions, fertility islands). Regeneration may occur in subsequent years, with such areas serving as seed banks from which colonisation takes place (Sect. 2.2.2.2). Also, where nutrients are limiting plant production, C_3 herbs predominate in fertility islands, characterised by a higher nutritious value than C_4 herbs. The latter predominate over large areas when resource availability is poor. Hence, spatial concentration improves plant nutritive values for pastoral use. In a similar way, Brouwer et al. (1993) hypothesise that spatial heterogeneity is helpful to farmers in semi-arid regions as it contributes to spreading of risks.

The other component of ecological sustainability can be formulated as "no increased land degradation", i.e. no negative impact on the water and nutrient balances. Even at low exploitation levels, agrosylvopastoral land use inevitably disturbs water and nutrient balances (which need not be closed in natural ecosystems). Reduced water and wind erosion allow a more effective use of the available resources for production of desirable products, while at the same time maintaining a certain equilibrium.

Table 5.6 indicates for which land-use systems, and in which situations, negative impacts are most severe. Risks for soil degradation vary according to climatic zone and soil type. Sandy-loam soils are particularly susceptible to physical degradation by crust formation, leading to high run-off. Although the associated soil and nutrient losses are limited (sheet erosion), severe erosion and soil losses occur down-stream, where run-off water collects. It leads to silting of dams, excessive peak flows and flooding of valleys, and insufficient recharge of groundwater. Indeed, these phenomena are widespread in virtually all semi-arid regions. The minimum vegetative biomass required to avoid crust formation and water erosion on loamy soils in the Sahel is about $0.7\,t\,ha^{-1}$ where rainfall is 250 mm, $2.0\,t\,ha^{-1}$ at 500 mm, and $2.5\,t\,ha^{-1}$ at 700 mm rainfall (Breman and de Ridder 1991). In the absence of woody plants, biomass is insufficient to provide such protection in units Pl and Sq (Fig. 4.30). Therefore, without woody plants such soils cannot be utilised (exploited) in a sustainable way.

Wind erosion causes losses of nutrients (by transport of fine soil particles). Most important are the abrasive effects of wind erosion on plant seedlings by wind-blown soil particles (Lyles and Woodruff 1960), and mechanical stress by air movements, influencing plant physiology (Stigter 1988). The minimum biomass required to limit wind erosion is less than the estimates given above for physical degradation of loamy soils.

Unfortunately, woody plant characteristics that are desirable to reduce water and wind erosion (Tables 5.4 and 5.5) often show severe competition with herbs (e.g. shallow roots, rapid growth), and should preferably not produce desirable products (e.g. be unpalatable) in order to maximise the return of nutrients and organic matter to the soil.

The reduction of water and wind erosion may take place directly or indirectly. Direct ways focus on reducing the impact of wind, rainfall, run-off and radiation. Here, woody plants are not necessarily more suitable than herbaceous plants. Advantages of herbaceous plants are their high cover of the soil, the small distance between soil surface and plant material, and their limited water use. However, on soils susceptible to physical degradation, additional protection by woody plants is essential (see above).

Indirect soil protection occurs by the improvement of the soil structure, of which maintenance of SOM contents is crucial (Sect. 4.6.1). For maintenance of soil structure Pieri (1989) considered $6\,g\,C\,kg^{-1}$ soil as a minimum SOM concent. Woody plants are more effective in maintaining high SOM contents than herbs. On the basis of assumptions on organic matter dynamics (Sect. 4.6), one can calculate that in the southern Sahel zone (500 mm annual rainfall), in order to maintain minimum SOM contents, are required at least 5.0 t ha^{-1} of herbaceous biomass without woody plants, or a 23% canopy cover. The latter value is obtained from Table 4.28, after adaptation of the estimates for croplands to the natural vegetation, where decomposition rates are lower. In order to maintain sufficiently high SOM contents organic matter should include poorly decomposable components (e.g. wood).

5.4.5 Sylvopastoralism

5.4.5.1 Pastoralism in Semi-Arid Regions

Where plant production is limited by water availability (situation II), most important for determining the potential of pastoralism is the fact that nutrient concentrations of forage (nutritive quality) remain high, contrary to situations where plant production is limited by nutrients (Penning de Vries and Djitèye 1991). This pattern is characteristic for (semi-)arid regions (Breman and de Wit 1983).

In the SSWA, average N concentrations in herbaceous biomass at the end of the growing season decrease from 30 g kg^{-1} in the northern Sahel to 5 g kg^{-1} in the southern Sudan zone. Net production, increasing in proportion with rainfall towards the south (Fig. 4.30), therefore shows an inverse relationship with nutrient concentrations, for soils with similar levels of soil fertility (Penning de Vries and Djitèye 1991). Hence, the zones with high pastoral potential have low plant productivity, and show great production variability due to rainfall variability. Whereas in situation II (Fig. 5.1) pastoral production is limited by forage quantity, of which a considerable proportion is also required for soil protection, in situation III forage quality is most limiting.

The importance of woody plants in sylvopastoral systems varies according to herbivores being grazers, for which herbaceous material constitutes the majority of the feed intake, or browsers, for which woody plant foliage is an important source of feed. Common grazers are cattle and sheep, common browsers are goats and camels. Grazers utilise woody plant forage in the dry season only, when herbage biomass of sufficient quality is no longer available. Whereas during the rainy season the proportion of woody plant forage in the daily menu of cattle does not exceed 10%, during the dry season this proportion may reach 20% in the Sahel zones and 60% in the Sudan zones (van Raay and de Leeuw 1974; Diallo 1978; Carew et al. 1980; Dicko-Touré 1980; Piot et al. 1980; ISRA 1983; Dicko and Sayers 1988; Leloup and Traoré 1989; Egging 1990). These data are obtained by measurements of the time spent consuming a certain type of forage, assuming that it reflects intake. The proportion of woody plant forage in the daily menu of sheep is probably similar to that of cattle, whereas for goats and camels it is much higher: in the dry season up to the entire goats' feed intake may consist of woody plant forage (Wilson 1969 and 1977; Carew et al. 1980; Piot et al. 1980; ISRA 1983; Cissé 1985).

Before assessment of the influence of woody plants on pasture production and the contribution of browse, it should be kept in mind that several factors limit the possibility to utilise browse material. Factors involved are accessibility, digestibility, palatability and phenology.

1. Most of the woody plants forage is beyond the reach of livestock. The maximum height to be reached by livestock varies from about 1 m for sheep and goats (although goats may climb trees), 1–2 m for cattle and more than 3 m for camels (ILCA 1975; Hiernaux 1983). The average maximum height that livestock can reach is 2 m. The proportion of foliage accessible in the layer of 0–2 m height varies from 25% in the Sahel zones to 15% in the Sudan zones (Le Houérou 1979; Hiernaux 1980; own unpubl. data). These estimates concern green leaves and fruits, the fallen fruits included. Variation in accessible forage is large: in woodlands in the Sudan zone it varies from less than 10% to more than 50% according to vegetation type (own unpubl. data). Vegetation changes caused by the drought periods (Sect. 2.2) have led to an increased proportion of browse being accessible. Likewise, intensive exploitation increases accessibility: in intensively exploited woody vegetations 70–90% of woody plant forage is accessible (ILCA 1975; Cornet and Poupon 1977; Grouzis 1979), but the quantity of accessible forage is not high since total productivity is low. Bush encroachment may occur (Sect. 2.4.3.2), leading to dense bush formations, impenetrable for most herbivores and involving relatively unpalatable species mainly. The influence of bush-fires is twofold: high accessibility after the fires (drop of leaves), and low availability during the rest of the dry season. On young fallowlands shrubs dominate, but the amount of available forage is not particularly high. Accessibility can be increased by deliberate lopping of woody plants, a common practice in semi-arid regions, threatening woody plant survival (Sect. 2.4.3.1).
2. The quality of browse biomass is characterised by high N concentrations (Tables 3.14–3.16). Intake of forage with high N concentrations should enable browsers to benefit more from herbaceous biomass with low N concentrations (Breman and de Ridder 1991). Assuming a minimum N concentration of total ingested feeds of 7.5 $g\,N\,kg^{-1}$ dry matter (maintenance level), 100 kg browse would enable intake of an additional 500 kg of herbaceous biomass. However, digestibility of browse material is low: the average digestibility of organic matter of woody plants foliage is only 500 $g\,kg^{-1}$, which is lower than that of green grasses. High tannin, lignin and silica concentrations in

woody plant foliage are the main underlying factors (Sect. 3.6). Low digestibility reduces effective N uptake from browse material. Therefore, the above calculation is probably not correct. Moreover, browsers are unlikely to ingest five times more herbaceous than browse material.

3. A third factor reducing browse uptake is its palatability. More woody species seem to be palatable to goats, sheep and camels than to cattle, but there are few unpalatable species (Table 3.17). Palatability refers to the combination of preference and voluntary intake. Preference of woody plants browse is related to factors such as leaf size, presence of thorns and plant morphology. The voluntary intake of a feed may be defined as the quantity ingested when animals are offered excessive feed (Minson 1990). It is strongly related to feed quality (Ketelaars and Tolkamp 1991). Important animal factors determining feed intake are the type of animal, its age, its appetite and the availability of water (Benjamin 1980). During drought years, browsers consume virtually all available browse.
4. Phenological patterns are a fourth factor limiting the possibility to utilise the browse biomass produced. At the end of the dry season, when forage is most scarce, the average proportion of annual browse production that is available varies from 25% in the northern Sahel to 5% in the southern Sudan zone (Table 3.13).

5.4.5.2 Woody Plants and Forage Availability

Where the quantity of animal feeds is limiting (situation II), woody plants can be beneficial for pastoral systems by their potential to increase water availability. Increase is most important in the Sudan zones (Table 4.8), for loamy and shallow soils where situation II occurs (Table 5.3: situation IIb). Here, most important is the process of soil structure improvement indirectly leading to increased water infiltration, run-off being the process limiting plant production in situation IIb. Desirable characteristics of woody plants are those increasing water availability (Table 5.4), and those related to survival under arid conditions. In order to improve soil structure (referring to SOM contents mainly), the woody plants should be productive and as little as possible exploited (Sect. 4.8.4).

In the above situation, benefits by woody plants are about 400 kg $ha^{-1} yr^{-1}$ increase of forage production on loamy soils in the Sudan zone (on shallow soil improvement is less). This value is based on a reduction of run-off from loamy soils from 65 to 50% due to the presence of woody plants. On loamy soils the average value for run-off at undegraded rangeland is 50% (Breman and de Ridder 1991). Woody cover to achieve such reduction of run-off is about 30%, corresponding to pre-drought canopy cover (Fig. 2.4). Run-off reduced to 50% leads to water availability increasing from 225 to 340 mm yr^{-1}, causing a shift from water-limited to nutrient-limited production (Fig. 5.1: situation IIIa), corresponding to a total plant production increase from 2.4 to 3.5 t $ha^{-1} yr^{-1}$ (Fig. 4.32). The additional plant production useful as animal feed (aboveground herbage production and fraction of browse that can be exploited in a sustainable way) is less than 40% (compare Figs. 4.30 and 4.31), corresponding to about 400 kg ha^{-1} feed production. Free-grazing animals will be able to utilise 30–50% of this additional production.

5.4.5.3 Woody Plants and Forage Quality

Feed quality limits animal production in situation III (Fig. 5.1). Improved forage quality by the presence of woody plants is related to the nutrient concentration effect of woody plants and secondly on improved water availability, leading to the following advantages, particularly in resource-poor environments:

- higher nutrient concentrations of browse (Tables 3.15, 3.16 and 3.17);
- the prolonged growing season under woody canopies (Sects. 4.4.3.2 and 4.8.3.3);
- the prolonged availability of woody plant forage in relation to herbaceous biomass (Sect. 3.5);
- the replacement of C_4 by C_3 species, the latter having higher nutrient concentrations (Sects. 4.8.2 and 5.4.2.3).

Other improvements require a relative increase of nutrient availability in relation to water availability. This can be achieved by reducing water availability or by increasing nutrient availability. The first process leads to increased forage quality on degraded soils showing increased run-off, but is of course undesirable from an ecological point of view. Here increased forage quality is at the expense of long-term sustainability.

Positive influences are most important in the Sudan zones and on loamy soils in the southern Sahel zone (Table 4.22), with beneficial characteristics of woody plants as indicated in Table 5.4 in relation to positive impacts on nutrient availability. Here, most important are the processes that reduce nutrient losses (Table 4.22). Improved soil structure is a key factor everywhere; as rainfall increases, deep and dense rooting becomes another important issue.

Figure 4.30 shows in which situations, and up to which woody cover, the presence of woody plants leads to improved feed availability for grazers, i.e. increased herbage production. The benefits are greatest in the northern Sudan zone, except for the geomorphological unit La. The greatest effect is observed on sandy soils in the northern Sudan zone, where the production of the herbage layer increases from 2.2 to 3.1 t ha^{-1}. Of the herbage production approximately 35% is available for pastoral use, due to various losses (Breman and de Ridder 1991). Moreover, during the dry season (8 months in the northern Sudan zone), a fraction of the herbaceous biomass is of insufficient quality to serve as animal feed. This fraction is 20% in a mixed vegetation, because of the higher quality of herbs under woody canopies, whereas it is 40% for C_4 herbs in the absence of woody plants. For grazers, not using browse material, the annual feed increase due to woody plants is therefore about $0.4\,t\,ha^{-1}$. Important conditions are low exploitation intensity of the woody plants, crown diameter about equal to trunk height, and a homogeneous distribution of the woody plants in the vegetation (Sect. 4.8.5).

For browsers the benefits due to woody plants are more pronounced. Browse production on sandy soils in the northern Sudan zone is about $1.2\,t\,ha^{-1}\,yr^{-1}$ (Table 3.8 and pre-drought canopy cover). However, protection of the fraction related to the internal N pool signifies that only half of it should be available (Fig. 4.30). Moreover, in order to obtain maximum beneficial effects of woody plants on nutrient availability, leading to improved herbage production as estimated above, the available browse from woody plants should not be exploited at all, or at very low levels only. The more browse, or other woody plant products, are exploited, the less are the benefits of woody plants on plant production factors. Based on the four factors mentioned above, limiting the use of woody plant forage (Sect. 5.4.5.1), the average fraction of browse that can be utilised during the dry season is estimated at 5% of the annual production. Supposing that at least this fraction may be exploited without risk for the beneficial effects, it is estimated that for pastoralism the sustainable benefit to be obtained from woody plants, in the vegetation on sandy soils in the northern Sudan zone, is 400 kg (herbs) + 60 (5% of browse) + 300 (additional herbage intake due to high browse quality) = $760\,kg\,ha^{-1}$. In that case feed quality is at the minimum level of $7.5\,g\,N\,kg^{-1}$, which permits animal production at maintenance levels only.

With the data presented in Section 4.8 (Fig. 4.30 in particular), similar estimates can be made for other situations (combinations of climatic zones and geomorphological units). Using insights presented in Section 5.2, an extrapolation to other semi-arid regions can be made as well.

5.4.5.4 Woody Plants and Forage Security

For pastoral systems, the variation in production levels is a problem in situations of limited forage quantities (situation II). Here, the reduced susceptibility of woody plants to drought (Sect. 5.4.4) leads to a lower reduction of woody plant production levels due to drought, than for herbaceous plants. Hardly any data are available for quantifying such effects. Table 3.5 provides some relations for foliage production in dry and average rainfall years, in the northern Sahel zone. Whereas herbage production in years with annual rainfall of 33 mm is virtually nil, woody plant foliage production varies from 25–50% of production levels in the previous year with average rainfall. The effects are highly variable between woody species, and also vary between different sites. Breman and de Ridder (1991) estimated browse production in a dry year on the basis of proportionality with the amount of infiltrated water. Sections 4.8.5 and 5.3.4 present a restricted support for this approach: in the vicinity of woody plants conditions of water-limited production persist up to higher water availability than in case of herbs. Expressed as a proportion of production in an average year, browse production in a dry year varies from 60% in situation IIa to 80% in situation IIb, against less than 40% for the herbage production in situation IIa to 70% in IIb. Since canopy cover is low in areas where plant production is limited by water (Table 5.3), browse as a proportion of total forage production is low. The advantage due to less impact by drought years is therefore little: in situation IIa herbage production in a dry year is more than 500 kg ha^{-1} below average, whereas browse availability is only 50 kg ha^{-1} below average, based on pre-drought canopy cover values. Browse availability can be increased fourfold by cutting and lopping of trees, which is of course not sustainable (Sect. 2.4.3) and reduces the beneficial impacts of woody plants. However, in situation IIb woody cover is higher (Table 5.3) and the contribution of browse to feed availability in dry years is more important.

For the majority of pastoral systems plant production is not limited by water availability and forage quality is the major problem (situation III). In mixed vegetations and under woody canopies in particular, plant production is more easily limited by water, thus reducing plant production in an absolute sense (Sect. 5.3). Nutrient concentrations of woody plants foliage and fruits show less variation between sites and years than for herbaceous biomass (Table 3.16). For instance, average N concentrations of foliage in the SSWA decreases from 27 to 22 g kg^{-1} from North to South, whereas those for herbage biomass in the same zone decreases from 25 to 5 g kg^{-1} (Sect. 3.6). Average N concentrations of old leaves are 40 to 80% of those of young leaves, but herbaceous biomass at the end of the growing season has N concentrations which are 30 to 60% of those in the early season. This has major advantages for enhancing security by the availability of high quality forage during the dry season. Benefits by woody plants are related to the gradual spatial concentration of nutrients, the efficient internal recycling under resource-poor conditions, nutrient and water stocks associated with woody plants, more widespread root systems, and genetic characteristics. Obviously, intensive exploitation disrupts these dynamics, and as soil fertility decreases, forage quality decreases (Sect. 3.6).

The sustainability objective for woody plants in sylvopastoral systems is not compatible with production and short-term security objectives. Whereas the first requires unpalatable woody plants, in order to have maximum benefits on soil structure and other water- and nutrient-improving processes, the use for pastoralism requires palatable species with maximum proportions of production being available for livestock feeding. From an ecological ànd productive point of view, it is most effective to utilise sites with optimal conditions for woody plants, i.e. where maximum potential for niche differentiation exists, for the establishment of fodder banks. This is the case in situation III, where subsoil water is available (Fig. 4.28), particularly on sandy soils with run-on or high rainfall and on loamy soils with relative high water availability, not yet causing anaerobic soil conditions. Here, selection of woody species and their rational management may lead to annual browse production levels of 2.5 t ha^{-1} in the southern Sahel zone and 7 t ha^{-1} in the southern Sudan zone. Management should include the selection of leguminous N_2-fixing woody plants with P supply to boost their N_2-fixing capacity.

5.4.5.5 Consequences

The improved animal production obtained from the benefits of woody plants as outlined above, can be evaluated in terms of the impact on livestock carrying capacity. The livestock carrying capacity can be expressed as the maximum stocking rate that:

a) allows a level of animal production corresponding to 50 kg annual liveweight gains of young heifers, considered minimum growth rates that enable pastoral systems to survive (Breman and de Ridder 1991), or
b) in addition to (a) allows ecological sustainability.

The maximum increased forage production of 400 kg ha^{-1} by reintroducing woody plants to pre-drought canopy cover on loamy soils in the Sudan zone to improve water infiltration from 35 to 50% in situation IIb (Sect. 5.4.5.2), doubles the livestock carrying capacity in dry years. This corresponds to an additional 0.07 TLU ha^{-1}, producing only 4 kg liveweight and 10 l milk annually.

The maximum increased forage production of 760 kg ha^{-1} by reintroducing woody plants to pre-drought canopy cover on sandy soils in the northern Sudan zone in situation III (Sect. 5.4.3.3), leads to a more important increase of animal production. An annual animal growth of 50 kg requires forage quality of at least 10 g N kg^{-1}, reducing the estimated available forage production by 50%. For a herd composition of grazers and browsers, the additional 380 kg corresponds to an additional 0.18 TLU ha^{-1}, producing 10 kg live weight and 22 l milk annually.

Obviously, animal production gains to be obtained from fodder banks as proposed (Sect. 5.4.5.4) are much higher.

For production objective (b), including the sustainability criterion, the livestock-carrying capacity is less improved, because the woody plants should provide maximum benefits for improved soil structure and water and nutrient availability. Wood exploitation is commonly proposed as an alternative to forage exploitation. Although wood has lower nutrient concentrations than foliage and fruits, the nutrient losses and the strongly reduced protective properties associated with lopped or cut woody plants are undesirable if sustainability is the main objective. Besides, wood litter plays an important role in maintaining SOM contents.

Specific recommendations are given in Section 5.4.8.

5.4.6 Agrosylviculture

5.4.6.1 Arable Farming in Semi-Arid Regions

Arable farming (crop production and horticulture) is not possible where plant production is limited by water (Fig. 5.1: situation II). Where crops are produced under arid conditions, one is dealing with patches of situation III: valleys, oases, depressions. Pumping of water also creates situations of plant production being limited by nutrients. Where land is fertilised, water may become more limiting due to *relatively* low rainfall, corresponding to increased risks of dry periods for well developed crops in the driest parts of situation III.

The potential role of woody plants to improve crop production, security or sustainability (Table 5.6) may be realised using woody plants with specific characteristics (Tables 5.4 and 5.5). For arable farming, in situation III mainly, the focus is on characteristics increasing nutrient availability and reducing water erosion, but increased water availability leading to a prolonged growing season is important as well (Table 5.4). Characteristics of associated crops also determine the potential of woody plants to have beneficial effects.

5.4.6.2 Prolonged Crop-Growing Season

Crop growth under conditions of nutrient limited plant production is also influenced by the length of the growing season. Woody plants improving water availability can increase the length of the crop-growing season by stimulating early germination and plant establishment and by a prolonged growth period following establishment.

Plant germination and establishment highly depend on unpredictable rainfall distribution. Patterns of plant germination and establishment under woody canopies are determined by the combination of germination rate and rainfall distribution: plants rapidly germinating under woody canopies may not survive a dry spell during the early growing season (Sect. 4.2.4). The reduction of wind erosion is also beneficial for plant establishment (Stigter, in press). Wind erosion is particularly important on sandy soils, which in the SSWA are found in the northern part mainly, where rainfall is relatively low. Here, possibilities for niche differentiation between woody plants and herbs are limited, because of homogeneous water infiltration on sandy soils (Sect. 2.3). Thus, where woody plants are beneficial for early plant establishment, competitive relationships predominate, affecting subsequent crop growth. Highest potential for woody plants to lead to a prolonged growing season that is beneficial for crops, occurs in rare situations of high groundwater table, e.g. in oases and wadis (Sect. 4.4; Smith et al., in press). Here the characteristic "stimulating stemflow" is most important (Table 5.4). The woody plants are preferably oriented perpendicular on the prevailing direction of the winds, acting as wind barriers effectively decreasing wind erosion.

Woody plants can cause a prolonged growth period by increased water availability as a result of improved water infiltration and/or an increased transpiration/evaporation ratio. The prolonged growth period is estimated at a maximum of 25% in comparison to the situation without woody plants (Sect. 4.4.3). Improved water infiltration is related to improved soil structure and SOM contents mainly; the processes involved also enhance nutrient availability. To increase the transpiration/evaporation ratio a sufficient overall water availability is required, because the absolute water availability for the crop decreases in spite of the prolonged growing season (Sect. 4.8.3.3). As a result, an increased transpiration/evaporation ratio due to woody plants can only be expected in areas with run-on (situation IIIb) and in situation IIIa where rainfall is high. For crops to benefit from the prolonged growing season, the following conditions should be met:

- soil moisture remains in the topsoil;
- radiation interception by woody plants does not exceed 50%;
- soil fertility under woody plants is sufficiently high;
- roots from woody plants are predominantly deep.

These conditions are rare in situations where the need for a prolonged growing season for crops is highest, i.e. the driest parts of situation III. On the contrary, these conditions are met in the most humid parts of semi-arid regions, where soils are deep and fertile. Here, the benefits due to woody plants are interesting in densely populated regions with a high demand for (fire-)wood and food, and sufficient labour available (Sect. 5.4.7).

5.4.6.3 Increased Nutrient Availability

Regarding the potential of woody plants to improve nutrient availability and SOM contents, most important are the processes of reducing losses, by water erosion and by leaching mainly (Table 4.22). The reduction of water erosion is related to the improved SOM content; processes involved are indicated in Table 4.22. For the reduction of wind erosion, most beneficial are woody plants with a low ratio of trunk height to crown diameter (shrubs are best) and with shallow roots (Table 5.4). These characteristics are not compatible with arable farming, except for the rare cases described in the preceding section. Leaching is reduced by a high root density and by deep rooting. Deep rooting also allows herbaceous plants to benefit from the increased nutrient availability.

Regarding ecological site conditions to maximise nutrient improving processes by woody plants, most beneficial are sandy and loamy soils (units Os+Ps and Pl), as can be observed in Table 4.34 and Fig. 4.30. The potential benefits are probably highest on sandy soils, because on croplands with loamy soils it is unlikely that woody plants can completely suppress high water erosion rates (Table 4.15), contrary to sandy soils. Average erosion rates on loamy soils with woody vegetations are much higher than on sandy soils (Breman and de Ridder 1991). Here, other soil and water conservation measures are required, and can be at least as effective as integrating woody plants.

In unit La no positive effects can be expected (Fig. 4.30d), whereas on shallow and rocky soils (unit Sq) positive effects are possible (Fig. 4.30c), but cropping on such soils requires soil conservation measures to avoid accelerated water erosion.

The maximum positive effect of woody plants on N availability for the herbage layer (i.e. crops) increases from 3 $kg\,ha^{-1}$ in the southern Sahel zone to 6 $kg\,ha^{-1}$ in the northern Sudan zone (Sect. 4.8.5). This corresponds to at most 200 to 400 $kg\,ha^{-1}$ additional cereal yields, provided that crops are predominantly shallow-rooting and of the C_3 photosynthetic pathway, and that all crop residues are recycled into the soil. For instance, deep-rooting millet shows less beneficial effects than shallow-rooting sorghum (ICRISAT 1993).

In the southern Sahel zone, this maximum effect can be expected on loamy soils only, since elsewhere woody cover is insufficient. In the Sudan zones the conditions for maximum effects are canopy cover of at most 35% and two leaf layers (Sect. 4.8.5.1), provided that the woody plants are regularly distributed (reducing shade effects), have a high trunk height to crown diameter ratio, and are not exploited. As soon as one or more of these conditions is not met, canopy cover should be less in order to maintain maximum benefits for crops. The condition of no exploitation of woody plants is essential (Table 4.35), but most unlikely to be met. This is probably why most agrosylvicultural systems do not reach the estimated improved crop production values. When continuously cropped, croplands with trees usually show average cereal production levels that are lower than those on croplands after a sufficiently long period of fallowing. On croplands in the SSWA, a maximum canopy cover of 20% with two leaf layers is probably most realistic.

In addition to the conditions given above, maximum beneficial effects on crop production are based on the assumption that 5% of woody plants and crops are N_2-fixing leguminous plants (Sect. 4.8.4.3). In view of the limited potential of woody plants to increase P availability by deep rooting and rhizosphere interactions (Sect. 4.5.4), and the lower enrichment factor for P than for N (Fig. 4.32), most agroforestry systems require additional P inputs in order to sustain increased crop yields.

In other semi-arid regions, situation III occurs with lower annual rainfall than in the SSWA (Sect. 5.2), with advantageous consequences as described above. Moreover, where in other semi-arid regions inherent soil fertility is higher, the beneficial effects of woody plants on crop production can be greater, particularly where P availability is higher (including possibilities for uptake from the subsoil). Theoretically, the beneficial effect can be two to three times higher (Sect. 5.3.4), but, since such effects are associated with higher woody cover, it will be difficult to reduce competition to such an extent that only crops can benefit from the increased nutrient availability. Reducing woody cover in order to reduce competition, as necessary in the southern Sudan zone (see above), should be evaluated against the reduced nutrient enrichment with reduced woody cover.

5.4.6.4 Light Competition

By light interception, situations are created where crops cannot make optimal use of the positive effects of woody plants on water and nutrient availability (Sect. 4.8.5.1; Fig. 4.32). In that case woody plants benefit most from the improved conditions, further suppressing crop production by light interception.

Light interception can be reduced by selection of woody species (Table 5.5) and by appropriate management. The number of leaf layers should not exceed two in order to limit the degree of light interception (Sect. 4.7.2). However, isolated trees on croplands in the Sudan zone easily reach four leaf layers (Table 3.10). Therefore, a high ratio of trunk height to crown diameter is required for maximum accessibility of diffuse and lateral light under the canopy, and homogeneous spreading of trees. The trees may be pruned to maintain an open canopy, allowing light to penetrate. Alternatively, the woody plants are concentrated on a limited surface, with biomass close to the surface (shrubs), such as in alleys with north-south direction, or along the contours if the objective of creating an anti-erosion barrier is more important, or perpendicular on the prevailing wind direction. When woody plants are concentrated, litter should be distributed over the entire cropland in order to evenly spread nutrients and organic matter.

Most crops grown in the SSWA being of the C_4 type, the general pattern in the parklands is that of crop yields under the tree canopies being lower than in the open field (Kessler and Boni 1991; Kater et al. 1992; Kessler 1992). Studies from other semi-arid areas show similar patterns, either in relation to isolated trees, or to wooded alleys or windbreaks (Singh et al. 1989; Malik and Sharma 1990; Lal 1991). A correlation was found between light availability under canopies of *Parkia biglobosa* and *Vitellaria paradoxa* and sorghum yields, in spite of improved nutrient availability under the tree canopies (Kessler 1992). It appears that, here, a reduction of radiation by about 50% leads to significant reductions in crop yields. These findings correspond to influences of shading by slash pine on partridge-pea production (Harshbarger and Perkins 1971). *Faidherhia albida* is unique because it sheds its leaves during the growing season. Advantages are not only limited light competition with crops during the rainy season, but also the fact that the tree utilises few nutrients and water during the growing season, thus limiting competition with crops. However, foliage production and tree growth in the dry season are limited by low water and nutrient availability in this period. The tree appears to depend on groundwater, but cannot be expected to obtain much nutrients from the subsoil. Therefore, build up of improved soil conditions is slow (e.g. by hydraulic lift). The beneficial effects of *Faidherbia albida* on soil fertility are not unique (Table 4.10), and associated with old trees mainly.

Besides reduced light availability, under woody canopies air humidity is commonly higher (Sect. 4.4.2.3), which provides suitable conditions for some diseases. Crops under the canopy of *Vitellaria paradoxa* are infested with the fungus *Spacelothera*, whereas no infestations occur in the open field (Kessler 1992). Temperatures lower under canopies than in the open field also lead to lower growth rates of C_4 crops. It is commonly observed that C_4 crops under dense canopies show retarded maturity (besides lower yields). Under shade conditions C_3 crops can more effectively benefit from the improved water and nutrient availability under canopies than C_4 crops. In the SSWA, crops grown under woody canopies during the rainy season are igname, peppers, tobacco and sweet potato. Farmers report that their productivity is highest under canopies (Kessler and Boni 1991). During the dry season, horticulture associated with woody plants may be particularly profitable, where irrigation enables plants to benefit from the high soil fertility in the topsoil under woody plants. Woody species should be selected that do not resprout when vegetable crops are irrigated. Such an integration of trees and crops is effective because water and nutrients are both found in the topsoil during crop growth and tree production, contrary to the *Faidherbia albida*–crop combination. On the other hand, most vegetables are C_3 plants, that can benefit from the shade of woody plants, even during the dry season. Here, *Faidherbia albida* may be of particular importance. In the preceding, it should be kept in mind that the relationship between soil fertility and light availability is complex (e.g. Eriksen and Whitney 1981). Fertilised tobacco and tea show better growth in full sunlight than in the shade, contrary to unfertilised crops (Murray and Nichols 1966).

5.4.6.5 Woody Plants and Food Security

In West Africa, fruits and leaves from woody plants are traditionally referred to as "hunger crops" (Raison 1988), because in the dry season and in dry years these products may be the only reliable food resources. Underlying reasons were elaborated in Sections 5.4.4 and 5.4.5.4. Fruits and leaves from woody plants also provide essential contributions in terms of food quality: certain vitamins and minerals are largely, or exclusively, obtained from tree crops (FAO 1989b; Bergeret and Ribot 1990). They are ingredients of most traditional meals.

Security is also enhanced by incomes derived from tree products. Kessler (1992) demonstrated how economic benefits resulting from the exploitation of tree crops are higher than the losses resulting from the reduced cereal yields due to shading by the woody plants on croplands. The economic analysis is, however, of doubtful validity, since cereals and tree crops are for home consumption mainly. The market would be saturated rapidly if farmers produced more tree crops in order to benefit from the positive economic trade-off between cereal loss and tree crops.

5.4.6.6 Woody Plants and Fertilisers

The influence of woody plants on soil fertility is more pronounced for N than for P (Fig. 4.13). Therefore, in vegetations with woody plants, and under woody canopies in particular, P is more likely than N to be, or to become, a limiting factor. Consequently, fertilising with P is most beneficial and can have synergistic effects by enhancing N_2 fixation. The returns to associated investments are obtained in the long term only. Before P application becomes effective, large P doses may be required to overcome the soil's capacity for P immobilisation, as encountered in several areas (e.g. Szott and Kass 1993).

In vegetations with woody plants and under woody canopies, applying fertilisers (organic or inorganic) is more effective because here, as compared to vegetations without woody plants or the open field, nutrient losses are less (Table 4.22). The processes leading to reduced nutrient losses do not show a negative feed back with improving soil fertility (Sect. 5.3.4), so that their impact increases with increasing soil fertility (Table 5.2). In other words: the combination of high soil fertility, resulting from natural or artificial factors, and the presence of woody plants, leads to synergistic effects due to an increased efficiency of the use of the available nutrients. For the SSWA this effect may be quantified as follows: average annual N availability for herbaceous production varies from $10\,kg\,ha^{-1}$ in the southern Sahel zone to $25\,kg\,ha^{-1}$ in the southern Sudan zone (Breman and de Ridder 1991). The increase of N availability due to woody plants is $3-6\,kg\,ha^{-1}$ (Sect. 5.4.4.3), corresponding to a maximum reduction of nutrient losses of about 25% (assuming that other processes have insignificant importance). Hence, crops on a cropland with woody plants receiving annual rates of $120\,kg\,N\,ha^{-1}$, may be able to utilise $30\,kg\,ha^{-1}$ more than would have been the case on a cropland without woody plants.

Where the improvement of nutrient availability under woody plants is a factor 2–3, plant production becomes limited by water (Fig. 4.32; Sect. 5.3). Therefore, increased nutrient availability due to the combined effect of fertilisation and the presence of woody plants can be more effectively utilised if nutrients and/or organic matter are distributed in the open field. This can be achieved by actively lopping woody plants and spreading vegetative material over the entire cropland, with dual benefits by reduced shading and improved soil fertility. Wood should be as much as possible left for decomposition. Poorly decomposable woody material is a crucial element for build up of SOM (Sect. 4.6).

In the process of agricultural intensification, where cropping systems include the use of fertilisers and make proper use of crop residues and soil conservation measures to maintain SOM contents, woody plants lose their additional value, as outlined above. In a general sense, woody plants lose their value as "fertility islands" in resource-rich environments (Fig. 5.2). Also, as crop yields increase, the absolute yield losses due to shading by woody plants increase. This actually is the major reason why trees are removed where cash crops are grown with the use of fertilisers. The removal of trees is, however, rarely justified because (a) alternative measures to maintain SOM contents are not taken, and (b) the value of woody and non-woody products of woody plants are neglected.

In most semi-arid regions, particularly in the SSWA, SOM contents are generally insufficient to allow efficient use of inorganic fertilisers (Pieri 1989), whereas the availability of rangelands in relation to croplands is insufficient to produce sufficient organic fertilisers to maintain SOM contents (Breman and Traoré 1987). Therefore, in the process of agricultural intensification woody plants are a crucial component by their soil protective functions mainly. Their role can be maximised by selection of woody species, by specific management and by concentrating efforts in specific areas, as outlined in previous sections.

5.4.7 Woody Plants and Socio-Economic Conditions

In the preceding, it has been assumed that the farmer is and remains primarily interested in sustained agricultural production. So crop and pasture production levels are generally not accepted to decline under the influence of woody plants. This assumption is not correct in densely populated (urban) areas, where the price of wood (construction wood mainly) has increased so much that it has become profitable for farmers to transform their cereal crop into one with rapidly growing trees (e.g. *Eucalyptus*). Also, it is well known that without regulations on land tenure and sufruct rights of woody plants, any management of woody plants aimed at maximising their beneficial effects as outlined above, cannot be expected. Such regulations are a sine qua non for any technical intervention proposed in this chapter. The complexity of these socio-economic factors may be illustrated by the case of semi-arid northern India, where the combination of unequal land distribution and labour shortage has led to large-scale *Eucalyptus* plantations. However, as the anticipated yields and profits from *Eucalyptus* farming dwindled very fast after 1987, many farmers uprooted their trees and returned to annual crops (Saxena 1992). Several other woody plant products have economic potential, often well-known in traditional land-use systems, but rarely is the economic potential such that large-scale tree plantations are profitable.

The optimal management of agroforestry systems requires a considerable amount of additional labour, as compared to land-use practices prevailing today in most semi-arid regions. This may be one reason of the preference for multi-purpose trees, although their benefit is always less than adding the maximum productive benefits for every separate goal (purpose). Labour is a production-limiting factor for most agricultural systems in semi-arid regions, the demand for which is highly concentrated during the onset and the end of the rainy period (Brossier and Jager 1983; Veeneklaas et al. 1991). Regular pruning of woody plants and spreading of vegetative biomass over croplands or of feeding foliage to livestock is highly labour-intensive, and will (partly) coincide with the periods of peak labour demands. Where sufficient labour is not available, it may be one reason for not optimising the role of woody plants on croplands or in rangelands. Although less productive, spatial segregation of agriculture and forestry is certainly less labour-intensive, but requires more external inputs to maintain production levels.

Examples of intensive tree management are from areas with relatively high population densities in semi-arid regions mainly. In the SSWA, examples are from the Mandara mountains in North Cameroon, where pruning of trees and live fences are common practices (Hallaire 1991), and from the Mossi plateau in Burkina Faso, where trees like *Parkia biglobosa* are planted and rationally pruned in order to reduce competition with crops and to improve fruit production (Timmer 1994). However, alley cropping and horticulture associated with woody plants require more labour than these indigenous systems. Such labour-intensive agroforestry systems may be most feasible around urban centres, in view of both profitable markets and labour availability.

The large-scale environmental benefits associated with the maintenance and proper management of woody plants in agrosylvopastoral systems may incite donors to subsidise activities leading to improved woody cover.

5.4.8 Conclusions and Recommendations

The following conclusions and recommendations are based on the previous sections, and are focused at improving the sustainability (long term security) of agrosylvopastoral land-use systems.

For sylvopastoral systems, three situations have been described where the integration of woody plants with livestock keeping has a synergistic effect.

1. In situations of water-limited production, where forage quantity is the limiting factor for pastoral systems (situation II), woody cover should be as high as climatic and site conditions allow. From a sustainability point of view, woody species should be as little as possible exploited. Their role in protecting the soil is crucial. Reforestation of situations IIb improves water availability (reducing run-off). It increases security and sustainability of pastoral land use, both on the spot and elsewhere, but does not substantially improve animal production. Controlled lopping of trees for forage is preferable to browsing by browsers, which generally appears difficult to control, particularly on communally owned rangelands.
2. In situations of nutrient limited production, where forage quality is the limiting factor for pastoral systems (situation IIIa), average woody cover should be at least 15–20%. Higher canopy cover leads to reduced animal productivity, following competition with herbs. The woody plants are preferably distributed homogeneously (in order to reduce shading), are not intensively exploited, and have a high ratio of trunk height to crown diameter. At run-on sites (situation IIIb) canopy cover is generally higher; in clay depressions fodder trees may be intensively exploited for pastoral use. For the overall forage availability, it is desirable to reduce canopy cover at run-on sites of loamy plains and shallow soils, in order to reduce competition with the herbage layer, but from a sustainability point of view this is not desirable.
3. The establishment of fodder banks at locations representing optimal growth conditions and niche differentiation for woody plants, is

generally more productive, and certainly more sustainable, than introducing or stimulating palatable woody species in semi-arid rangelands.

For agrosylvicultural systems, three situations have been described where the integration of woody plants with cropping has a synergistic effect.

1. On sandy soils, in the driest parts of situation III, or at run-on sites in situation II where nutrients have become the limiting factor for plant production (Fig. 5.1), woody plants oriented as windbreaks improve germination and establishment of crops by reducing the impact of wind. One condition is that the woody plants can effectively use subsoil groundwater reserves. One good example is that of the windbreaks of the Maggia valley in Niger. Here, prevailing winds from one direction, subsoil groundwater reserves, deep sandy soils, and relatively high population densities, offer optimal conditions for the establishment of windbreaks. However, in semi-arid regions such conditions are rare. In a general sense, as crop productivity increases, the profitability of windbreaks to protect croplands increases. The possibility to obtain high yields (in terms of biomass or incomes) are a condition for the establishment and management of windbreaks and vice versa (Breman 1987). For maximum benefits of windbreaks for crop production, selection and management of woody plants should be oriented at minimising shallow rooting, and maximum protective biomass with low growth rates.
2. In the more humid parts of situation III, where run-off and leaching are significant processes, evenly distributed woody plants can lead to improved nutrient availability for crops. In the SSWA, maximum benefits correspond to additional cereal yields of at most 400 kg ha^{-1} yr^{-1}; in other semi-arid regions, characterised by higher inherent soil fertility, benefits may be slightly higher. Benefits are highest with C_3 crops (e.g. vegetables), which are more shade-tolerant than C_4 crops. Selection of woody species and management of the woody plants should be oriented at the combination of minimum light reduction and maximum woody plant production. Here, maintaining a high ratio of trunk height to crown diameter is probably most important. Every single exploitation of the woody plants greatly reduces the beneficial effects of the woody plants on plant production factors.
3. Woody plants can lead to a prolonged growing season for crops, but the conditions that favour this effect and that also permit crops to benefit by increased production, are rare in semi-arid regions. In fact, most suitable are relatively humid conditions, and the presence of fertile and deep soils.

Woody plants can play a significant role in the transition phase of agrosylvopastoral systems in semi-arid regions from extensive systems (characterised by a high land-production ratio, and a low labour-production ratio) to intensified systems (with opposite characteristics). Woody plants provide buffering functions, stabilising ecosystem dynamics, and allowing effective use of additional nutrient and water inputs, or allowing effective use of these resources where they occur naturally. In most semi-arid regions, the use of external inputs requires subsidies, because of the combination of high costs and doubtful results (Breman 1990). The maintenance of woody plants reduces the necessity of such subsidies, provided that their management is aimed at maximising the beneficial role of woody plants for soil structure and organic matter. However, it should be recognised that high labour inputs are required to maximise positive effects of woody plants. Once the agricultural system has reached higher production levels in a sustainable way, woody plants loose their relative advantages.

In most semi-arid regions the management involved in the transition process is not easy: the absence of the necessary technical know-how, the prevailing poor socio-economic conditions, labour shortage at crucial moments, land tenure insecurity are among the most important bottlenecks. So far woody plants have been predominantly used for productive purposes. Substantial changes are required to change the focus to protective and supportive functions. It cannot be overemphasised that the use of woody plants in agrosylvopastoral land use has never been simple, and should include much more insight and diversity to allow optimal use. Experimenting with woody plants to maximise production, the risks for adverse effects are indeed great. For instance, woody plants can be used to further deplete soils, from greater depths and during prolonged periods within a year. In that case, short-term productive objectives prevail, which may be rational from the farmers' point of view. Recognising that local communities are often not in the position to choose for long-term objectives, support is required to make optimum use of woody plants, as one element of a sustainable land-use system, which in the long term will be beneficial for the population as a whole.

References

Adams ME (1967) A study of the ecology of *Acacia mellifera*, *A. seyal* and *Balanites aegyptiaca* in relation to land-clearing. J Appl Ecol 4: 221–237

Adams MA, Attiwill PM (1984) Role of *Acacia* spp. in nutrient balance and cycling in regenerating *Eucalyptus regnans* F. Muell. forests. II. Field studies of acetylene reduction. Aust J Bot 32: 217–223

Afolayan TA (1978) Savanna burning in Kainji Lake National Park, Nigeria. East Afr Wildl J 16: 245–255

Afolayan TA (1979) Change in percentage ground cover of perennial grasses under different burning regimes. Vegetatio 39: 35–41

Ahuja LD, Verma CM, Sharma SK, Lamba TR (1978) Range management studies on the contribution of ground storey (grass) in afforested areas in arid regions. Ann Arid Zone 17: 304–310

Alberda Th et al. (1977) Crop photosynthesis: methods and compilation of data obtained with a mobile field equipment. Pudoc, Wageningen, Agric Res Rep 865: 46

Alexander I (1989) Mycorrhizas in tropical forests. In: Proctor J (ed) Mineral nutrients in tropical forest and savanna ecosystems. Blackwell, Oxford, Spec Publ Br Ecol Soc 9: 169–188

Alexandre DY (1990) Importance quantitative et qualitative des échanges d'eau entre racines d'espèces différentes. Commun Sém sur la physiologie des ligneux en zone aride, 27 Mars–7 Avril 1990, Nancy. ORSTOM, Ouagadougou

Alexandre DY, Ouedraogo SJ (1991) Variations de la morphologie racinaire de *Faidherbia albida* en relation avec le sol et effets agronomiques de l'arbre: quelques observations au Burkina Faso. Note présentée à l'atelier *Faidherbia albida*, 22–26 Avril 1991, Niamey. ORSTOM, Ouagadougou

Allen JC (1985) Soil response to forest clearing in the United States and the tropics: geological and biological factors. Biotropica 17: 15–27

Anderson JM, Swift MJ (1983) Decomposition in tropical forests. In: Sutton SL, Whitmore TC, Chadwick AC (eds) Tropical rain forest: ecology and management. Blackwell, Oxford, Spec Publ Br Ecol Soc 2: 287–309

Anonymous (1980) A handbook of research. Charlesville Pastoral Lab, Dep Primary Ind, Queensland, 89 pp

Archer S (1989) Have southern Texas savannas been converted to woodlands in recent history? Am Nat 134: 545–561

Archer S, Scifres CJ, Bassham CR, Maggio R (1988) Autogenic succession in a subtropical savanna: conversion of grassland to thorn woodland. Ecol Monogr 58: 111–127

Areola O (1980) Quelques questions et problèmes sur l'étude des jachères de savanes. Environ Afr 13: 55–67

Arshad MA (1982) Influence of the termite *Macrotermes michaelseni* (Sjöst) on soil fertility and vegetation in a semi-arid savanna ecosystem. Agro-Ecosystems 8: 47–58

Atkinson D (1980) The distribution and effectiveness of the roots of tree crops. Hortic Rev 2: 425–490

Attiwill P (1981) Energy, nutrient flow, and biomass. In: Productivity in perpetuity. Proc Aust For Nutr Workshop, 10–14 August 1981, Canberra. CSIRO (Commonwealth Scientific and Industrial Research Organization) Division of Forest Research, Canberra, pp 131–144

Aubréville A (1963) Principes d'une systématique des formations végétales tropicales. Adansonia 5: 153–196

Audry P, Rossetti C (1962) Observations sur le sol et la végétation en Mauretanie du sud-est et sur la bordure adjacente du Mali 1959 et 1961. Prospection écologique/etudes en Afrique occidentale. FAO, Rome, 190 pp

Aweto AO (1981) Secondary succession and soil fertility restoration in south-western Nigeria. I. Succession, II. Soil fertility restoration. J Ecol 69: 601–614

Baker JM, Bavel CHM (1986) Resistance of plant roots to water loss. Agron J 78: 641–644

Barnes DL (1979) Cattle ranching in the semi-arid savannas of East and Southern Africa. In: Walker BH (ed) Management of semi-arid ecosystems. Elsevier, Amsterdam, Dev Agric Managed-For Ecol 7: 9–54

Barry JP, Boudet G, Bourgeot A, Celles JC, Coulibaly AM, Leprun JC, Manière R (1983) Etude des potentialités pastorales et de leur évolution en milieu sahélien au Mali. ACC – GRIZA-LAT, Groupe de Recherches Interdisciplinaires en Zones Arides, ORSTOM, Paris, 116 pp

Barth RC, Klemmedson JO (1978) Shrub-induced spatial patterns of dry matter, nitrogen, and organic carbon. Soil Sci Soc Am J 42: 804–809

Batanouny KH, Abdel Wahab AM (1973) Eco-physiological studies on desert plants. VIII. Root penetration of *Leptadenia pyrotechnica* (Forsk.) Decne. in relation to its water balance. Oecologia (Berlin) 11: 151–161

Beadle NCW (1964) Nitrogen economy in arid and semi-arid plant communities. III. The symbiotic nitrogen fixing organisms. Proc Linn Soc NSW 89: 273–286

Beadle NCW, White GJ (1968) The mineral content of the trunks of some Australian woody plants. Proc Ecol Soc Aust 3: 55–60

Beale IF (1973) Tree density effects on yields of herbage and tree components in south west Queensland mulga (*Acacia aneura* F. Muell.) scrub. Trop Grassl 7: 135–142

Beer J (1988) Litter production and nutrient cycling in coffee (*Coffea arabica*) or cacao (*Theobroma cacao*) plantations with shade trees. Agrofor Syst 7: 103–114

Behnke RH, Scoones I (1991) Rethinking range ecology: implications for rangeland management in Africa. Overview of paper presentations and discussions at the Technical Meeting on savanna development and pasture production, 19–21 November 1990, Woburn, UK. Overseas Dev Inst Int Inst Environ Dev (ODI), London, 38 pp

Belsky AJ, Amundson RG, Duxbury JM, Riha SJ, Ali AR, Mwonga SM (1989) The effects of trees on their physical, chemical, and biological environments in a semi-arid savanna in Kenya. J Appl Ecol 26: 1005–1024

Benjamin RW (1980) The use of forage shrubs in the Norte Chico region of Chile. In: Le Houérou HN (ed) Browse in Africa: the current state of knowledge. Papers presented at the Int Symp on Browse in Africa, Addis Ababa, 8–12 April 1980. ILCA, Addis Ababa, pp 299–303

Benjaminsen TA (1988) Ressursendringer i Gourma, Mali. Årsaker og konsekvenser. Hovedoppgave til cand. scient.-graden. Ressursgeografi. Geogr Inst Univ Oslo, 116 pp

Berendse F (1990) Organic matter accumulation and nitrogen mineralisation during secondary succession in heathland ecosystems. J Ecol 78: 413–427

Berendse F, Bobbink R, Rouwenhorst G (1989) A comparative study on nutrient cycling in wet heathland ecosystems. I. Litter decomposition and nutrient mineralization. Oecologia (Berlin) 78: 338–348

Bergeret A, Ribot JC (1990) L'arbre nourricier en pays sahélien. Editions de la Maison des Sciences de l'Homme, Paris, 237 pp

Berish CW, Ewel JJ (1988) Root development in simple and complex tropical successional ecosystems. Plant Soil 106: 73–84

Bernhard-Reversat F (1982) Biogeochemical cycle of nitrogen in a semi-arid savanna. Oikos 38: 321–332

Bernhard-Reversat F, Poupon H (1980) Nitrogen cycling in a soil-tree system in a Sahelian savanna. Example of *Acacia senegal*. In: Rosswall T (ed) Nitrogen cycling in West African ecosystems, SCOPE/UNEP International Nitrogen Unit, R Swed Acad Sci Stockholm, pp 363–369

Bie S de (1991) Wildlife resources of the West African savanna. Wageningen Agric Univ Pap 91(2): 226

Bille JC (1977) Etude de la production primaire nette d'un écosystème sahélien. ORSTOM, Paris, Trav Doc ORSTOM 65: 82

Bille JC (1980) Measuring the primary palatable production of browse plants. In: Le Houérou HN (ed) Browse in Africa: the current state of knowledge. Papers presented at the Int Symp on Browse in Africa, Addis Ababa, 8–12 April 1980. ILCA, Addis Ababa, pp 185–196

Birch (1964) Mineralisation of plant nitrogen following alternate wet and dry conditions. Plant Soil 20: 43–49

Birk EM, Simpson RW (1980) Steady state and the continuous input model of litter accumulation and decomposition in Australian Eucalypt forests. Ecology 61: 481–485

Bodine MC, Ueckert DN (1975) Effect of desert termites on herbage and litter in a shortgrass ecosystem in west Texas. J Range Manage 28: 353–358

Boer F de, Kessler JJ (1994) Livestock husbandry and the environment in the Province of Sissili, Burkina Faso. Trop Resour Manage Pap no 4 Dep Nature Conserv, Wageningen Agric Univ, Wageningen, 106 pp

Boerner REJ (1982) Fire and nutrient cycling in temperate ecosystems. Bioscience 32: 187–192

Bosch DJH, Wijk JJP van (1970) Die invloed van bosveldbome op die produktiwiteit van *Panicum maximum*: 'n voorlopige verslag. Grassl Soc South Afr Proc 5: 69–74

Boudet G (1961) Etude botanique et agrostologique de la Haute Vallée du Niger. Rev Elev Med Vet Pays Trop 15: 1

Boudet G (1971) Esquisse pastorale et esquisse de transhumance de la région du Gourma, le Mali. IEMVT, Maisons-Alfort, 131 pp

Boudet G (1972) Etude agrostologique de la cinquième région du Mali (région de Mopti). IEMVT, Maisons-Alfort, Etud Agrostol 37: 309

Boudet G (1978) Manuel sur les pâturages tropicaux et les cultures fourragères. Manuels et Précis d'Elevage 4, IEMVT, Maisons-Alfort, 202 pp

Boudet G (1979) Etude de l'évolution d'un systéme d'exploitation sahélien au Mali. ACC/IEMVT, Maisons-Alfort, 48 pp

Boudet G, Crue J (1966) Etude des pâturages naturels du Ferlo-Oriental (Sénégal). IEMVT, Maisons-Alfort, Etud Agrostol 13, 69 pp

Boudet G, Duverger E (1961) Etude des pâturages naturels Sahéliens: e Hodh (Mauretanie). Vigot Frères, Paris, 161 pp

Boudet GC, Toutain B (1980) The integration of browse plants within pastoral and agropastoral systems in Africa. In: Le Houérou HN (ed) Browse in Africa: the current state of knowledge. Papers presented at the Int Symp on Browse in Africa, Addis Ababa, 8–12 April 1980. ILCA, Addis Ababa, pp 427–432

Bourlière F (1983) Tropical savannas: ecosystems of the world. Elsevier, Amsterdam

Boutrais J (1980) L'arbre et le boeuf en zone soudano-guinéenne. Cah ORSTOM, Ser Sci Hum 17: 235–246

Braun Wilke RH (1982) Net primary productivity and nitrogen and carbon distribution in two xerophytic communities of central-west Argentina. Plant Soil 67: 315–323

Breman H (1981) Le trajet Nouakchott-Kankossa: une première reconnaissance. Projet APPAM, ITC, Enschede

Breman H (1987) The struggle of the green against the yellow dragon. The Chinese approach to desertification control and its usefulness for the Sahel. CABO, Wageningen, 62 pp

Breman H (1990) No sustainability without external inputs. In: Beyond adjustment. Africa Seminar, Maastricht, The Netherlands. Ministry of Foreign Affairs, 's-Gravenhage, pp 124–133

Breman H (1992) Desertification control, the West African case; prevention is better than cure. Biotropica 24: 328–334

Breman H, Cissé AM (1979) Dynamics of Sahelian pastures in relation to drought and grazing. Oecologia (Berlin) 28: 310–315

Breman H, Diarra L (1994) Méthodologie pour le suivi léger des ressources pastorales au Sahel. Rapp Cent Rech Agrobiol (CABO-DLO), Wageningen (in prep)

Breman H, Ridder N de (eds) (1991) Manuel sur les pâturages des pays sahéliens. Karthala, Paris, 480 pp

Breman H, Traoré N (eds) (1986) Analyse des conditions de l'élevage et propositions de politiques et de programmes. République du Niger. OCDE/CILSS/Club du Sahel, Paris, 194 pp (Sahel D(86) 284)

Breman H, Traoré N (eds) (1987) Analyse des conditions de l'élevage et de propositions politiques et de programmes. République du Mali. OCDE/CILSS/Club du Sahel, Paris, 243 pp (Sahel D(86) 302)

Breman H, Wit CT de (1983) Rangeland productivity and exploitation in the Sahel. Science 221: 1341–1347

Breman H, Diallo A, Traoré G (1979) Compte rendu des analyses de la digestibilité des espèces fourragères du Sahel. Rapp interne PPS, Bamako

Brinckman WL, Leeuw PN de (1984) The nutritive value of browse and its importance in traditional pastoralism. Misc Rep, ILCA, Addis Ababa, pp 190–206

Brookman-Amissah J, Hall JB, Swaine MD, Attakorah JY (1980) A re-assessment of a fire protection experiment in north-eastern Ghana savanna. J Appl Ecol 17: 85–99

Brossier J, Jager B (1983) La programmation linéaire appliquée à deux U.P.A. de la région de Fonsébougou, Mali. Coefficients techniques et résultats d'ordinateur. Documents complémentaires. KIT, Amsterdam

Brouwer J, Geiger SC, Vandenbeldt RJ (1991) Variability in the growth of *Acacia albida*: the termite connection? Paper presented at Regional *Acacia albida* workshop, 22–26 April 1991, Niamey

Brouwer J, Fussell LK, Herrmann L (1993) Soil and crop growth micro-variability in the West African semi-arid tropics: a possible risk-reducing factor for subsistence farmers. Agric Ecosyst Environ 45: 229–238

Bruijnzeel LA (1989) Nutrient cycling in moist tropical forests: the hydrological framework. In: Proctor J (ed) Mineral nutrients in tropical forest and savanna ecosystems. Blackwell, Oxford Spec Publ, Br Ecol Soc 9: 383–415

Budelman A (1988) The decomposition of the leaf mulches of *Leucaena leucocephala*, *Gliricidia sepium* and *Flemingia macrophylla* under humid tropical conditions. Agrofor Syst 7: 33–45

Budelman A (1989) The performance of selected leaf mulches in temperature reduction and moisture conservation in the upper soil stratum. Agrofor Syst 8: 53–66

Budelman A, Pol F van der (1992) Farming systems research and the guest for a sustainable agriculture. Agrofor Syst 19: 187–207

Budelman A, Zander PM (1990) Land-use by immigrant Baoulé farmers in the Taï region, south-west Côte d'Ivoire (Ivory Coast). Agrofor Syst 11: 101–123

Buffington LC, Herbel CH (1965) Vegetational changes on a semi-desert grassland range from 1850 to 1963. Ecol Monogr 35: 139–164

Burnham CP (1989) Pedological processes and nutrient supply from parent material in tropical soils. In: Proctor J (ed) Mineral nutrients in tropical forest and savanna ecosystems, Blackwell, Oxford, Spec Publ Br Ecol Soc 9: 27–41

Burrows WH (1976) Aspects of nutrient cycling in semi-arid mallee and mulga communities. PhD Thesis, Aust National University (ANU), Canberra

Buxton RD (1981) Changes in composition and activities of termite communities in relation to changing rainfall. Oecologia (Berlin) 51: 371–378

Cameron DM, Rance SJ, Jones RM, Charles-Edwards DA, Barnes A (1989) Project STAG: an experimental study in agroforestry. Aust J Agric Res 40: 699–714

Carew BAR, Mosi AK, Mba AU, Egbunike GN (1980) The potential of browse plants in the nutrition of small ruminants in the humid forest and derived savanna zones of Nigeria. In: Le Houérou HN (ed) Browse in Africa: the current state of knowledge. Papers presented at the Int Symp on Browse in Africa, Addis Ababa, 8–12 April 1980. ILCA, Addis Ababa, pp 307–311

Catchpoole DW, Blair G (1990) Forage tree legumes. II. Investigation of nitrogen transfer to an associated grass using split-root technique. Aust J Agric Res 41: 531–537

Catinot R (1983) La stratégie du bois de feu dans les pays en voie de développement. Rev Energie 356: 516–529

César J, Menaut JC (1974) Analyse d'un écosystème tropical humide: la savane de Lamto, Côte d'Ivoire. II. Le peuplement végétal. Station d'écologie tropical de Lamto, Côte d'Ivoire

Charley JL (1971) The role of shrubs in nutrient cycling. In: McKell CM, Blaisdell JP, Goodin JR (eds) Wildland shrubs – their biology and utilization. US For Serv Gen, Tech Rep INT-1: 182–203

Charley JL, West NE (1975) Plant-induced soil chemical patterns in some shrub-dominated semi-desert ecosystems of Utah. J Ecol 63: 945–963

Charley JL, West NE (1977) Micro-patterns of nitrogen mineralization activity in soils of some shrub-dominated semi-desert ecosystems of Utah. Soil Biol Biochem 9: 357–365

Charney LG (1975) Dynamics of desert and drought in the Sahel. Q J R Meteorol Soc 101: 193–202

Charreau C, Vidal P (1965) Influence de l'*Acacia albida* Del. sur le sol, nutrition minérale et rendements des mils *Pennisetum* au Sénégal. Agron Trop (Paris) 20: 600–626

Chaturvedi AN, Sharma SC, Srivastava R (1988) Water consumption and biomass production of some forest tree species. Int Tree Crops J 5: 71–76

Christie EK (1975) A note on the significance of *Eucalyptus populnea* for buffel grass production in infertile semi-arid rangelands. Trop Grassl 9: 243–246

CILSS (1981) Analyse du secteur forestier et propositions. Le Niger. CILSS/Club du Sahel, Paris

CILSS (1982) Analyse du secteur forestier et propositions Le Mali. CILSS/Club du Sahel, Paris

Cissé AM (1986) Dynamique de la strate herbacée des pâturages de la zone sud-sahélienne. Thèse, Univ Agrono, Wageningen, 212 pp

Cissé L (1986) Etude des effets d'apports de matière organique sur les bilans hydriques et minéraux et la production du mil et de l'arachide sur un sol sableux dégradé du Centre-Nord de Sénégal. Thèse de Doctorat, Inst Natl Polytech Louraine, 184 pp

Cissé MI (1980) Production fourragère de quelques arbres sahéliens: 1. Effets de divers régimes d'effeuillage sur la production foliaire de quelques buissons fourragers de la zone soudano-sahélienne. 2. Relations entre la biomasse foliaire maximale et divers paramètres pfysiques. Programme des zones arides, Mali, ILCA/CIPEA, Bamako

Cissé MI (1981) Variations saisonnières de biomasse foliaire chez quelques ligneux fourragers sahéliens. Programme des zones aride et semi-aride, ILCA/CIPEA, Bamako, Doc Progr AZ72

Cissé MI (1983a) Disponibilités en gousses d'*Acacia tortilis* pour la nutrition des petits ruminants dans les ranches Masai de Mbirikani et Kimana. Programme des zones aride et semi-aride, ILCA/CIPEA, Bamako, Doc Progr AZ78

Cissé MI (1983b) Bilan de cinq années de mise en défens du peuplement ligneux de divers milieux au ranch de Niono. Programme des zones arides et semi-arides, CIPEA, Bamako

Cissé MI (1984) Synthèse des essais d'ébranchages pratiqués sur quelques arbres fourragers sahéliens de 1978 à 1983. Programme des zones aride et semi-aride, ILCA/CIPEA, Bamako, Doc Progr AZ103

Cissé MI (1985) Contribution des peuplements ligneux à l'alimentation des petits ruminants en zone semi-aride du Mali central. In: Wilson RT, Bourzat D (eds) Les petits ruminants dans l'agriculture africaine. CIPEA, Addis Ababa, pp 64–74

Cissé MI, Wilson RT (1984) Statut et utilisation de *Pterocarpus lucens* Lepr. dans les écosystèmes sahéliens. Programme des zones aride et semi-aride, ILCA/CIPEA, Bamako, Doc Progr AZ101d

Clanet JC, Gillet H (1980) Le *Commiphora africana*, véritable arbre fourrager sahélien. Colloque sur les fourrages ligneux en Afrique, Addis Ababa, 8–12 Avril 1980. ILCA/CIPEA, Addis Ababa

Clément J (1982) Estimation des volumes et de la productivité des formations mixtes forestières et graminéennes tropicales. Données concernant les pays de l'Afrique francophone au nord de l'équateur et recommandations pour la conduite de nouvelles études. Rev Bois For Trop 198: 35–58

Cline-Cole RA, Falola JA, Main HAC, Mortimore MJ, Nichol JE, O'Reilly FD (1988) Woodfuel in Kano, Nigeria: the urban-rural conflict. ODI, London, Soc For Network Pap 76, 16 pp

Coleman DC, Oades JM, Uehara G (eds) (1989) Dynamics of soil organic matter in tropical ecosystems. NifTAL project, Univ Hawaii Press, Honolulu, 249 pp

Connor DJ (1983) Plant stress factors and their influence on production of agroforestry plant associations. In: Huxley PA (ed) Plant research and agroforestry. ICRAF, Nairobi, pp 249–256

Conway GR (1987) The properties of agroecosystems. Agric Syst 24: 95–117

Cooper SM, Owen-Smith N (1985) Condensed tannins deter feeding by browsing ruminants in a South African savanna. Oecologia (Berlin) 67: 142–146

Coppock DL, Reed JD, Atsedu M, Assefa M (1991) Evaluation of *Acacia* forages for improved livestock feeding in Ethiopia. IVth Int Rangeland Congr, April 1991, CIRAD, Montpellier

Corlett JE, Ong CK, Black CR (1989) Microclimate modification in intercropping and alley cropping. In: Reifsnyder WS, Darnhofer TO (eds) Meteorology and agroforestry. ICRAF, Nairobi, pp 419–430

Cornet A, Poupon H (1977) Description des facteurs du milieu et de la végétation dans cinq parcelles situées le long d'un gradient climatique en zone sahélienne au Sénégal. Bull Inst Fondam Afr Noire (IFAN) T 39, Ser A 2: 241–302

Corré WJ (1983) Growth and morphogenesis of sun and shade plants. I. The influence of light intensity. Acta Bot Neerl 32: 49–62

Corré WJ (1984) Growth and morphogenesis of sun and shade plants. IV. Competition between sun and shade plants in different light environments. Acta Bot Neerl 33: 25–38

Crane WJB, Raison RJ (1980) Removal of phosphorus in logs when harvesting *Eucalyptus delegatensis* and *Pinus radiata* forests on short and long rotations. Aust For 43: 253–260

CRTO (1988) Etude de l'occupation des sols – Province de la Tapoa. Cent Régional de Télédétection, Ouagadougou

CTFT (1988) *Faidherbia albida* (Del.) A. Chev. Monographie. Cent Tech For Trop, Nogent-sur-Marne, 72 pp

Daccarett M, Blydenstein J (1968) La influencia deá rboles leguminosos y no leguminosos sobre et forraje que crece bajo ellos. Turrialba 18: 405–408

Dancette C, Poulain JF (1969) Influence of *Acacia albida* on pedoclimatic factors and crop yields. Afr Soils 14: 143–184

Darnhofer T, Gatama D, Huxley P, Akunda E (1989) The rainfall distribution at a tree/crop interface. In: Reifsnyder WS, Darnhofer TO (eds) Meteorology and agroforestry. ICRAF, Nairobi, pp 371–382

Davies SJJF (1976) Studies of the flowering season and fruit production of some arid zone shrubs and trees in western Australia. J Ecol 64: 665–687

Davies WJ, Rhizopoulou S, Sanderson R, Taylor G, Metcalfe JC, Zhang J (1989) Water relations and growth of roots and leaves of woody plants. In: Pereira JS, Landsberg JJ (eds) Biomass production by fast-growing trees. Kluwer, Dordrecht, pp 13–26

Dekker AJFM (1980) Etudes préliminaire de la végétation du Parc National de la Boucle du Baoulé et quelques réserves adjacentes au Mali. Rapp Tech Projet RURGS. Département d'Aménagement de la Nature, Univ Agron Wageningen, Wageningen (verslag no 496)

De Winter J, Fischer M, Hellemans P, Doulaye I, Gambo S, Bagoudou M (1989) Etude: Aménagement et protection des forêts naturelles dans la région de Niamey. Rapp de la deuxième phase. Dtsch Forstinventur Serv (DFS), Feldkirchen

Diagayeté M (1981) Untersuchungen zur Erweiterung der Kentnisse über den Futterwert westafrikanischer Futterpflanzen. Dissertation, Universität Hohenheim

Diallo A (1978) Transhumance: comportement, nutrition et productivité d'un troupeau zébus de Diafarabé. Thèse, Centre Pédagogique Supérieure, Bamako

Dicko MS, Sayers R (1988) Recherches sur le système agropastoral de production de la zone semi-aride du Niger: étude de la composante animale. CIPEA, Niger, 152 pp

Dicko-Touré MS (1980) The contribution of browse to cattle fodder in the sedentary system of the Office du Niger. In: Le Houérou HN (ed) Browse in Africa: the current state of knowledge. Papers presented at the Int Symp on Browse in Africa, Addis Ababa, 8–12 April 1980. ILCA, Addis Ababa, pp 313–320

Dommergues YR (1987) The role of biological nitrogen fixation in agroforestry. In: Steppler HA, Nair PKR (eds) Agroforestry, a decade of development. ICRAF, Nairobi, pp 245–271

Donaldson CH (1967) The immediate effects of the 1964/1966 drought on the vegetation of specific study areas in the Vryburg district. Grassl Soc South Afr Proc 2: 137–141

Donaldson CH, Kelk DM (1970) An investigation of the veld problems of the Molopo area: I. Early findings. Grassl Soc South Afr Proc 5: 50–57

Donaldson CH, Kelk DM (1973) The development of semi-arid bushveld to its full potential. 3. Molopo. Grassl Soc South Afr Proc 8: 143–146

Doyle CJ, Evans J, Rossiter J (1986) Agroforestry: an economic appraisal of the benefits of intercropping trees with grassland in lowland Britain. Agric Syst 21: 1–32

Drechsel P, Glaser B, Zech W (1991) Effect of four multipurpose tree species on soil amelioration during tree fallow in central Togo. Agrofor Syst 16: 193–202

Dregne HE (1988) Irrigation development in northwestern Xinjiang. Chin J Arid Land Res 1: 185–190

Ducousso M (1991) Importance des symbioses racinaires pour l'utilisation des Acacias d'Afrique de l'Ouest. CIRAD/ISRA, Dakar, 205 pp

Duivenbooden N van (1985) The effect of grazing on the growth of subshrubs and the water balance in the northwestern coastal zone of Egypt. Cent Agrobiol Res (CABO-DLO), Wageningen

Duivenbooden N van, Gosseye PA, Keulen H van (eds) (1991) Competing for limited resources: the case of the fifth region of Mali. Part 2: Plant, livestock and fish production. Cent Agrobiol Res (CABO-DLO), Wageningen, 242 pp

Dunham KM (1989) Litterfall, nutrient-fall and production in an *Acacia albida* woodland in Zimbabwe. J Trop Ecol 5: 227–238

Dye PJ, Spear PT (1982) The effects of bush clearing and rainfall variability on grass yield and composition in south-west Zimbabwe. Zimbabwe J Agric Res 20: 103–118

Eastham J, Rose CW (1988) Pasture evapotranspiration under varying tree-planting density in an agroforestry experiment. Agric Water Manage 15: 87–105

Eastham J, Rose CW (1990) Tree/pasture interactions at a range of tree densities in an agroforestry experiment. I. Rooting patterns. Aust J Agric Res 41: 683–695

Eastham J, Rose CW, Cameron DM, Rance SJ, Talsma T, Charles-Edwards DA (1990) Tree/pasture interactions at a a range of tree densities in an agroforestry experiment. II. Water uptake in relation to rooting patterns. Aust J Agric Res 41: 697–707

Ebersohn JP, Lucas P (1965) Trees and soil nutrients in south-western Queensland. Queensl J Agric Anim Sci 22: 431–435

Egging P (1990) La capacité de charge du Département de Tô. Une étude de la situation fourragère dans la savane du Centre-Sud du Burkina Faso. Département de Fourrages Tropicaux, Univ Agron Wageningen, Wageningen, 116 pp

Egunjobi JK, Bada SO (1979) Biomass and nutrient distribution in stands of *Pinus caribaea* L. in the dry forest zone of Nigeria. Biotropica 11: 130–135

Elberse WTh, Breman H (1990) Germination and establishment of Sahelian rangeland species. II. Effects of water availability. Oecologia (Berlin) 85: 32–40

Elkins NZ, Sabol GV, Ward TJ, Whitford WG (1986) The influence of subterranean termites on the hydrological characteristics of a Chihuahuan desert ecosystem. Oecologia (Berlin) 68: 521–528

Eriksen FI, Whitney AS (1981) The effects of light intensity on the growth of some tropical forage species. I. Interaction of light intensity and nitrogen fertilization on six forage grasses. Agron J 73: 427–433

Evans J (1979) The effects of leaf position and leaf age in foliar analysis of *Gmelina arborea*. Plant Soil 52: 547–552

Everett R, Sharrow S, Thran D (1986) Soil nutrient distribution under and adjacent to single leaf *Pinyon* crowns. Soil Sci Soc Am J 50: 788–792

Ewel JJ (1986) Designing agricultural ecosystems for the humid tropics. Annu Rev Ecol Syst 17: 245–271

Eyog-Matig O (1982) Contribution à la sylviculture de l'*Eucalyptus* au Nord-Cameroun. Thèse de 3ème cycle. Univ P Sabatier, Toulouse, 111 pp

Fall ST (1991) Digestibilité in vitro et dégradabilité in situ dans le rumen de ligneux fourragers disponibles sur pâturages naturels au Sénégal. Premiers résultats. Rev Elev Med Vet Pays Trop 44: 345–354

FAO (1984) Agroclimatological data for Africa. 1. Countries north of the equator. FAO, Rome, FAO Plant Prod Protect Ser 22(1)

FAO (1986) Brise-vent et rideaux abris avec référence particulière aux zones sèches. FAO, Rome, Cah FAO Conserv 15: 385

FAO (1989a) Etude sur les volumes et la productivité des peuplements forestiers tropicaux. 1. Formations forestières sèches. FAO, Rome, Etud FAO For 51/1

FAO (1989b) Forestry and food security. FAO, Rome, FAO For Pap 90: 128

Faria SM de, Lewis GP, Sprent JI, Sutherland JM (1989) Occurrence of nodulation in the Leguminosae. New Phytol 111: 607–619

Felker P (ed) (1986) Establishment and productivity of tree plantings in semiarid regions. Spec Issue For Ecol Manage 16: 1–444

Felker P, Clark PR (1982) Position of mesquite (*Prosopis* spp.) nodulation and nitrogen fixation (acetylene reduction) in 3-m-long phraetophytically simulated soil columns. Plant Soil 64: 297–305

Fenner M (1980) Some measurements on the water relations on Baobab trees. Biotropica 12: 205–209

Foran BD, Bastin G, Remenga E, Hyde KW (1982) The response to season, exclosure, and distance from water of three central Australian pasture types grazed by cattle. Aust Range J 4: 5–15

Ford ED, Deans JD (1978) The effects of canopy structure on stemflow, throughfall and interception loss in a young Sitka spruce plantation. J Appl Ecol 15: 905–917

Fotius G, Valenza J (1966) Etude des pâturages naturels du Ferlo-Oriental (République du Sénégal). IEMVT, Maisons-Alfort, Etud Agrostol 13: 180

Friedel MH (1991) Range condition assessment and the concept of thresholds: a viewpoint. J Range Manage 44: 422–427

Frost PGH (1985a) The response of savanna organisms to fire. In: Tothill JC, Mott JJ (eds) Ecology and management of the world's savannas. Aust Acad Sci, Canberra, pp 232–237

Frost PGH (1985b) Organic matter and nutrient dynamics in a broadleaved African savanna. In: Tothill JC, Mott JJ (eds) Ecology and management of the world's savannas. Aust Acad Sci, Canberra, pp 200–206

Gallais J (1967) Le sens de la diversité des paysages agraires. In: Gallais J (ed) Le Delta intérieur du Niger, vol 1. IFAN, Dakar, Mem Inst Fondam Afr Noire 79: 284–286

Garcia-Moya E, McKell CM (1970) Contribution of shrubs to the nitrogen economy of a desert-wash plant community. Ecology 51: 81–88

Garner W, Steinberger Y (1989) A proposed mechanism for the formation of "fertile islands" in the desert ecosystem. J Arid Environ 16: 257–262

Garrison GA (1953) Effect of clipping on some range shrubs. J Range Manage 6: 309–317

Gaston A (1967) Etude agrostologique du Kanem, préfecture du Kanem au Sud du 16ème parallèle et préfecture du Lac. IEMVT, Maisons-Alfort, Etud Agrostol 19: 147 pp

Gaston A (1974) Etude agrostologique des pâturages du projet Assale – Serbewel. IEMVT, Maisons-Alfort, Etud Agrostol 41: 143 pp

Gaston A (1975) Pâturages de Kanem. Effets de la sécheresse de 1973 sur les pâturages du Kanem – Lac. Comparaison avec les études de 1964 et 1966. IEMVT, Maisons-Alfort

Geerling C (1973) The vegetation of Yankari game reserve: its utilization and condition. Dep For, Univ Ibadan, Bull 3: 49

Geerling C (1976) Vegetation map of Borgu game reserve, Nigeria. A report prepared for the Kainji Lake Research Project. FAO, Rome

Geerling C (1987) Guide de terrain des ligneux sahéliens et soudano-guinéens. Agric Univ Wageningen Pap 87-4: 340

Geerling C, Diakité MD (eds) (1988) Rapport final du projet Recherche pour l'utilisation rationnelle du gibier au Sahel (RURGS). Département d'Aménagement de la Nature, Univ Agron Wageningen, Wageningen, 93 pp

Gerakis PA, Tsangarakis CZ (1970) The influence of *Acacia senegal* on the fertility of a sand sheet soil in the central Sudan. Plant Soil 33: 81–86

Gersper PL, Holowaychuk N (1971) Some effects of stem flow from forest canopy trees on chemical properties of soils. Ecology 52: 691–702

Gijsbers I, Kessler JJ, Knevel M (1994) Regeneration dynamics of the parklands in the northern Sudan zone in Burkina Faso. For Ecol Manage 65: 1–12

Gill HS, Abrol IP, Samra JS (1987) Nutrient recycling through litter production in young plantations of *Acacia nilotica* and *Eucalyptus tereticornis* in a highly alkaline soil. For Ecol Manage 22: 57–69

Giller KE, Wilson KJ (1991) Nitrogen fixation in tropical cropping systems. CAB Int, Oxon, 320 pp

Glover PE (1950) Rain-water penetration in British Somaliland soils. East Afr Agric J 16: 26–33

Glover PE (1951) The root systems of some British Somaliland plants-IV. East Afr Agric J 17: 38–50

Glover PE, Glover J, Gwynne MD (1962) Light rainfall and plant survival in eastern Africa. II. Dry grassland vegetation. J Ecol 50: 199–206

Gorham E (1961) Factors influencing supply of major ions to inland waters with special reference to the atmosphere. Geol Soc Am Bull 72: 795–840

Gosz JR (1980) Nutrient budget studies for forests along an elevational gradient in New Mexico. Ecology 61: 515–525

Goudriaan J (1977) Crop meteorology: a simulation study. Pudoc, Wageningen, 249 pp

Goudriaan J (1989) Simulation of micrometeorology of crops, some methods and their problems, and a few results. Agric For Meteorol 47: 239–258

Granier P (1977) Rapport d'activités agro-pastorales en République du Niger. IEMVT, Maisons-Alfort

Gray JT (1983) Nutrient use by evergreen and deciduous shrubs in southern California. I. Community nutrient cycling and nutrient-use efficiency. J Ecol 71: 21–41

Greenland DJ (1985) Nitrogen and food production in the tropics. In: Kang BT, Heide J van der (eds) Nitrogen management in farming systems in humid and subhumid tropics. Inst Soil Fertil (IB-DLO), Haren, pp 9–38

Grier CC, Running SW (1977) Leaf area of mature northwestern coniferous forests: relation to site water balance. Ecology 58: 893–899

Grossman D, Grunow JO, Theron GK (1980) Biomass cycles, accumulation rates and nutritional characteristics of grass layer plants in canopied and uncanopied subhabitats of Burkea savanna. Soc South Afr Proc Grassl 15: 157–161

Grouzis M (1979) Structure, composition floristique et dynamique de la production de matière sèche de formations végétales sahéliennes (Mare d'Oursi – Haute Volta). ACC Lutte contre l'aridité dans l'Oudalan. DGRST-ORSTOM, Ouagadougou

Grouzis M, Sicot M (1980) A method for the phenological study of browse populations in the Sahel: the influence of some ecological factors. In: Le Houérou HN (ed) Browse in Africa: the current state of knowledge. Papers presented at the Int Symp on Browse in Africa, Addis Ababa, 8–12 April 1980. ILCA, Addis Ababa, pp 233–240

Grouzis M, Nizinski G, Fournier C (1990) L'arbre et l'herbe au Sahel. Séminaire physiologie des arbres et arbustes des zones arides, 27 Mars–6 Avril 1990, Nancy

Grove TS (1988) Growth responses of trees and understorey to applied nitrogen and phosphorus in *Eucalyptus diversicolor* forest. For Ecol Manage 23: 87–103

Grubb PJ (1989) The role of mineral nutrients in the tropics: a plant ecologist's view. In: Proctor J (ed) Mineral nutrients in tropical forest and savanna ecosystems. Blackwell, Oxford, Spec Publ Br Ecol Soc 9: 417–439

Guinaudeau C (1987) Les brise-vent au Sénégal. Les techniques d'implantation, méthodes et outils de sensibilisation, programme de formation. FAO, Dakar

Gupta JP, Ramakrishna YS, Ramana Rao BV, Rao GGSN (1984) Role of shelterbelts in arid zone. Indian Farming 34: 29–30

Guyot G (1989) Les effets aerodynamiques et micro-climatiques des brise-vent et des aménagements régionaux. In: Reifsnyder WS, Darnhofer TO (eds) Meteorology and agroforestry. ICRAF, Nairobi, pp 485–520

Hairiah K, Noordwijk M van (1986) Root studies on a tropical ultisol in relation to nitrogen management. Report of field work at IITA's high rainfall substation at Onne (Port Harcourt, Nigeria) in 1985. Inst Soil Fertil (IB-DLO), Haren, 7: 121

Hallaire A (1991) Paysans montagnards du Nord-Cameroun. Les monts Madara. ORSTOM, Paris, 253 pp

Halvorson WL, Patten DT (1975) Productivity and flowering of winter ephemerals in relation to Sonoran desert shrubs. Am Midl Nat 93: 311–319

Hanley TA, Robbins CT, Hagerman AE, McArthur C (1992) Predicting digestible protein and digestible dry matter in tannin-containing forages consumed by ruminants. Ecology 73: 537–541

Harshbarger TJ, Perkins CJ (1971) Effect of shade on growth and seed production of partridge pea. J Wildl Manage 35: 25–32

Hellmers H, Horton JS, Juhren G, O'Keefe J (1955) Root systems of some chaparral plants in southern California. Ecology 36: 667–678

Helvey JD, Patric JH (1965) Canopy and litter interception of rainfall by hardwoods of eastern United States. Water Resour Res 1: 193–206

Hiernaux P (1980) Inventory of the browse potential of bushes, trees and shrubs in an area of the Sahel in Mali: method and initial results. In: Le Houérou HN (ed) Browse in Africa: the current state of knowledge. Papers presented at the Int Symp on Browse in Africa, Addis Ababa, 8–12 April 1980. ILCA, Addis Ababa, pp 197–204

Hiernaux P (1983) Les ressources végétales et leur productivité. In: Wilson RT, Leeuw PN de, Haan C de (eds) Recherches sur les systèmes des zones arides du Mali: résultats préliminaires. CIPEA, Bamako, Rapp Rech 5

Hiernaux P, Diarra L, Maiga A (1990) Dynamique de la végétation sahélienne après sécheresse. Un bilan de suivi des sites pastoraux du Gourma en 1989. Programme des zones arides et semi-arides, CIPEA, Bamako

Hingston FJ, Malajczuk N, Grove TS (1982) Acetylene reduction (N2-fixation) by Jarrah forest legumes following fire and phosphate application. J Appl Ecol 19: 631–645

Hirst SM (1975) Savanna ecosystem project – progress report 1974/1975. South Afr Nat Sci Progr 3: 27

Hoffmann O (1985) Pratiques pastorales et dynamique du couvert végétal en pays Lobi. ORSTOM, Paris, Trav Doc ORSTOM 189

Högberg P (1986) Nitrogen fixation and nutrient relations in savanna woodland trees (Tanzania). J Appl Ecol 23: 675–688

Högberg P (1989) Root symbiosis of trees in savannas. In: Proctor J (ed) Mineral nutrients in tropical forest and savanna ecosystems. Blackwell, Oxford, Spec Publ Br Ecol Soc 9: 121–136

Hopkins B (1965) Forest and savanna; an introduction to tropical plant ecology with special reference to West Africa. Heinemann, Ibadan

IAEA (1975) Root activity patterns of some tree crops. Int Atomic Energy Agency, Vienna, Tech Rep Ser 170

ICRISAT (1990) ICRISAT West African Programs. Annual Report 1989. ICRISAT, Niamey

ICRISAT (1991) ICRISAT West African Programs. Annual Report 1990. ICRISAT, Niamey

ICRISAT (1993) Shade increases crop yields. Spore 44: 10

ILCA (1975) Evaluation and mapping of tropical African rangelands. Proc Sem, Bamako, Mali, 3–8 March 1975, ILCA, Addis Ababa, 399 pp

ISRA (1983) Rapport annuel sur les recherches de physiologie 1982. Inst Sénégal Rech Agricoles (ISRA) Vét, Dakar, 70 pp

IWACO (1990) Etude des quatres bassins versants expérimentaux au Burkina Faso. I. Synthèse des études. IWACO, Bureaux d'études en eau et environnement, Ouagadougou

Jackson IJ (1975) Relationships between rainfall parameters and interception by tropical forest. J Hydrol 24: 215–238

Jackson JE (1989) Tree and crop selection and management to optimize overall system productivity, especially light utilization, in agroforestry. In: Reifsnyder WS, Darnhofer TO (eds) Meteorology and agroforestry. ICRAF, Nairobi, pp 163–173

Jackson JK, Taylor GF, Condé-Wane C (1983) Gestions des ressources forestières naturelles dans la région du Sahel. OCDE/CILSS, Paris, Rapp D(83)232

Jackson JE, Hamer PJC, Jackson B (1989) Water balance and soil water relations studies in a mixed tree/grass/bare soil system. In: Reifsnyder WS, Darnhofer TO (eds) Meteorology and agroforestry. ICRAF, Nairobi, pp 431–442

Jaksic FM, Fuentes ER (1980) Why are native herbs in the Chilean matorral more abundant beneath bushes: microclimate or grazing? J Ecol 68: 665–669

Jameson DA (1966) Pinyon – juniper litter reduces growth of blue gama. J Range Manage 19: 214–217

Jameson DA (1967) The relationship of tree overstory and herbaceous understory vegetation. J Range Manage 20: 247–249

Janssen BH (1984) A simple method for calculating decomposition and accumulation of young soil organic matter. Plant Soil 76: 297–304

Janssen BH (1986) Organische stof en bodemvruchtbaarheid (Organic matter and soil fertility). Lecture notes (in Dutch), Dep Soil Fertil, Wageningen Agric Univ, Wageningen

Janssen BH, Noij IGAM, Wesselink LG, Grinsven JJM van (1990) Simulation of the dynamics of nutrients and moisture in tropical ecosystems. Fert Res 26: 145–156

Janzen DH (1974) Ecology of plants in the tropics. Edward Arnold, London, Stud Biol 58: 66

Jarvis PG (1985) Transpiration and assimilation of tree and agricultural crops: the "omega factor". In: Cannell MGR, Jackson JE (eds) Attributes of trees as crop plants. Inst Terrestrial Ecol, pp 460–480

Jenkinson DS, Ayanaba A (1977) Decomposition of carbon-14-labeled plant material under tropical conditions. Soil Sci Soc Am J 41: 912–915

Joffre R, Rambal S (1988) Soil water improvement by trees in the rangelands of southern Spain. Acta Oecol Oecol Plant 9: 405–422

Jones E (1960) Contribution of rainwater to the nutrient economy of soil in northern Nigeria. Nature (London) 188: 432

Jonsson K, Fidjeland L, Maghembe JA, Högberg P (1988) The vertical distribution of fine roots of five tree species and maize in Morogoro, Tanzania. Agrofor Syst 6: 63–69

Jung G (1969) Cycles biogéochimiques dans un écosystème de région tropicale sèche *Acacia albida* (Del.) sol ferrugineux tropical peu lessivé (Dior). Oecol Plant 4: 195–210

Kang BT (1993) Alley cropping: past achievements and future directions. Agrofor Sys 23: 141–155

Kang BT, Reynolds L (1986) Alley farming in the humid and subhumid tropics. Proc Workshop Ibadan, Nigeria, 10–14 March 1986. IDRC, Ottawa

Kater LJM, Kante S, Budelman A (1992) Karité (*Vitellaria paradoxa*) and néré (*Parkia biglobosa*) associated with crops in South Mali. Agrofor Syst. 18: 89–105

Kaul RN, Ganguli R (1963) Fodder potential of *Ziziphus* in the scrub grazing lands of arid zone. Indian For 89: 623–630

Keita MN (1982) Les disponibilités de bois de feu en région sahélienne de l'Afrique occidentale – situation et perspectives. FAO, Rome

Kellman M (1979) Soil enrichment by neotropical savanna trees. J Ecol 67: 565–577

Kellman M (1985) Nutrient retention by savanna ecosystems. III. Response to artificial loading. J Ecol 73: 963–972

Kellman M, Sanmugadas K (1985) Nutrient retention by savanna ecosystems. I. Retention in the absence of fire. J Ecol 73: 935–951

Kellman M, Hudson J, Sanmugadas K (1982) Temporal variability in atmospheric nutrient influx to a tropical ecosystem. Biotropica 14: 1–9

Kellman M, Miyanishi K, Hiebert P (1985) Nutrient retention by savanna ecosystems. II. Retention after fire. J Ecol 73: 953–962

Kelly RD (1977) The significance of the woody component of semi-arid savanna vegetation in relation to meat production. Grassl Soc South Afr Proc 12: 105–108

Kelly RD, Walker BH (1976) The effects of different forms of land-use on the ecology of a semi-arid region in south-eastern Rhodesia. J Ecol 64: 553–576

Kemp PR (1983) Phenological patterns of Chihuahuan desert plant in relation to the timing of water availability. J Ecol 73: 427–436

Kemp RH (1962) Growth and regeneration of open savanna woodland in northern Nigeria. Commonw For Rev 42: 200–206

Kennard DG, Walker BH (1973) Relationships between tree canopy cover and *Panicum maximum* in the vicinity of Fort Victoria. Rhod J Agric Res 11: 145–153

Kennenni L, Maarel E van der (1990) Population ecology of *Acacia tortilis* in the semi-arid region of the Sudan. J Vegetation Sci 1: 419–424

Kerfoot O (1963) The root systems of tropical forest trees. Emp For Rev 42: 19–26
Kessler JJ (1992) The influence of karité (*Vitellaria paradoxa*) and néré (*Parkia biglobosa*) on sorghum production in Burkina Faso. Agrofor Syst 14: 97–118
Kessler JJ, Boni J (1991) L'agroforesterie au Burkina Faso. Bilan et analyse. Nature Conservation Dep, Wageningen Agric Univ, Wageningen, Resour Manage Pap 1: 144
Kessler JJ, Breman H (1991) The potential of agroforestry to increase primary production in the Sahelian and Sudanian zones of West Africa. Agrofor Syst 13: 41–62
Kessler JJ, Ohler FMJ (1983) Interventions dans les pays sahéliens: une approche écologique. Cent Agrobiol Res (CABO-DLO), Wageningen, 72 pp
Ketelaars JJMH, Tolkamp BJ (1991) Toward a new theory of feed intake regulation in ruminants. Doctoral Thesis, Wageningen Agric Univ, Wageningen, 254 pp
Keulen H van (1975) Simulation of water use and herbage growth in arid regions. Simulation Monographs series. Pudoc, Wageningen, 176 pp
Keulen H van, Breman H (1990) Agricultural development in the West African Sahelian region: a cure against land hunger? Agric Ecosyst Environ 32: 177–197
Keulen H van, Wolf J (eds) (1986) Modelling of agricultural production: weather, soils and crops. Simulation Monogr Ser, Pudoc, Wageningen, 479 pp
King KFS, Chandler MT (1978) The wasted lands; the programme of work of the International Council for Research in Agroforestry. ICRAF, Nairobi
Klein HD, Keita I, Mesnil JG (1981) Trois années d'activités agropastorales en République du Niger – 1977 à 1980. IEMVT, Maisons-Alfort, 290 pp
Knevel M (1993) Solitary trees: light interception and shade patterns. Dep Theor Prod Ecol, Wageningen Agric Univ, Wageningen, 92 pp
Knoop WT, Walker BH (1985) Interactions of woody and herbaceous vegetation in a southern African savanna. J Ecol 73: 235–253
Koechlin B, Rambal S, Debussche M (1986) Rôle des arbres pionniers sur la teneur en eau du sol en surface de friches de la région méditerranéenne. Acta Oecol Oecol Plant 7: 177–190
Kramer PJ (1983) Water relations of plants. Academic Press, New York, 489 pp
Kumar R (1983) Chemical and biochemical nature of fodder tree leaf tannins. J Agric Food Chem 31: 1364–1366
Kummerow J, Mangan R (1981) Root systems in *Quercus dumosa* Nutt.- dominated chaparral in southern California. Acta Oecol Oecol Plant 2: 177–188
Kummerow J, Krause D, Jow W (1977) Root systems of chaparral shrubs. Oecologia (Berlin) 29: 163–177
Laar HH van, Goudriaan J, Keulen H van (eds) (1992) Simulation of crop growth for potential and water-limited production situations (as applied to spring wheat). Cent Agrobiol Res (CABO-DLO), Wageningen, Simulation Rep 27: 72
Laishram ID, Yadava PS (1988) Lignin and nitrogen in the decomposition of leaf litter in a sub-tropical forest ecosystem at Shiroy hills in north-eastern India. Plant Soil 106: 59–64
Lal R (1975) Role of tropical mulching techniques in tropical soil and water management. IITA, Ibadan, Tech Bull 1: 48 pp
Lal R (1976) Soil erosion problems on an alfisol in Western Nigeria and their control. IITA, Ibadan, Monogr 1: 208
Lal R (1987) Termites. In: Lal R (ed) Tropical ecology and physical edaphology. Wiley, New York, pp 337–421
Lal R (1989) Agroforestry systems and soil surface management of a tropical alfisol: I: Soil moisture and crop yields. Agrofor Syst 8: 7–29
Lal R (1991) Myths and scientific realities of agroforestry as a strategy for sustainable management for soils in the tropics. In: Stewart BA (ed) Advances in soil science, vol 15. Springer, Berlin Heidelberg New York, pp 91–137
Lambert JDH, Arnason JT, Gale JL (1980) Leaf-litter and changing nutrient levels in a seasonally dry tropical hardwood forest, Belize, C.A. Plant Soil 55: 429–443

Lamers B, Roorda T (1991) La mise en défens de Zana Mogho. Dép d'Aménagement de la Nature, Univ Agron Wageningen, Wageningen

Lamotte M (1979) The structure and functioning of the savanna ecosystems of Lamto (Ivory Coast). In: Tropical grazing land ecosystems. UNESCO/UNEP/FAO, UNESCO Nat Resour Res 16: 511–561

Lamprey HF, Halevy G, Makacha S (1974) Interactions between *Acacia*, bruchid seed beetles and large herbivores. East Afr Wildl J 12: 81–85

Landsberg W, McMurtrie R (1984) Water use by isolated trees. Agric Water Manage 8: 223–242

Langkamp PJ, Swinden LB, Dalling MJ (1979) Nitrogen fixation (acetylene reduction) by *Acacia pellita* on areas restored after mining at Groote Eylandt, Northern Territory. Aust J Bot 27: 353–361

Langkamp PJ, Farnell GK, Dalling MJ (1982) Nutrient cycling in a stand of *Acacia holosericea* A Cunn ex G Don. I Measurements of precipitation interception, seasonal acetylene reduction, plant growth and nitrogen requirement. Aust J Bot 30: 87–106

Lavelle P, Martin A, Blanchart E, Gilot C, Melendez G, Pashanasi B (1990) Conservation de la fertilité des sols de savane par la gestion de l'activité de la macrofaune du sol. In: Pieri C, Bodichon J, Fontanel P, Forgue J-M (eds) Savanes d'Afrique, terres fertiles? Comment produire plus et de façon durable en zone de savanes au sud du Sahara. Actes des rencontres internationales: Montpellier, 10–14 Décembre 1990. La Documentation Française, Paris, pp 371–399

Lawson GW, Jeník J (1967) Observations on microclimate and vegetation interrelationships in the Accra Plains (Ghana). J Ecol 55: 773–785

Lay DW (1965) Effects of periodic clipping on yield of some common browse species. J Range Manage 18: 181–184

Lebreton P (1982) Tanins ou alcaloides: deux tactiques phytochimiques de dissuasion des herbivores. Terre Vie 36: 539–572

Leenaars J (1990) Regeneration of the nitrogen availability. Fallow lands of Oula, Burkina Faso. Dep Nature Conserv, Wageningen Agric Univ, Wageningen

Leeuw PN de (1975) Species preferences of domestic ruminants grazing Nigerian savanna. In: Ajayi SS, Geerling C, Laveren LP van, Hall JB (eds) Wildlife management in savannah woodland. Proc of the Ibadan/Garoua Int Symp on Wildlife Management, 22–26 Sept 1975, Ibadan, pp 207–224

Leeuw LM de (1981) Waardering van houtigen als voedselgewas voor het wild in West Afrika. Dep Nature Conserv, Wageningen Agric Univ, Wageningen

Le Houérou HN (1979) Le rôle des arbres et arbustes dans les pâturages sahéliens. In: Le rôle des arbres au Sahel. C R Colloq Dakar, 5–10 Nov 1979. IDRC, Ottawa

Le Houérou HN (1980a) The role of browse in the Sahelian and Sudanian zones. In: Le Houérou HN (ed) Browse in Africa: the current state of knowledge. Papers presented at the Int Symp on Browse in Africa, Addis Ababa, 8–12 April 1980. ILCA, Addis Ababa, pp 83–102

Le Houérou HN (1980b) Chemical composition and nutritive value of browse in tropical West Africa. In: Le Houérou HN (ed) Browse in Africa: the current state of knowledge. Papers presented at the Int Symp on Browse in Africa, Addis Ababa, 8–12 April 1980. ILCA, Addis Ababa, pp 261–290

Le Houérou HN (1989) The grazing land ecosystems of the African Sahel. Springer, Berlin Heidelberg New York, Ecol Stud 75: 282

Leigh JH, Mulham WE (1970) The effect of defoliation on the persistence of *Atriplex vesicaria*. Aust J Agric Res 22: 239–245

Leloup S, Traoré M (1989) La situation fourragère dans le sud-est du Mali, une étude agro-écologique. DRSPR, Volet Fonsébougou, Sikasso and Royal Institute of the Tropics, KIT, Amsterdam, 94 pp

Lepage M (1974) Influence de la sécheresse sur le peuplement en termites. Terre Vie 28: 76–94

Liebermann D (1982) Seasonality and phenology in a dry tropical forest in Ghana. J Ecol 70: 791–806

Lieth H, Whittaker RH (1975) Primary productivity of the biosphere. Ecological Studies 14. Springer, Berlin Heidelberg New York, 339 pp

Lohan OP, Lall D, Pal RN, Neg SS (1980) Note on tannins in tree fodders. Indian J Anim Sci 50: 881–883

Lövenstein HM, Berliner PR, Keulen H van (1991) Runoff agroforestry in arid lands. For Ecol Manage 45: 59–70

Lövenstein H, Rabbinge R, Keulen H van (1992) Water-limited crop production. In: World food production textbook. Section 2. Biophysical factors of agricultural production. Open Univ, Amsterdam, pp 55–74

Lowry JB (1990) Toxic factors and problems: methods of alleviating them in animals. In: Devendra C (ed) Shrubs and tree fodders for farm animals. IDRC, Ottawa, pp 76–88

Lowry JB, Lowry JBC, Jones R (1988) Enhanced grass growth below canopy of *Abizia lebbeck*. Nitrogen Fixing Tree Res Rep 6: 45–47

Lundgren B (1982) Introduction. Agrofor Syst 1: 3–6

Lunt HA (1934) Distribution of soil moisture under isolated forest trees. J Agric Res 49: 695–703

Lyles L, Woodruff NP (1960) Abrasive action of wind blown soil on plant seedlings. Agron J 52: 533–536

Mai Moussa KA, William JH, Stern RD (1994) Effet de la grandeur du tronc sur le gradient de fertilité aux environs de *Faidherbia albida*. In: C R Symp Int Parcs Agrofor, 25–28 Oct 1993, Ouagadougou. ICRAF/IRBET/CILSS/LTC (in press)

Malik RS, Sharma SK (1990) Moisture extraction and crop yield as a function of distance from a row of *Eucalyptus tereticornis*. Agrofor Syst 12: 187–195

Marchal JY (1983) Yatenga, nord Haute Volta: la dynamique d'un espace rural soudano-sahélien. ORSTOM, Paris, Trav Doc ORSTOM 167: 873

Mariaux A (1975) Essai de dendroclimatologie en climat sahélien sur *Acacia raddiana*. Rev Bois For Trop 163: 27–35

Martin GG (1988) Productivity, stability, sustainability, equitability and autonomy as properties for agroecosystem assessment. Agric Syst 26: 291–316

Mathieu P (1976) Influence des apports atmosphériques et du pluviolessivage forestier sur la qualité des eaux de deux bassins versants en Côte d'Ivoire. Cah ORSTOM Ser Geol 8: 11–32

Maydell H-J von (1986) Trees and shrubs of the Sahel. Their characteristics and uses. GTZ, Eschborn, 525 pp

McKey D, Waterman PG, Mbi CN, Gartlan JS, Struhsaker TT (1978) Phenolic content of vegetation in two African rain forests: ecological implications. Science 202: 61–64

McNaughton SJ (1983) Compensatory plant growth as a response to herbivory. Oikos 40: 329–336

McTainsh GH, Walker PH (1982) Nature and distribution of Harmattan dust. Z Geomorphol 26: 417–435

Menaut JC (1983) The vegetation of African savannas. In: Bourlière F (ed) Tropical savannas. Elsevier, Amsterdam, pp 109–149

Menaut JC, Barbault R, Lavelle P, Lepage M (1985) African savannas: biological systems of humification and mineralization. In: Tothill JC, Mott JJ (eds) Ecology and management of the world's savannas. Aust Acad Sci, Canberra, pp 14–33

Menke JW, Trlica MJ (1981) Carbohydrate reserve, phenology and growth cycles of nine Colorado range species. J Range Manage 34: 269–277

Miehe S (1990) Inventur und Überwachung der Vegetation auf Weideparzellen in Widou Thiengoly. GTZ, Eschborn, 95 pp

Miettinen P, Karsisto M, Musa MG (1992) Nodulation of nine nitrogen-fixing tree species grown in central Sudan. For Ecol Manage 48: 107–119

Miller HG (1989) Internal and external cycling of nutrients in forest stands. In: Pereira JS, Landsberg JJ (eds) Biomass production by fast-growing trees. Kluwer, Dordrecht, pp 73–80

Minson DJ (1990) Forage in ruminant nutrition. Academic Press, London, 483 pp

Mohren GMJ (1987) Simulation of forest growth, applied to Douglas fir stands in the Netherlands. Thesis, Dep For, Wageningen Agric Univ, Wageningen

Montagnini F, Sancho F (1990) Impacts of native trees on tropical soils: a study in the Atlantic lowlands of Costa Rica. Ambio 19: 386–390

Montana C, Cavagnaro B, Briones O (1991) Water use of a simulated rainfall by coexisting shrub and grasses in a semi-arid rangeland. IVth Int Rangeland Congr, April 1991, Montana, Montpellier

Moral R del, Muller CH (1970) The allelopathic effects of *Eucalyptus camaldulensis*. Am Midl Nat 83: 254–282

Mosnier M (1961) Pâturages naturels sahéliens: Région de Kaedi (Mauritania). IEMVT, Maisons-Alfort

Mott JJ, Williams J, Andrew MH, Gillison AN (1982) Structure and functional aspects of Australian savanna ecosystems. Int Pap, CSIRO, Melbourne, 62 pp

Mueller-Harvey I, Juo ASR, Wild A (1985) Soil organic C, N, S and P after forest clearance in Nigeria: mineralization rates and spatial variability. J Soil Sci 36: 585–591

Muoghalu JI, Isichei AO (1991) Effect of tree canopy cover on the yield, crude protein and fibre content of forb species in Nigerian Guinea savanna. Vegetatio 95: 167–175

Murphy AH, Crampton B (1964) Quality and yield of forage as affected by chemical removal of blue oak (*Quercus douglasii*). J Range Manage 17: 142–144

Murray DB, Nichols R (1966) Light, shade and growth of some tropical plants. In: Bainbridge R, Evans GC, Rackham O (eds) Light as an ecological factor. Blackwell, Oxford, Br Ecol Soc Sym 6: 249–263

Nair PKR (1984) Soil productivity aspects of agroforestry. Science and practice of agroforestry 1. ICRAF, Nairobi

Nakano K, Syahbuddin (1989) Nutrient dynamics in forest fallows in South-East Asia. In: Proctor J (ed) Mineral nutrients in tropical forest and savanna ecosystems. Blackwell, Oxford, Spec Publ Br Ecol Soc 9: 325–336

Neumann F, Pietrowicz P (1989) Light and water availability in fields with and without trees. In: Reifsnyder WS, Darnhofer TO (eds) Meteorology and agroforestry. ICRAF, Nairobi, pp 401–406

Nicholson SE (1982) The Sahel: a climatic perspective. OCDE/CILSS/Club du Sahel, Paris, Sahel D(82)187: 80

Nix HA (1983) Methods for coping with variation in climate and soils. In: Proc Agrores Aust semi-arid tropics, Symp on the role of Res Agric Dev. 21–25 March 1983, Darwin, NT Australia, CSIRO

Noij IGAM, Janssen BH, Wesselink LG, Grinsven JJM van (1990) Modeling nutrient and moisture cycling in tropical forests. Dep Trop Agric, Wageningen Agric Univ, Wageningen, Tech Ser 5: 195 pp

Noordwijk M van (1984) Ecology textbook for the Sudan. Ecol Uitgeverij, Amsterdam, 280 pp

Noordwijk M van (1989a) Tree roots as components of agroforestry systems. Reader for IBSRAM/ICRAF workshop on research in soil management and agroforestry. Inst Soil Fertil (IB-DLO), Haren, Note 217, 46 pp

Noordwijk M van (1989b) Rooting depth in cropping systems in the humid tropics in relation to nutrient use efficiency. In: Heide J van der (ed) Nutrient management for food crop production in tropical farming systems. Inst Soil Fertil (IB-DLO), Haren, pp 129–144

Noordwijk M van, Widianto, Heinen M, Hariah K (1991) Old tree root channels in acid soils in the humid tropics: important for crop root penetration, water infiltration and nitrogen management. Plant Soil 134: 37–44

Noy Meir I (1979) Structure and function of desert ecosystems. Isr J Bot 28: 1–19
Nwoboshi LC (1972) Differential influences of two exotic forest tree species on a soil. J West Afr Sci Assoc 4: 35–47
Nyami DO (1991) Phosphorus uptake by different tree species on low phosphorus soils. Int Tree Crops J 7: 77–82
Nye PH, Greenland DJ (1960) The soil under cultivation. Commonw Agric Bur Soils, Harpenden, CAB Farnham Royal, Bucks, Tech Commun 51: 155
Nyerges AE (1989) Coppice swidden fallows in tropical deciduous forest: biological, technological, and sociocultural determinants of secondary forest successions. Hum Ecol 17: 379–400
Ohler FMJ (1982) De karité of shea notenboom. Dep Trop Agric, Wageningen Agric Univ, Wageningen
Ohler FMJ (1983) De braaklanden nabij het National Park Boucle de Baoulé. Dep Nature Conserv, Wageningen Agric Univ, Wageningen, Student Rep 660, 90 pp
Ohler FMJ (1985) The fuelwood production of wooded savanna fallows in the Sudan zone of Mali. Agrofor Syst 3: 15–23
Okeke AI, Omaliko CPE (1991) Nutrient accretion to the soil via litterfall and throughfall in *Acioa barteri* stands at Ozala, Nigeria. Agrofor Syst 16: 223–229
Olson JS (1963) Energy storage and the balance of producers and decomposers in ecological systems. Ecology 44: 322–331
Ong CK, Rao MR, Mathuva M (1992) Trees and crops, competition for resources above and below the ground. Agrofor Today 4: 4–5
Ouédraogo M (1986) Etude de la relation herbe-arbre dans une savane naturelle soudanienne: phytosociologie et phytomasse maximale dans divers faciès de savane plus ou moins denses de la réserve de Nazinga (Burkina Faso). Mémoire Ingénieur Développement Rural. IDR, Univ Ouagadougou, Ouagadougou
Ovalle C, Avendaño J (1987) Interactions de la strate ligneuse avec la strate herbacée dans les formations d'*Acacia caven* (Mol.) Hook. et Arn. au Chile. I. Influence de l'arbre sur la composition floristique, la production et la phénologie de la strate herbacée. Acta Oecol Oecol Plant 8: 385–404
Ovalle C, Avendaño J (1988) Interactions de la strate ligneuse avec la strate herbacée dans les formations d'*Acacia caven* (Mol.) Hook. et Arn. au Chile. II. Influence de l'arbre sur quelques éléments du milieu: microclimat et sol. Acta Oecol Oecol Plant 9: 113–134
Palm CA, Sanchez PA (1991) Nitrogen release from the leaves of some tropical legumes as affected by their lignin and polyphenolic contents. Soil Biol Biochem 23: 83–88
Panda SK, Panda NC, Sahu BK (1983) Effect of tree-leaf tannin on dry matter intake by goats. Indian Vet J 60: 660–664
Papadakis J (1975) Climates of the world and their potentialities. Libro de Edicion Argentina, Buenos Aires, 200 pp
Parker GG (1983) Throughfall and stemflow in the forest nutrient cycle. Adv Ecol Res 13: 57–133
Parker VT, Muller CH (1982) Vegetational and environmental changes beneath isolated live oak trees (*Quercus agrifolia*) in a California annual grassland. Am Midl Nat 107: 69–81
Parnot J (1988) Suivi des feux de brousse au Burkina Faso 1986/1987. Cent Régional de Télédétection, Ouagadougou
Parton WJ, Anderson DW, Cole CV, Stewart JWB (1983) Simulation of soil organic matter formations and mineralization in semiarid agroecosystems. In: Lowrance RR, Todd RL, Asmusen LE, Leonard RA (eds) Nutrient cycling in agricultural ecosystems. Coll Agric Exp Stat, Univ Georgia, Athens, Spec Publ 23: 533–550
Pellew RA (1980) The production and consumption of Acacia browse and its potential for animal protein production. In: Le Houérou HN (ed) Browse in Africa: the current

state of knowledge. Papers presented at the Int Symp on Browse in Africa, Addis Ababa, 8–12 April 1980. ILCA, Addis Ababa, pp 223–232

Peltier R, Eyog-Matig O (1989) Un essai sylvo-pastoral au Nord-Cameroun. Rev Bois For Trop 221: 3–23

Penning de Vries FWT, Djitèye MA (eds) (1991) La productivité des pâturages sahéliens. Une etude des sols, des végeptations et de l'exploitation de cette ressource naturelle, 2nd edn. Pudoc, Wageningen, Agric Res Rep 918: 523

Pereira JS, Pallardy S (1989) Water stress limitations to tree productivity. In: Pereira JS, Landsberg JJ (eds) Biomass production by fast-growing trees. Kluwer, Dordrecht, pp 37–56

Perry RA (1970) The effects on grass and browse production of various treatments on a mulga community in central Australia. Proc XIth Int Grassl Congr, 13–23 April, Surfers Paradise, Queensland, pp 63–66

Pianka ER (1970) On r- and K-selection. Am Nat 52: 592–597

Pichot J (1975) Rôle de la matiére organique dans la fertilité du sol. Agron Trop (Paris) 30: 170–175

Pieri C (1989) Fertilité des terres de savanes. Bilan de trente ans de recherche et de développement agricol au Sud du Sahara. CIRAD, Montpellier

Piot J (1969) Végétaux ligneux et pâturages des savanes de l'Adamoua au Cameroun. Rev Elev Med Vet Pays Trop 22: 541–559

Piot J (1970) Pâturage aérien au Cameroun. Utilisation des ligneux par les bovins. Rev Elev Med Vet Pays Trop 23: 503–514

Piot J (1980) Management and utilization methods for ligneous forages: natural stands and artificial plantations. In: Le Houérou HN (ed) Browse in Africa: the current state of knowledge. Papers presented at the Int Symp on Browse in Africa, Addis Ababa, 8–12 April 1980. ILCA, Addis Ababa, pp 339–349

Piot J, Nébout JP, Nanot R, Toutain B (1980) Utilisation des ligneux sahéliens par les herbivores domestiques, étude quantitative dans la zone Sud de la Mare d'Oursi (Haute-Volta). IEMVT, Maisons-Alfort

PIRL (1988) Rapport de synthèse, première phase, les formations végétales. Projet Inventaire des Ressources Ligneuses au Mali, Bamako

Poel P van der, Kaya B (1990) Synthèse des activités de recherche de la DRSPR sur l'aménagement anti-érosif (1986–1989). Inst R Trop, KIT, Amsterdam

Poore MED, Fries C (1985) The ecological effects of *Eucalyptus*. FAO, Rome, FAO For Pap 59, 87 pp

Post WM, Emanuel WR, Zinke PJ, Stangenberger AG (1982) Soil carbon pools and world life zones. Nature (London) 298: 156–159

Poupon H (1976) La biomasse et l'évolution de sa répartition au cours de la croissance d'*Acacia senegal* dans une savane sahélienne (Sénégal). Bois For Trop 166: 23–38

Poupon H (1977a) Evolution d'un peuplement d'*Acacia senegal* (L.) Willd. dans une savane sahélienne de 1972 à 1976. Cah ORSTOM Ser Biol 12: 283–291

Poupon H (1977b) Recherches écologiques sur une savane sahélienne du Ferlo septentrional, Sénégal: premières données sur *Commiphora africana* (Rich.) Engl. Terre Vie 31: 127–162

Poupon H (1979) Etude de la phénologie de la strate ligneuse à Fété Olé (Sénégal septentrional) de 1971 à 1977. Bull Inst Fondam Afr Noire (IFAN) 41: 44–91

Poupon H (1980) Structure et dynamique de la strate ligneuse d'une steppe sahélienne au Nord du Sénégal. ORSTOM, Paris, Trav Doc ORSTOM 115, 351 pp

Poupon H, Bille JC (1974) Recherches écologiques sur une savane sahélienne du Ferlo septentrional (Sénégal). Influence de la sécheresse de l'année 1972–1973 sur la strate ligneuse. Terre Vie 28: 49–75

Powell J, Box TW (1967) Mechanical control and fertilization as brush management practices affect forage production in South Texas. J Range Manage 20: 227–236

Pratchett D (1983) Botswana: recent range research findings. World Anim Rev 46: 18–25
Prebble RE, Stirk GB (1980) Throughfall and stemflow on silverleaf ironbark (*Eucalyptus melanophloia*) trees. Aust J Ecol 5: 419–427
Pressland AJ (1973) Rainfall partitioning by an arid woodland (*Acacia aneura* F. Muell) in south-west Queensland. Aust J Bot 21: 235–245
Pressland AJ, Fisher IH (1982) The place of catchment hydrology and productivity in catchment stability. Conf Agric Engin, Armidale, 22–24 Augus 1982, Inst Engin, Australia
Pritchett WL (1979) Properties and management of forest soils. Wiley, New York
Pullan RA (1977) Farmed parkland in West Africa. Savanna 3: 119–151
Raay HGT van, Leeuw PN de (1974) Fodder resources and grazing management in a savanna environment: an ecosystem approach. Occas Pap, Inst Soc Stud, the Hague, 31 pp
Radersma S (1994) The influence of parkland trees on the soil and the herbaceous layer in the southern Sahel; with special attention to phosphorus. Projet Production Soudano-Sahélienne (PSS), Niono/Wageningen, MSc Thesis, Dep Trop Crop Sci, Agric Univ, Wageningen, 98 pp
Radwanski SA (1969) Improvement of red acid sands by Neem tree (*Azadirachta indica*) in Sokoto, north-western state of Nigeria. J Appl Ecol 6: 507–511
Raison J-P (1988) Les parcs en Afrique: état des connaissances et perspectives de recherches. Encycl Tech Agric Afr Trop, 79 pp
Raison RJ (1979) Modification of the soil environment by vegetation fires, with particular reference to nitrogen transformations: a review. Plant Soil 51: 73–108
Ram SC, Ramakrishnan PS (1988) Hydrology and soil fertility of degraded grasslands at Cherrapunji in northeastern India. Environ Conserv 15: 29–35
RAMS (1980) Rural assessment and manpower surveys – Mauretania. Oasis Agriculture. USAID, Washington
Ramsay JM, Innes RR (1963) Some quantitative observations on the effects of fire on the Guinea savanna vegetation of northern Ghana over a period of eleven years. Sols Afr 8: 41–85
Reifsnyder WS (1989) Control of solar radiation in agroforestry practice. In: Reifsnyder WS, Darnhofer TO (eds) Meteorology and agroforestry. ICRAF, Nairobi, pp 141–157
Renard C, Vandenbeldt RJ (1990) Bordures d'*Andropogon gayanus* Kunth comme moyen de lutte contre l'érosion éolienne au Sahel. Agron Trop (Paris) 45: 227–231
Renes G, Coulibaly S (1988) Etude de la capacité de régénération naturelle et de la productivité des forêts naturelles au Burkina Faso. Minist Environ Tourisme, Ouagadougou, 40 pp
Reuler H van, Janssen BH (1989) Nutritional constraints in secondary vegetation and upland rice in south-west Ivory Coast. In: Proctor J (ed) Mineral nutrients in tropical forest and savanna ecosystems, Blackwell, Oxford, Spec Publ Br Ecol Soc 9: 371–382
Richards BN, Charley JL (1983) Mineral cycling processes and system stability in the Eucalypt forest. For Ecol Manage 7: 31–47
Richards JH, Caldwell MM (1987) Hydraulic lift: substantial nocturnal water transport between soil layers by *Artemisia tridentata* roots. Oecologia (Berlin) 73: 486–489
Ridder N de, Keulen H van (1990) Some aspects of the role of organic matter in sustainable intensified arable farming systems in the West-African semi-arid tropics (SAT). Fert Res 26: 299–310
Robins E (1994) Researching tree management strategies in Thiougou village, central plateau, Burkina Faso. Int Symp Agrofor Parklands, 28–28 Oct 1993, Ouagadougou, Burkina Faso (in press)
Robinson PJ (1986) The dependence of crop production on trees and forest land. In: Prinsley RT, Swift MJ (eds) Amelioration of soil by trees, a review of current concepts and practices. Commonw Sci Counc, London, pp 104–120

Roose E (1981) Dynamique actuelle de sols ferralitiques et ferrugineux tropicaux d'Afrique Occidentale. ORSTOM, Paris, Trav Doc ORSTOM 130: 569
Roy MM (1991) Some tropical fodder trees for sustained fodder and firewood availability during lean period. J Trop For 7: 196–205
Rutherford MC (1979) Above ground biomass subdivisions in woody species of the savanna ecosystem project study area, Nylsvley. South Afr Natl Sci Progr Rep 36: 33 pp
Ruyssen B (1957) Le Karité au Soudan. 3. Caractère de la production. Agron Trop (Paris) 12: 279–306
Sabiiti EN, Cobbina J (1992) Initial agronomic evaluation of *Parkia biglobosa* in the humid zone of Nigeria. Agrofor Syst 17: 271–279
Sanchez PA, Palm CA, Davey CB, Szott LT, Russell CE (1985) Tree crops as soil improvers in the humid tropics? In: Cannell MGR, Jackson JE (eds) Attributes of trees as crop plants. Inst Terrest Ecol, pp 327–358
Sanford WW, Usman S, Obot EO, Isichei AO, Wari M (1982) Relationship of woody plants to herbaceous production in Nigerian savanna. Trop Agric (Trinidad) 59: 315–318
Satoo T (1982) Forest biomass. Junk, the Hague, 152 pp
Savenije H (1993) Sylvicultural management practices in the Sudan and Sahel zone with an emphasis on sylvopastoral vegetations. A compilation of literature. Antenne Sahélienne de l'Univ Agron Wageningen, Wageningen, Doc Proj 17: 109
Saxena NC (1992) *Eucalyptus* planting as a response to farm management problems faced by "on-site" and "off-site" farmers. Agrofor Syst 19: 159–172
Soni P, Vasistha HB (1991) Understorey vegetation in *Eucalyptus* plantations – a review. J Trop For 7: 15–26
Soumaré A, Groot JJR, Koné D, Radersma S (1994) Structure spatiale du système racinaire de deux arbres du Sahel: *Acacia seyal* et *Sclerocarya birrea*. Rapp PSS no 5, IER, Bamako/AB-DLO, Wageningen & Haren/DAN-UAW, Wageningen
Spitters CJT, Keulen H van, Kraalingen DWG van (1989) A simple and universal crop growth simulator: SUCROS87. In: Rabbinge R, Ward SA, Laar HH van (eds) Simulation and systems management in crop protection. Pudoc, Wageningen, Simulation Monogr Ser 32: 147–181
Sprent JI (1986) Potentials for nitrogen-fixing legume trees in the tropics. Int Tree Crops J 4: 49–55
Sprent JI, Sutherland JM (1990) Nitrogen-fixing woody legumes. Nitrogen-Fixing Tree Res Rep 8: 17–23
Squire RO, Marks GC, Craig FG (1978) Root development in a *Pinus radiata* D. Don plantation in relation to site index, fertilizing and soil bulk density. Aust For Res 8: 103–114
Stahr K, Bleich KE, Hebel A (1993) The influence of organic matter and dust deposition on site characteristics and their microvariability. Rep Univ Hohenheim, Stuttgart
Sterck FJ, Zandt GJAW van der, Oldeman RAA (1991) Architectural development of *Faidherbia albida* (Del.) A. Chev. In: Physiologie des arbres et arbustes en zones arides et semi-arides. Groupe d'Etude de l'Arbre, Paris, pp 249–256
Stigter CJ (1985) Wind protection in traditional microclimate management and manipulation – examples from East Africa. Progr Biometeorol 2: 145–154
Stigter CJ (1988) Microclimate management and manipulation in agroforestry. In: Wiersum KF (ed) Viewpoints on agroforestry. Dep For, Wageningen Agric Univ, Wageningen
Stigter CJ (1994) Essentials and cases of wind protection from scattered trees and shelterbelts. Int Symp Agrofor Parklands, 25–28 Oct 1993, Ouagadougou, ICRAF/IRBET/CILSS/LTC (in press)
Stocker O (1970) Water and photosynthesis relations of desert plants in the Mauretanian Sahara. 1. Raingreen and evergreen trees. Flora (Jena) 159: 539–572

Stoorvogel JJ, Smaling EMA (1990) Assessment of soil nutrient depletion in sub-saharan Africa: 1983–2000. Vol 1: Main report. The Winand Staring Centre, Wageningen, Rep 28: 137
Story R (1967) Pasture patterns and associated soil water in partially cleared woodland. Aust J Bot 15: 175–187
Strojan CL, Turner FB, Castetter R (1979) Litter fall from shrubs in the northern Mojave desert. Ecology 60: 891–900
Stromgaard P (1985) Biomass, growth, and burning of woodland in a shifting cultivation area of south central Africa. For Ecol Manage 12: 163–178
Stromgaard P (1986) Early secondary succession on abandoned shifting cultivator's plots in the miombo of south-central Africa. Biotropica 18: 97–106
Stuart Chapin F (1980) The mineral nutrition of wild plants. Annu Rev Ecol Syst 11: 233–260
Stuart-Hill GC, Tainton NM (1989) The competitive interaction between *Acacia karroo* and the herbaceous layer and how this is influenced by defoliation. J Appl Ecol 26: 285–298
Suresh KK, Vinaya Rai RS (1988) Allelopathic exclusion of understorey by a few multipurpose trees. Int Tree Crops J 5: 143–151
Sutcliffe JF (1979) Plants and water. Arnold, London, Stud Biol 14: 122
Swift MJ, Russel-Smith A, Perfect TJ (1981) Decomposition and mineral-nutrient dynamics of plant litter in regeneration bush-fallow in sub-humid tropical Nigeria. J Ecol 69: 981–995
Switzer GL, Nelson LE (1972) Nutrient accumulation and cycling in loblolly pine (*Pinus taeda* L.) plantation ecosystems: the first twenty years. Soil Sci Soc Am Proc 36: 143–147
Szott LT, Kass DCL (1993) Fertilizers in agroforestry systems. Agrofor Syst 23: 157–176
Szott LT, Fernandes ECM, Sanchez PA (1991) Soil-plant interactions in agroforestry systems. For Ecol Manage 45: 127–152
Tanner CB, Sinclair TR (1983) Efficient use of water in crop production: research or re-search. In: Taylor HM, Jordan WR, Sinclair TR (eds) Limitations to efficient water use in crop production. ASA-CSSA-SSSA, Madison, pp 1–27
Scanlan JC, Burrows WH (1990) Woody overstorey impact on herbaceous understorey in *Eucalyptus* spp. communities in central Queensland. Aust J Ecol 15: 191–197
Schlesinger WH (1985) Decomposition of chaparral shrub foliage. Ecology 66: 1353–1359
Schlesinger WH, Gill DS (1980) Biomass, production, and changes in the availability of light, water, and nutrients during the development of pure stands of the chaparral shrub, *Ceanothus megacarpus*, after fire. Ecology 61: 781–789
Scholte PT (1989) Vegetation soil relations in an area with sealed chronic luvisols, Kenya. Arid Soil Res Rehabilit 3: 337–348
Scott RM (1962) Exchangeable bases of mature, well-drained soils in relation to rainfall in East Africa. J Soil Sci 13: 1–9
Scott-Wendt J, Chase RG, Hossner LR (1988) Soil chemical variability in sandy ustalfs in semiarid Niger, West Africa. Soil Sci 145: 414–419
Sedogo TP (1988) Les pâturages naturels de Gampèla: quelques aspects des relations herbes/arbres-herbes/arbustes. Mémoire Ingénieur Développement Rural. IDR, Univ Ouagadougou, Ouagadougou
Seghieri J (1990) Dynamique saisonnière d'une savane soudano-sahélienne au nord-Cameroun. Thése, Acad Montpellier, Univ Montpellier II, 200 pp
Seif el Din A, Obeid M (1971) Ecological studies of the vegetation of Sudan. IV. The effect of simulated grazing on the growth of *Acacia senegal* (L.) seedlings. J Appl Ecol 8: 211–216
Seligman NG, Benjamin RW, Forti M (1991) Growth responses of three forage shrub species to intra- and inter-specific competition. IVth Int Rangeland Congr, April 1991, CIRAD, Montpellier

Shaikh A, Arnould E, Christophersen K, Hagen R, Tabor J, Warshall P (1988) Opportunities for sustained development. Successful natural resources management in the Sahel. Vol III: Financial analysis of interventions. US-AID, Office Tech Resour and Sahel Office, Bur Afr, E/DI, Washington, 39 pp

Shankarnarayan KA, Harsh LN, Kathju S (1989) Agroforestry in the arid zones of India. In: Nair PKR (ed) Agroforestry systems in the tropics. ICRAF/Kluwer, Dordrecht

Shmida A, Burgess TL (1988) Plant growth-form strategies and vegetation types in arid environments. In: Werger MJA, Aart PJM, van der, During HJ, Verhoeven JTA (eds) Plant form and vegetation structure. Acad, the Hague, pp 211–241

Sidibé M (1976) Contribution à l'étude comparative de quelques modes de traitement appliquées sur les pâturages naturels du CNRZ de Sotuba. Mém DEA, CPS/EN Sup, Bamako

Sieder P (1983) Grossflächen aufforstungen in Obervolta. Ertragskundliche und betriebswirtschaftliche Ergebnisse beim grossflächigen Anbau von Exoten in der Sudan Zone Westafrikas. Forst Holzwirt 5: 112–118

Singh RP, Ong CK, Saharan N (1989) Above and below ground interactions in alley-cropping in semi-arid India. Agrofor Syst 9: 259–274

Sipkens L (1984) Agroforestry and soil fertility: finding facts. Dep Soil Sci Plant Nutr, Wageningen Agric Univ, Wageningen, 55 pp

Skarpe C, Bergström R (1986) Nutrient content and digestibility of forage plants in relation to plant phenology and rainfall in the Kalahari, Botswana. J Arid Environ 11: 147–164

Slatyer RO (1961) Methodology of a water balance study conducted on a desert woodland (*Acacia aneura* F. Muell.) community in central Australia. Proc Symp on plant water relationships in arid and semi-arid conditions. UNESCO, Paris

Slatyer RO (1965) Measurements of precipitation interception by an arid zone plant community (*Acacia aneura* F. Muell.). Arid Zone Res 25: 181–192

Smith DM (1986) The practice of silviculture. Wiley, New York, 71 pp

Smith DM, Jarvis PG, Odongo JCW (1994) Water and energy relations of windbreak tree in the Sahel. Int Symp Agrofor Parklands, 25–28 Oct 1993, Ouagadougou, ICRAF/IRBET/CILSS/LTC (in press)

Smith TM, Goodman PS (1986) The effect of competition on the structure and dynamics of *Acacia* savannas in southern Africa. J Ecol 74: 1031–1044

Soest PJ van (1982) Nutritional ecology of the ruminant: ruminant metabolism, nutritional strategies, the cellulolytic fermentation and the chemistry of forages and plant fibers. O & B Books, Corvallis, 374 pp

Taylor RD, Walker BH (1978) Comparisons of vegetation use and herbivore biomass on a Rhodesian game and cattle ranch. J Appl Ecol 15: 565–581

Teklehaimanot Z, Jarvis PG, Ledger DC (1991) Rainfall interception and boundary layer conductance in relation to tree spacing. J Hydrol 123: 261–278

Thalen DCP (1979) Ecology and utilization of desert shrub rangelands in Iraq. Junk, the Hague

Tiedemann AR, Klemmedson JO (1973) Nutrient availability in desert grassland soils under mesquite (*Prosopis juliflora*) trees and adjacent open areas. Soil Sci Soc Am Proc 37: 107–110

Tiedemann AR, Klemmedson JO (1977) Effect of mesquite trees on vegetation and soils in the desert grassland. J Range Manage 30: 361–367

Tiedemann AR, Helvey JD, Anderson TD (1980) Effects of chemical defoliation of an *Abies grandis* habitat on amounts and chemistry of throughfall and stemflow. J Environ Qual 9: 320–328

Tilman D (1985) The resource-ratio hypothesis of plant succession. Am Nat 125: 827–852

Timmer L (1994) Inventaire des pratiques de gestion de *Parkia biglobosa* au Burkina Faso. Dép For, Univ Agron Wageningen, Wageningen,120 pp

Togola M, Cissé MI, Breman H (1975) Evolution de la végétation du ranch de Niono depuis 1969. In: Inventaire et cartographie des pâturages tropicaux africains. Actes Colloq, 3–8 Mars, Bamako, Mali. CIPEA, Addis Ababa, pp 195–201

Toky OP, Bisht RP (1988) Observations on the rooting patterns of some agroforestry trees in an arid region of north-western India. Agrofor Syst 8: 245–259

Tolman M (1989) Transpiration des essences principales dans la période mars-mai 1989, dans quatre bassins versants. IWACO, Ouagadougou, Burkina Faso/Univ Groningue, Groningen

Tolsma DJ (1989) On the ecology of savanna ecosystems in south-eastern Botswana. Free Univ, Amsterdam, 243 pp

Tongway DJ, Ludwig JA (1990) Vegetation and soil patterning in semi-arid mulga lands of eastern Australia. Aust J Ecol 15: 23–34

Topps JH (1992) Potential, composition and use of legume shrubs and trees as fodders for livestock in the tropics. Review. J Agric Sci 118: 1–8

Torres F (1983) Role of woody perennials in animal agroforestry. Agrofor Syst 1: 131–163

Tothill JC, Mott JJ (1985) Ecology and management of the world's savannas. Aust Acad Sci, Canberra, 384 pp

Toutain B (1978) ACC lutte contre l'aridité dans l'Oudalan (Haute-Volta). Rapp Campagne. IEMVT, Maisons-Alfort, 67 pp

Toutain B (1980) The role of browse plants in animal production in the Sudanian zone of West Africa. In: Le Houérou HN (ed) Browse in Africa: the current state of knowledge. Papers presented at the Int Symp on Browse in Africa, Addis Ababa, 8–12 April 1980. ILCA, Addis Ababa, pp 103–108

Toutain B, Piot J (1980) Mises en défens et possibilités de régénération des ressources fourragères sahéliennes. Etudes expérimentales dans le bassin de la Mare d'Oursi (Haute-Volta). IEMVT, Maisons-Alfort

Toutain B, Wispelaere G de (1977) Pâturage de l'ORD du Sahel et de la zone de délestage au nord-est de Fada N'Gourma, Haute-Volta. Tome I. Les p̂turages naturels et leur mise en valeur. IEMVT, Maisons-Alfort, Etude Agrostol 51: 95

Trapnell CG, Friend MT, Chamberlain GT, Birch HF (1975) The effects of fire and termites on a Zambian woodland soil. J Ecol 64: 577–588

Trochain JL (1940) Contribution à l'étude de la végétation du Sénégal. IFAN, Dakar, Mém 2, 434 pp

Tunstall BR, Walker J (1975) The effect of woodland disturbance on soil water. Proc Ecol Soc Aust 9: 49–57

Turner J (1981) Nutrient supply in relation to immobilization in biomass and nutrient removal in harvesting. In: Proc Aust For Nutr Workshop: productivity in perpetuity. 10–14 August, Canberra, CSIRO (Commonwealth Scientific and Industrial Research Organization), Division of Forest Research, Canberra, pp 263–275

Tybirk K (1991) Régénération des légumineuses ligneuses du Sahel. Bot Inst Aarhus Univ, AAU Rep 28: 86

Ullmann I (1985) Diurnal courses of transpiration and stomatal conductance of Sahelian and Saharian Acacias in the dry season. Flora (Jena) 176: 383–409

UNESCO-IPAL (1982) Resource management plan for the Rendill area of northern Kenya. UNESCO – Integrated Project on Arid Lands, Marsabit

Upadhyay VP, Singh JS (1989) Patterns of nutrient immobilization and release in decomposing forest litter in central Himalaya, India. J Ecol 77: 127–146

Vallis I, Jones RJ (1973) Net mineralization of nitrogen in leaves and leaf litter of *Desmodium intortum* and *Phaseolus atropurpureus* mixed with soil. Soil Biol Biochem 5: 391–398

Veeneklaas FR, Cissé S, Gosseye PA, Duivenbooden N van, Keulen H van (eds) (1991) Competing for limited resources: the case of the fifth region of Mali. Report 4: Development scenarios. CABO, Wageningen & ESPR, Mopti, 144 pp + annexes

Veenendaal EM (1991) Growth responses to shading of seedlings of savanna grasses in Botswana with different C_4 photosynthetic pathways? In: Adaptive strategies of grasses in a semi-arid savanna in Botswana. Doctoral Thesis, Free Univ, Amsterdam, pp 101–126

Vegten JA van (1983) Thornbush invasion in a savanna ecosystem in eastern Botswana. Vegetatio 56: 3–7

Verberne ELJ, Hassink J, Willigen P de, Groot JJR, Veen JA van (1990) Modelling organic matter dynamics in different soils. Neth J Agric Sci 38: 221–238

Verinumbe I (1987) Crop production on soil under some forest plantations in the Sahel. Agrofor Syst 5: 185–188

Vetaas OR (1992) Micro-site effects of trees and shrubs in dry savannas. J Vegetat Sci 3: 337–344

Vitousek PM, Sanford RL Jr (1986) Nutrient cycling in moist tropical forest. Am Rev Ecol Syst 17: 137–167

Voigt GK (1960) Distribution of rainfall under forest stands. For Sci 6: 2–10

Walker BH (1980) A review of browse and its role in livestock production in Southern Africa. In: Le Houérou HN (ed) Browse in Africa: the current state of knowledge. Papers presented at the Int Symp on Browse in Africa, Addis Ababa, 8–12 April 1980. ILCA, Addis Ababa, pp 7–24

Walker BH, Noy-Meir I (1980) Aspects of the stability and resilience of savanna ecosystems. In: Huntley BJ, Walker BH (eds) Ecology of tropical savannas. Springer, Berlin Heidelberg New York, pp 556–590

Walker BH, Ludwig D, Holling CS, Peterman RM (1981) Stability of semi-arid savanna grazing systems. J Ecol 69: 473–498

Walter H (1973) Vegetation of the earth in relation to climate and the eco-physiological conditions. Springer, Berlin Heidelberg New York, p 237

Walker J, Moore RM, Robertson JA (1972) Herbage response to tree and shrub thinning in *Eucalyptus populnea* shrub woodlands. Aust J Agric Res 23: 405–410

Walker J, Robertson JA, Penridge LK (1986) Herbage response to tree thinning in a *Eucalyptus crebra* woodland. Aust J Ecol 11: 135–140

Waterman PG, Ross JAM, McKey DB (1984) Factors affecting levels of some phenolic compounds, digestibility, and nitrogen content of the mature leaves of *Barteria fistulosa* (Passifloraceae). J Chem Ecol 10: 387–401

Weltzin JF, Coughenor MB (1990) Savanna tree influence on understory vegetation and soil nutrients in northwestern Kenya. J Vegetat Sci 1: 325–334

Werger MJA (1978) Primary production ecology in southern Africa. In: Werger MJA (ed) Biogeography and ecology of southern Africa. Junk, the Hague, pp 623–651

Werger MJA (1983) Tropical grasslands, savannas, woodlands: natural and manmade. In: Holzer W, Werger MJA, Ihusima I (eds) Man's impact on vegetation. Junk, the Hague, pp 107–137

Westoby M, Walker B, Noy-Meir I (1989) Opportunistic management for rangelands not at equilibrium. J Range Manage 42: 266–274

White F (1983) The vegetation of Africa. UNESCO, Paris, 356 pp

White LP (1970) Brousse tigrée patterns in southern Niger. J Ecol 58: 549–553

Whitmore TC (1990) An introduction to tropical rain forest. Clarendon Press, Oxford, 226 pp

Whittaker RH (1970) Communities and ecosystems. MacMillan, New York, 162 pp

Whittaker RH, Niering WA, Crisp MD (1979) Structure, pattern and diversity of a mallee community in New South Wales. Vegetatio 39: 65–76

Wickens GE (1980) Alternative uses of browse species. In: Le Houérou HN (ed) Browse in Africa: the current state of knowledge. Papers presented at the Int Symp on Browse in Africa, Addis Ababa, 8–12 April 1980. ILCA, Addis Ababa, pp 155–182

Wiersum KF (1984) Surface erosion under various tropical agroforestry systems. In: O'Luoghlin CL, Pearce AJ (eds) Symposium on effects of forest land use on erosion and slope stability. Environ Policy Inst, East-West Center, Honolulu, pp 231–239

Wijngaarden W van (1985) Elephants, trees, grass, grazers: relationships between climate, soils, vegetation and large herbivores in a semi-arid savanna ecosystem (Tsavo, Kenya). ITC, Enschede, ITC Publ 4, 159 pp

Wildin JH (1986) Trees in forage systems. Queensl Dep Primary Ind, Rockhampton

Willard EE, McKell CM (1978) Response of shrubs to simulated browsing. J Wildl Manage 42: 514–519

Williams RT (1981) Biology. In: Evans G (ed) Rural research in northern Australia. Commonw Counc Rural Res Extension, Commonw Aust, Canberra, pp 23–36

Wilson AD (1969) A review of browse in the nutrition of grazing animals. J Range Manage 22: 23–28

Wilson AD (1977) The digestibility and voluntary intake of the leaves of trees and shrubs by sheep and goats. Aust J Agric Res 28: 501–508

Wilson AD, Tongway DJ, Tupper GJ (1988) Factors contributing to differences in forage yield in the semi-arid woodlands. Aust Range J 10: 13–17

Wilson JR, Catchpoole VR, Weier KL (1986) Stimulation of growth and nitrogen uptake by shading a rundown green panic pasture on Brigalow clay soil. Trop Grassl 20: 134–143

Wilson RT (1980) Fuelwood consumption in a central Malian town and its effects on browse availability. In: Le Houérou HN (ed) Browse in Africa: the current state of knowledge. Papers presented at the Int Symp on Browse in Africa, Addis Ababa, 8–12 April 1980. ILCA, Addis Ababa, pp 473–476

Winterbottom R, Hazelwood PT (1987) Agroforestry and sustainable development: making the connection. Ambio 16: 100–110

Wispelaere G de (1980) Les photographies aériennes témoins de la dégradation du couvert ligneux dans un géosystème sahélien sénégalais. Influence de la proximité d'un forage. Cah ORSTOM Ser Sci Hum 17: 155–166

Wispelaere G de (1983) Systèmes de production d'élevage au Sénégal. Compte rendu de fin d'études. IEMVT, Maisons-Alfort

Wit CT de (1958) Transpiration and crop yields. Pudoc, Wageningen, Agric Res Rep 64(6): 88

World Bank (1985) Desertification in the Sahelian and Sudanian zones of West Africa. The World Bank, Washington DC, 60 pp

Wotton NJ, Sparrow AD (1991) Spatial distribution and competition for soil moisture in populations of the semi-arid Australian shrub *Maireana sedifolia*. IVth Int Rangeland Congr, April 1991, CIRAD, Montpellier

Yavitt JB, Smith EL (1982) Spatial patterns of mesquite and associated herbaceous species in an Arizona desert grassland. Am Midl Nat 109: 89–93

Young A (1989) Agroforestry for soil conservation. CAB Int, Wallingford Sci Practice Agrofor 4, 276 pp

Young JA, Evans RA, Easi DA (1984) Stem flow on western Juniper (*Juniperus occidentalis*) trees. Weed Sci 32: 320–327

Zech W, Maydell HJ von (1981) Futterwert forstlicher Pflanzen im Sahel. Forschungsber Bundesminist Wirtsch Zusammenarb 7, Weltforum, München, 167 pp

Zinke PJ (1962) The pattern of influence of individual forest trees on soil properties. Ecology 43: 130–133

Zoungrana I, Yelemou B, Hien F (1994) Etude des relations Neem-sol-sorgho dans les systèmes agroforestiers du Bulkiemde (Burkina Fasa). In: C R Symp Int Parcs Agrofor, 25–28 Oct 1993, Ouagadougou. ICRAF/IRBET/CILSS/LTC (in press)

Subject Index

Springer-Verlag and the Environment

We at Springer-Verlag firmly believe that an international science publisher has a special obligation to the environment, and our corporate policies consistently reflect this conviction.

We also expect our business partners – paper mills, printers, packaging manufacturers, etc. – to commit themselves to using environmentally friendly materials and production processes.

The paper in this book is made from low- or no-chlorine pulp and is acid free, in conformance with international standards for paper permanency.

GPSR Compliance

The European Union's (EU) General Product Safety Regulation (GPSR) is a set of rules that requires consumer products to be safe and our obligations to ensure this.

If you have any concerns about our products, you can contact us on ProductSafety@springernature.com

In case Publisher is established outside the EU, the EU authorized representative is:

Springer Nature Customer Service Center GmbH
Europaplatz 3
69115 Heidelberg, Germany

Batch number: 10844669

Printed by Printforce, the Netherlands